Imaging Modalities for Biological and Preclinical Research: A Compendium, Volume 2

Parts II–IV: *In vivo* preclinical imaging, multimodality imaging and outlook

IPEM–IOP Series in Physics and Engineering in Medicine and Biology

About the Series

Series in Physics and Engineering in Medicine and Biology will allow IPEM to enhance its mission to 'advance physics and engineering applied to medicine and biology for the public good.'

Focusing on key areas including, but not limited to:

- clinical engineering
- diagnostic radiology
- informatics and computing
- magnetic resonance imaging
- nuclear medicine
- physiological measurement
- radiation protection
- radiotherapy
- rehabilitation engineering
- ultrasound and non-ionising radiation.

A number of IPEM–IOP titles are published as part of the EUTEMPE Network Series for Medical Physics Experts.

Imaging Modalities for Biological and Preclinical Research: A Compendium, Volume 2

Parts II–IV: *In vivo* preclinical imaging, multimodality imaging and outlook

Edited by
Andreas Walter
Chair of COMULIS, Austrian BioImaging/CMI, Vienna BioCenter Core Facilities, Vienna, Austria

Julia G Mannheim
Werner Siemens Imaging Center, Department of Preclinical Imaging and Radiopharmacy, Eberhard Karls University Tübingen, Tübingen, Germany

Carmel J Caruana
Department of Medical Physics, Faculty of Health Sciences, University of Malta, Msida, Malta

IOP Publishing, Bristol, UK

Multimedia content and supplementary material are available for this book at http://iopscience.iop.org/book/978-0-7503-3747-2.

ISBN 978-0-7503-3747-2 (ebook)
ISBN 978-0-7503-3745-8 (print)
ISBN 978-0-7503-3748-9 (myPrint)
ISBN 978-0-7503-3746-5 (mobi)

DOI 10.1088/978-0-7503-3747-2

Version: 20210501

IOP ebooks

British Library Cataloguing-in-Publication Data: A catalogue record for this book is available from the British Library.

Published by IOP Publishing, wholly owned by The Institute of Physics, London

IOP Publishing, Temple Circus, Temple Way, Bristol, BS1 6HG, UK

US Office: IOP Publishing, Inc., 190 North Independence Mall West, Suite 601, Philadelphia, PA 19106, USA

Contents

Section 8 Magnetic particle imaging

Section 9 Computed tomography

Section 10 Nuclear imaging

Section 4 Multimodal image processing

Section Editor: Nataša Sladoje

Preface

Imaging is a technology platform that is vital for the advancement of the life sciences. The ongoing emergence of novel biological and preclinical imaging techniques is driving the *ex vivo* and *in vivo* characterisation of biophysical properties of tissue and the visualisation of real-time molecular mechanisms within their tissue-specific context with unprecedented resolution and penetration depths. Indeed, knowledge of the multitude of available imaging modalities and their strengths and limitations has become essential, particularly for the mechanistic understanding of diseases and tissue regeneration. Unfortunately, knowledge of the various imaging modalities is dispersed in many books and acquiring the necessary background information required to be able to choose the appropriate technique for a specific research task is time consuming for the student, lecturer and researcher.

Imaging Modalities for Biological and Preclinical Research: A Compendium is perhaps the first ever attempt to produce a comprehensive collection of the imaging techniques that are used in the biological and preclinical areas. It was the brainchild of the members of the European Cooperation in Science and Technology (COST) consortium '*Correlated Multimodal Imaging in Life Sciences*' (COMULIS, www.comulis.eu). COMULIS members and researchers in the preclinical and life sciences come from diverse scientific backgrounds including advanced light and electron microscopy, biology, radiochemistry, image analysis, small-animal imaging and medical physics. The aim of COMULIS is to bring together these diverse scientists with their specific expertise to harness the power of multimodality imaging, promote and disseminate its benefits through showcase pipelines, train a new generation of young scientists for whom multimodality imaging research is the new normal, and pave the way for its technological advancement and implementation as a versatile tool in biological and preclinical research. Getting such diverse interests to work together and teaching scientists for whom the multimodality imaging approach is an integral component of their research tool-chest will only be possible if we produce the necessary learning resources to facilitate the transition.

This compendium will provide the means of exploring the immense potential of bioimaging for acquiring complementary, multiscale information of structure, function, dynamics and molecular composition. To highlight the importance and the many benefits of multimodality imaging approaches, we specifically included a section about correlative multimodality imaging and multimodality image processing. The compendium reflects the highly dynamic nature of the field and provides the reader with an overview of each imaging modality covering physical principles, typical setups, biomedical relevance, sample preparation, parameters of image quality, and data processing and visualisation. Each chapter concludes with a paragraph discussing strengths and limitations as well as future developments of each technique. In this way, the readers are equipped with a comprehensive resource to properly choose the most promising imaging technique for their specific biomedical research question—or use it as a reference work for basic and advanced imaging techniques and their principles.

The authors who contributed to the individual chapters have many years of experience within their respective field, and we sincerely would like to thank them for finding the time within their busy schedules to dedicate to this important educational initiative and to share their expertise with the readers of this compendium. We are sure it has not been easy, and we deeply appreciate their efforts and time. Furthermore, we sincerely would like to thank all reviewers, who contributed to this book by revising and amending the chapters. Last but not least, we do hope that the users of this compendium will enjoy browsing through the various imaging modalities as much as we have enjoyed the process of bringing them together.

Dr Andreas Walter
Chair of COMULIS
Austrian BioImaging/CMI, Vienna BioCenter Core Facilities, Vienna, Austria

Dr Julia G Mannheim
Werner Siemens Imaging Center, Department of Preclinical Imaging and Radiopharmacy, Eberhard Karls University Tübingen, Tübingen, Germany

Professor Carmel J Caruana PhD FIPEM
Department of Medical Physics, Faculty of Health Sciences, University of Malta, Malta

Acknowledgements

This publication is based upon work from the COST Action COMULIS (CA17121), supported by COST (European Cooperation in Science and Technology). COST is a funding agency for research and innovation networks. COST Actions help connect research initiatives across Europe and enable scientists to grow their ideas by sharing them with their peers. This boosts their research, career and innovation. We would like to thank all members of the COST Action COMULIS that help to make COMULIS a success. COMULIS (Correlated Multimodality Imaging in Life Sciences) specifically aims at fueling urgently needed collaborations in the field of correlated multimodality imaging (CMI), promoting and disseminating its benefits through showcase pipelines, and paving the way for its technological advancement and implementation as a versatile tool in biological and preclinical research.

We sincerely thank both the authors and the reviewers for finding the time within their busy schedules to contribute to this comprehensive imaging compendium and to revise the chapters. We would like to thank Prof. Dr. Natasa Sladoje specifically for her superb coordination and editing of the section on multimodal image processing as part of Volume II.

Co-funded by the Horizon 2020 programme of the European Union

Editor biographies

Andreas Walter

Dr Andreas Walter studied physics and biophysics at the University of Heidelberg and in Sankt Petersburg on a Baden-Wuerttemberg Scholarship. He spent his research career developing and applying various imaging techniques for life sciences, including advanced fluorescence, super-resolution, cryo-light, transmission electron, scanning electron and soft X-ray microscopy. After his tenure as a guest scientist at the European Molecular Biology Laboratory, he did his PhD at the Max Planck Institute of Biophysics on method development in cryo-electron microscopy. He finished his studies with distinction. He continued his research in imaging and worked as a postdoc and DFG fellow at the University of California San Francisco and at the Lawrence Berkeley National Laboratory to correlate soft X-ray tomography and cryo-light microscopy. He is also a fellow of the German Scholars Organisation. Dr Walter was appointed Director of Austrian BioImaging/CMI in October 2016 (www.bioimaging-austria.at). His goal is to advance correlated multimodality imaging (CMI) as a valuable tool in biomedical research. He is also the main proposer and Chair of the EU-funded COST Action COMULIS (CA17121) that aims at promoting the benefits of CMI through showcase pipelines and paving the way for its technological advancement and implementation as a versatile tool in biological and preclinical research (www.comulis.eu).

Julia G Mannheim

Julia Mannheim is a group leader for PET and Multimodal Imaging Science in the Department of Preclinical Imaging and Radiopharmacy at the Eberhard Karls University Tübingen, Germany, alongside being the Deputy Director of the Department since November 2014. She is an expert in PET quantification and reproducibility/reliability of the acquired data. Her focus is currently on the standardisation of preclinical imaging to enhance the scientific integrity and reliability of the acquired data.

Carmel J Caruana

Carmel J Caruana is Professor and Head of the Department of Medical Physics of the Faculty of Health Sciences at the University of Malta where he lectures in X-ray imaging, fluoroscopy, magnetic resonance imaging, computed tomography, ultrasound and nuclear medicine imaging including PET/CT and PET/MRI. Carmel has over 30 years' experience in the teaching of all imaging modalities to all healthcare professions and at all levels including the specialty level. He is very much involved in curriculum development for Medical Physicists at the European and international levels and was previously the Chair of the Education and Training Committee of the European Federation of Organisations for Medical Physics. In the last few years, he has been encouraging Medical Physicists to expand their scope of practice to pre-clinical and biological imaging.

List of contributors

Mandy Ahlborg
Institute of Medical Engineering, Universität zu Lübeck, Ratzeburger Allee 160, 23562 Lübeck, Germany
Fraunhofer Research Institution for Individualized and Cell-Based Medical Engineering IMTE, Mönkhofer Weg 239 a, 23562 Lübeck, Germany

Fatih Alioglu
Laboratory of MacroMolecular Cancer Therapeutics (MMCT), Center of Pharmaceutical Sciences, Department of Pharmaceutical Chemistry, University of Vienna, Althanstrasse 14, A-1090 Vienna, Austria

Marco Andreana
Austrian BioImaging/CMI, Medical University of Vienna-Center for Medical Physics and Biomedical Engineering, Waehringer Guertel 18-20/4L A-1090 Vienna, Austria

Pedro Assunção
Polytechnic of Leiria / Instituto de Telecomunicações, Portugal, Campus 2, ESTG, Alto do Vieiro, Leiria, Portugal

Anna Bakenecker
Institute of Medical Engineering, Universität zu Lübeck, Ratzeburger Allee 160, 23562 Lübeck, Germany
Fraunhofer Research Institution for Individualized and Cell-Based Medical Engineering IMTE, Mönkhofer Weg 239 a, 23562 Lübeck, Germany

André Behrends
Institute of Medical Engineering, Universität zu Lübeck, Ratzeburger Allee 160, 23562 Lübeck, Germany
Fraunhofer Research Institution for Individualized and Cell-Based Medical Engineering IMTE, Mönkhofer Weg 239 a, 23562 Lübeck, Germany

Monique R Bernsen
AMIE Core Facility, Erasmus MC - University Medical Hospital Dr. Molewaterplein 40, 3015 GD Rotterdam, The Netherlands

Yvonne Blancke Soares
Institute of Medical Engineering, Universität zu Lübeck, Ratzeburger Allee 160, 23562 Lübeck, Germany

Carles Bosch
Sensory Circuits and Neurotechnology Lab, The Francis Crick Institute, London, UK

Kevin Brindle
Cancer Research UK Cambridge Institute, University of Cambridge, Li Ka Shing Centre, Robinson Way, Cambridge CB2 0RE, United Kingdom

Katja Bühler
Biomedical Image Informatics, Austrian Bioimaging/CMI, VRVis Zentrum für Virtual Reality und Visualisierung Forschungs-GmbH, Donau-City-Straße 11, 1220 Vienna, Austria

Thorsten M Buzug
Institute of Medical Engineering, Universität zu Lübeck, Ratzeburger Allee 160, 23562 Lübeck, Germany
Fraunhofer Research Institution for Individualized and Cell-Based Medical Engineering IMTE, Mönkhofer Weg 239 a, 23562 Lübeck, Germany

Alexander Bykov
Optoelectronic and Measurements Techniques, University of Oulu, Oulu, Finland

Xin Chen
Institute of Medical Engineering, Universität zu Lübeck, Ratzeburger Allee 160, 23562 Lübeck, Germany

Lucy Collinson
Francis Crick Institute, London, UK

Arnaud Comment
General Electric Healthcare, Chalfont St Giles, Buckinghamshire HP8 4SP, UK

Francisco Cunha
Polytechnic of Leiria / Instituto de Telecomunicações, Portugal, Campus 2, ESTG, Alto do Vieiro, Leiria, Portugal

Pieter A A De Beule
International Iberian Nanotechnology Laboratory, Braga, Portugal

Michal M Dykas
Correscopy, Frysztak, Poland

Viktoria Ehret
Austrian BioImaging/CMI, Department of Biomedical Imaging and Image guided Therapy, Division of Molecular and Structural Imaging, Medical University of Vienna

Martin Falk
Department of Science and Technology, Linköping University ITN/MIT, SE-601 74 Norrköping, Sweden

Sérgio Faria
Polytechnic of Leiria / Instituto de Telecomunicações, Portugal, Campus 2, ESTG, Alto do Vieiro, Leiria, Portugal

Thomas Friedrich
Institute of Medical Engineering, Universität zu Lübeck, Ratzeburger Allee 160, 23562 Lübeck, Germany
Fraunhofer Research Institution for Individualized and Cell-Based Medical Engineering IMTE, Mönkhofer Weg 239 a, 23562 Lübeck, Germany

Natasa Gaspar
Department of Radiology and Nuclear Medicine, Erasmus Medical Center, Wythemaweg 80, 3015CE Rotterdam, The Netherlands

Adam Gaunt
Cancer Research UK Cambridge Institute, University of Cambridge, Li Ka Shing Centre, Robinson Way, Cambridge CB2 0RE, United Kingdom

Maria Gomez-Lazaro
i3S - Instituto de Investigação e Inovação em Saúde, INEB -Instituto de Engenharia Biomédica, Rua Alfredo Allen, 208. 4200-135, Porto, Portugal

Ana I Gómez-Varela
International Iberian Nanotechnology Laboratory, Braga, Portugal
Department of Applied Physics, University of Santiago de Compostela, Santiago de Compostela, Spain

Ksenija Gräfe
Institute of Medical Engineering, Universität zu Lübeck, Ratzeburger Allee 160, 23562 Lübeck, Germany

Felix Gremse
Experimental Molecular Imaging, RWTH Aachen University Clinic, Forckenbeckstr. 55, 52074 Aachen, Germany

Jan Grimm
Molecular Pharmacology Program, Memorial Sloan Kettering Cancer Center, 417 E 68th Street, New York, NY 10065, USA

Jaromir Gumulec
Department of Pathological Physiology, Faculty of Medicine, Masaryk University, Kamenice 5, Brno, CZ-62500, Czech Republic

David Haberthür
Institute of Anatomy, University of Bern, Baltzerstrasse 2, 3012 Bern, Switzerland

Joost C Haeck
AMIE Core Facility, Erasmus MC - University Medical Hospital Dr. Molewaterplein 40, 3015 GD Rotterdam, The Netherlands

Andreas Haghofer
FH Upper Austria, Hagenberg Campus, Bioinformatics Research Group, Hagenberg, Austria
Department of Computer Science, Johannes Kepler University Linz, Linz, Austria

Thomas Helbich
Austrian BioImaging/CMI, Department of Biomedical Imaging and Image guided Therapy, Division of Molecular and Structural Imaging, Medical University of Vienna, Austria

Ruslan Hlushchuk
Institute of Anatomy, University of Bern, Baltzerstrasse 2, 3012 Bern, Switzerland

Hannah Janout
FH Upper Austria, Hagenberg Campus, Bioinformatics Research Group, Hagenberg, Austria

Martin Jones
Francis Crick Institute, London, UK

Matthia A Karreman
Heidelberg University Hospital and German Cancer Research Center (DKFZ) Heidelberg, Germany

Prateek Katiyar
Werner Siemens Imaging Center, Department of Preclinical Imaging and Radiopharmacy, Eberhard Karls University Tübingen, Tübingen, Germany

Felix Kreis
Institute for Biomedical Engineering, University and ETH Zurich, Switzerland., Gloriastrasse 35, 8092 Zürich, Switzerland

Claudia Kuntner-Hannes
Austrian BioImaging/CMI, Department of Biomedical Imaging and Image guided Therapy, Medical University Vienna, Vienna, Austria

Claire Lefort
Institut de recherche XLIM, CNRS, Université de Limoges, UMR 7252, 123 avenue Albert Thomas, 87000 Limoges, France

Joakim Lindblad
Centre for Image Analysis, Dept. of Information Technology, Uppsala University, Box 337, SE-751 05 Uppsala, Sweden

Kerstin Lüdtke-Buzug
Institute of Medical Engineering, Universität zu Lübeck, Ratzeburger Allee 160, 23562 Lübeck, Germany

Ankit Malhotra
Institute of Medical Engineering, Universität zu Lübeck, Ratzeburger Allee 160, 23562 Lübeck, Germany
Fraunhofer Research Institution for Individualized and Cell-Based Medical Engineering IMTE, Mönkhofer Weg 239 a, 23562 Lübeck, Germany

Julia G Mannheim
Werner Siemens Imaging Center, Department of Preclinical Imaging and Radiopharmacy, Eberhard Karls University Tübingen, Tübingen, Germany
Cluster of Excellence iFIT (EXC 2180) 'Image Guided and Functionally Instructed Tumor Therapies', University of Tübingen, Tübingen, Germany

Wendy McDougald
The University of Edinburgh, Queen's Medical Research Institute, Edinburgh BioQuarter, 47 Little France Crescent, EH16 4TJ, Edinburgh, UK

Benedict McLarney
Molecular Pharmacology Program, Memorial Sloan Kettering Cancer Center, 417 E 68th Street, New York, NY 10065, USA

Igor Meglinski
Optoelectronic and Measurements Techniques Unit, University of Oulu, Oulu, Finland
College of Engineering and Physical Sciences, Aston University, Birmingham B4 7ET, UK

Laura Mezzanotte
Department of Radiology and Nuclear Medicine, Erasmus Medical Center, Wythemaweg 80, 3015CE Rotterdam, The Netherlands

Adelaide Miranda
International Iberian Nanotechnology Laboratory, Braga, Portugal

Carmel M Moran
Centre for Cardiovascular Science, Edinburgh Preclinical Imaging, University of Edinburgh, Queen's Medical Research Institute, 47 Little France Crescent, Edinburgh, EH16 4TJ, UK

Catarina C Moura
International Iberian Nanotechnology Laboratory, Braga, Portugal

Arrate Munoz-Baruttia
Universidad Carlos III de Madrid, Spain

Alexander Neumann
Institute of Medical Engineering, Universität zu Lübeck, Ratzeburger Allee 160, 23562 Lübeck, Germany

Robert Nuster
Institute of Physics, University of Graz, Universitätsplatz 5, 8010 Graz, Austria

Manfred Ogris
Austrian BioImaging/CMI, Laboratory of MacroMolecular Cancer Therapeutics (MMCT), Center of Pharmaceutical Sciences, Department of Pharmaceutical Chemistry, University of Vienna, Althanstrasse 14, A-1090 Vienna, Austria

Alexandra Pacureanu
Sensory Circuits and Neurotechnology Lab, The Francis Crick Institute, London, UK
Department of Neuroscience, Physiology and Pharmacology, University College London, UK
ESRF, the European Synchrotron, Grenoble, France

Patricia Pais-Roldán
Institute of Neuroscience and Medicine - 4, Medical Imaging Physics (Forschungszentrum Juelich, Germany), Wilhelm-Johnen-Straße, 52428 Jülich, Germany

Karla Palma-Alejandro
Center for Advanced Preclinical imaging (CAPI), First Faculty of Medicine, Charles University, Prague, Czech Republic

Guenther Paltauf
Institute of Physics, University of Graz, Universitätsplatz 5, 8010 Graz, Austria

João Parracho
Polytechnic of Leiria / Instituto de Telecomunicações, Portugal, Campus 2, ESTG, Alto do Vieiro, Leiria, Portugal

Perrine Paul-Gilloteaux
Université de Nantes, CHU Nantes, Inserm, CNRS, SFR Santé, Inserm UMS 016, CNRS UMS 3556, F-44000 Nantes, France
Université de Nantes, CNRS, INSERM, l'institut du thorax, F-44000 Nantes, France

Mailyn Pérez-Liva
Université de Paris, PARCC, INSERM, F-75006 Paris, France
Grupo de Física Nuclear and IPARCOS, U. Complutense Madrid, CEI Moncloa, Madrid, Spain

Teodora-Adriana Perles-Barbacaru
Center for Magnetic Resonance in Biology and Medicine, UMR CNRS 7339, Aix-Marseille University, 27, bd Jean Moulin, 13385 Marseille, France

Alain Pitiot
Ilixa Ltd, London, UK
Ludwig Boltzmann Institute for Experimental and ClinicalTraumatology, Vienna, Austria

Aileen Prosche
Laboratory of MacroMolecular Cancer Therapeutics (MMCT), Center of Pharmaceutical Sciences, Department of Pharmaceutical Chemistry, University of Vienna, Althanstrasse 14, A-1090 Vienna, Austria

Stefanie Rosenhain
Experimental Molecular Imaging, RWTH Aachen University Clinic, Forckenbeckstr. 55, 52074 Aachen, Germany

Haider Sami
Laboratory of MacroMolecular Cancer Therapeutics (MMCT), Center of Pharmaceutical Sciences, Department of Pharmaceutical Chemistry, University of Vienna, Althanstrasse 14, A-1090 Vienna, Austria

Michal Sarna
Department of Biophysics, Faculty of Biochemistry, Biophysics and Biotechnology, Jagiellonian University, Krakow, Poland

Andreas T Schaefer
Sensory Circuits and Neurotechnology Lab, The Francis Crick Institute, London, UK
Department of Neuroscience, Physiology and Pharmacology, University College London, UK

Andreas M Schmid
Werner Siemens Imaging Center, Department of Preclinical Imaging and Radiopharmacy, Eberhard Karls University Tübingen, Tübingen, Germany
Cluster of Excellence iFIT (EXC 2180) 'Image Guided and Functionally Instructed Tumor Therapies', University of Tübingen, Tübingen, Germany

Julia Scholda
Laboratory of MacroMolecular Cancer Therapeutics (MMCT), Center of Pharmaceutical Sciences, Department of Pharmaceutical Chemistry, University of Vienna, Althanstrasse 14, A-1090 Vienna, Austria

Jonas Schumacher
Institute of Medical Engineering, Universität zu Lübeck, Ratzeburger Allee 160, 23562 Lübeck, Germany
Fraunhofer Research Institution for Individualized and Cell-Based Medical Engineering IMTE, Mönkhofer Weg 239 a, 23562 Lübeck, Germany

Ludek Sefc
Center for Advanced Preclinical imaging (CAPI), First Faculty of Medicine, Charles University, Prague, Czech Republic

Karla-Nikita Singeorzan
Laboratory of MacroMolecular Cancer Therapeutics (MMCT), Center of Pharmaceutical Sciences, Department of Pharmaceutical Chemistry, University of Vienna, Althanstrasse 14, A-1090 Vienna, Austria

Magdalena Skubal
Molecular Pharmacology Program, Memorial Sloan Kettering Cancer Center, 417 E 68th Street, New York, NY 10065, USA

Nataša Sladoje
Centre for Image Analysis, Department of Information Technology, Uppsala University, Box 337, SE-751 05 Uppsala, Sweden

Noeska Smit
Department of Informatics, University of Bergen, HIB, Post Office Box 7803, 5020, Bergen, Norway
Mohn Medical Imaging and Visualization Centre, Haukeland University Hospital, Post Office Box 1400, N – 5021, Bergen, Norway

Jan Stelzner
Institute of Medical Engineering, Universität zu Lübeck, Ratzeburger Allee 160, 23562 Lübeck, Germany

Andreas Stylianou
Cancer Biophysics Laboratory, University of Cyprus, Nicosia, Cyprus
European University, Nicosia, Cyprus

Adriana A S Tavares
The University of Edinburgh, Queen's Medical Research Institute, 47 Little France Crescent, EH16 4TJ, Edinburgh, UK

Bertrand Tavitian
Université de Paris, PARCC, INSERM, F-75006 Paris, France
Radiology Department, AP-HP, Hôpital européen Georges Pompidou, F-75015 Paris, France

Lucas Thomaz
Polytechnic of Leiria / Instituto de Telecomunicações, Portugal, Campus 2, ESTG, Alto do Vieiro, Leiria, Portugal

Adrian J W Thomson
Centre for Cardiovascular Science, Edinburgh Preclinical Imaging, University of Edinburgh, Queen's Medical Research Institute, 47 Little France Crescent, Edinburgh, EH16 4TJ, UK

Angelika Unterhuber
Austrian BioImaging/CMI, Medical University of Vienna- Center for Medical Physics and Biomedical Engineering, Waehringer Guertel 18-20/4L- A-1090 Vienna, Austria

Sandra T van Tiel
AMIE Core Facility, Erasmus MC - University Medical Hospital Dr. Molewaterplein 40, 3015 GD Rotterdam, The Netherlands

Christian Vanhove
Ghent University, Institute of Biomedical Engineering and Technology, Corneel Heymanslaan 10, 9000 Gent, Belgium

Paul Verkade
School of Biochemistry, University of Bristol, Bristol, UK, Biomedical Sciences Building, University Walk, BS8 1TD, Bristol, UK

Tomas Vicar
Department of Biomedical Engineering, Faculty of Electrical Engineering and Communication, Brno University of Technology, Technicka 3058/10, Brno, CZ-61600, Czech Republic

Anna Vilanova
Department of Mathematics and Computer Science, Eindhoven University of Technology, PO Box 513, 5600 MB Eindhoven, The Netherlands

Angele Viola
Center for Magnetic Resonance in Biology and Medicine, UMR CNRS 7339, Aix-Marseille University, 27, bd Jean Moulin, 13385 Marseille, France

Wolfgang Wadsak
Medical University of Vienna, Department of Biomedical Imaging and Image-guided Therapy, Division of Nuclear Medicine, Waehringer Guertel 18-20, 1090 Vienna
CBmed - Center for Biomarker Research in Medicine, Stiftingtalstr. 5, 8010 Graz, Austria

Andreas Walter
Austrian BioImaging/CMI, Vienna BioCenter Core Facilities, Dr Bohr Gasse 3, 1030 Vienna, Austria

Thomas Wanek
Austrian BioImaging/CMI, Department of Biomedical Imaging and Image guided Therapy, Medical University Vienna, Vienna, Austria

Franz Wegner
Department of Radiology and Nuclear Medicine, University Hospital Schleswig-Holstein, Campus Lübeck, Ratzeburger Allee 160, 23538 Lübeck, Germany

Huimin Wei
Institute of Medical Engineering, Universität zu Lübeck, Ratzeburger Allee 160, 23562 Lübeck, Germany

Stephan Winkler
Austrian BioImaging/CMI, FH Upper Austria, Hagenberg Campus, Bioinformatics Research Group, Hagenberg, Austria
Johannes Kepler University Linz, Department of Computer Science, Linz, Austria

Xin Yu
Athinoula A. Martinos Center for Biomedical Imaging, Department of Radiology, Harvard Medical School, Massachusetts General Hospital, Charlestown, Massachusetts, USA

Part II

In vivo imaging

Section 1

Intravital microscopy

IOP Publishing

Imaging Modalities for Biological and Preclinical Research: A Compendium, Volume 2

Parts II-IV: *In vivo* preclinical imaging, multimodality imaging and outlook

Andreas Walter, Julia G Mannheim and Carmel J Caruana

Chapter II.1

Intravital microscopy

Matthia A Karreman

Intravital microscopy uniquely allows visualising an event of interest in the complex microenvironment of a living organism. Various animal models and microscopy setups can be used enabling one to study neurological, pathological, developmental and oncological processes at high spatio- and temporal resolution. Here, examples include tumour growth and metastasis, infection and associated immune-response, developmental processes, and cell lineage tracking. Intravital microscopy relies on imaging fluorescence markers or reporters, sometimes in combination with label-free imaging of the microenvironment. Scattering of light in the tissue influences the image quality and makes it necessary to graft optical windows to image deeper organs in e.g. mice. A rise in interest in intravital microscopy has led to the development of sophisticated software for image analysis and correction, and for advanced techniques that improve image quality. This chapter focuses on current approaches, new developments, and the wide range of possibilities of intravital microscopy in biomedical research.

1 Introduction

Although many advanced *in vitro* models, including 3D culture, co-culture, microfluidic systems, and organoid models, are employed in the field of medical and developmental biology, these models often do not sufficiently reflect the complexity of a living organism. Intravital imaging enables monitoring critical biological events in space and time in living tissue in animal model systems [1–5]. This chapter focuses on high resolution of living whole organisms, allowing monitoring and tracking biological events over time [6–8]. Low resolution intravital imaging techniques, such as IVIS or PET, will not be discussed here. Studying living animals enables including factors that may influence the event of interest but that are hard to mimic *in vitro*; immune-response, blood or lymphatic flow, chemical gradients, influence of the

doi:10.1088/978-0-7503-3747-2ch1

specific microenvironment and other factors. However, intravital imaging comes with its own challenges. This chapter provides a brief overview of currently used methods, possible applications, challenges and opportunities.

2 Principles and setups

2.1 Physical principles

Intravital optical imaging demands illuminating the region of interest with visible or near-infrared light, and image contrast is generated by chemical labels, fluorescent proteins, and/or label-free, based on tissue-intrinsic properties [3, 7]. One of the main challenges for intravital imaging is preserving image quality when imaging deep into the tissue. This does not pose a problem when using transparent organisms, such as the zebrafish and Drosophila embryos, however, when imaging higher organisms, such as mice or rats, the image quality will be affected by light absorption and scattering. To overcome these issues, sophisticated imaging setups (exemplified below) have been developed and optimised to enable deep-tissue imaging. Moreover, an optical window can be engrafted to expose the organ of interest [1, 2, 9] (see section 3.2).

2.2 Typical setups and state-of-the-art

- **Single-photon laser scanning microscopy**
  For small or transparent organisms, such as *C. elegans*, chicken embryo or the zebrafish embryo, it is possible to use a confocal laser scanning microscopy (CLSM) setup (figure 1(a)). Here, the animal can be kept alive and stabilised into low-melting point agarose [8]. Pharmaceutical treatments and/or (tumour cell) injections can be performed and the animal can be monitored live for several hours. However, the visible light used in CLSM is scattered or adsorbed in non-transparent tissues (e.g. zebrafish brain or mouse tissues), and therefore can then only be used for superficial imaging (<100 μm). Since in CLSM, images are built up from a focused laser beam scanning the sample point-by-point, acquisition speed may not be sufficient to monitor rapid dynamic processes taking place *in vivo*. Here, a spinning disk confocal provides higher frame-rates by illuminating the sample through a spinning disk with arrays of micro-lenses, which generate multiple parallel focused laser beams scanning the sample. A second disk, spinning at the same rate, contains micro-pinholes that are aligned to the lenses. Spinning disk microscopy enables one to simultaneously scan multiple positions of the sample, greatly speeding up the acquisition rate. The pinhole used in CLSM limits the out-of-focus light on the detector. An Airyscan allows detecting a larger part of the emitted light using a detector composed of different elements that are arranged like a compound eye. Each element captures a different part of the Airy disk, and their contribution to the image can be set enabling collecting light more efficiently and improving signal-to-noise, acquisition speed and sensitivity.

- **Multi-photon laser scanning microscopy (2/3PEM):**
In multi-photon microscopy, a pulsed infrared laser is used to excite the fluorophores in the sample. Here, excitation requires the absorption of multiple low-energy photons, typically two (two-photon microscopy) or three (figure 1(b)). Importantly, this principle requires a high density of photons, which is only achieved in the focal point, reducing out-of-focus signal and phototoxicity. Moreover, infrared photons have the capacity to penetrate much deeper into the tissue due to their longer wavelengths, which allows for non-invasive deep tissue imaging [1–8]. Finally, specific elements of tissue can be visualised label-free based on their capacity for second- or third harmonic generation (SHG or THG), inducing, respectively, an exact doubling or tripling of the excitation wavelength. SHG can occur at e.g. collagen and striated muscle, whereas THG excitation can occur at interfaces between water and lipids (such as membranes) or water and large protein clusters [6–8].

- **Light sheet microscopy:**
In contrast to laser scanning microscopy, in light-sheet microscopy the excitation and detection is performed by separate, perpendicularly positioned objectives. The excitation laser-beam is shaped as a sheet and illuminates a single plane of the sample, limiting out-of-focus fluorescence and phototoxicity (figure 1(c)). The setup combines high-speed and long-term volume imaging and, owing to high-sensitivity cameras, can be used at unprecedented acquisition rates or to follow developmental processes over time at high resolution [3, 7, 8]. One challenge for this method is the large quantity and complexity of the data generated for each experiment, and novel methods and software are under development to overcome this.

- **Super-resolution intravital microscopy**
Super-resolution approaches, such as localisation microscopy (LoM), stimulated emission depletion microscopy (STED) and structured illumination microscopy (SIM), have revolutionised the life sciences. Applications *in vivo* have been demonstrated but are hindered by either slow acquisition rate (as with LoM) or difficulty in overlapping the stimulation and depletion beam (as with STED). Recently, however, SIM was applied to show structural dynamics of dendrites in zebrafish and mouse brain [1, 2, 4, 9]. SIM allows for a two-fold improvement in image resolution and, importantly, is the fastest super-resolution technique, making it an excellent tool for high spatio-temporal resolution imaging (figure 1(d)).

- **Biomechanics**
In recent years, there has been an increasing interest in measuring mechanical forces, such as stiffness, strength of binding and flow, in living tissues. Optical tweezers have been used in the zebrafish embryo to measure the haemodynamic forces in zebrafish [8]. Here, microscopic particles can be trapped in the focal point of a laser beam, allowing manipulation of movement and

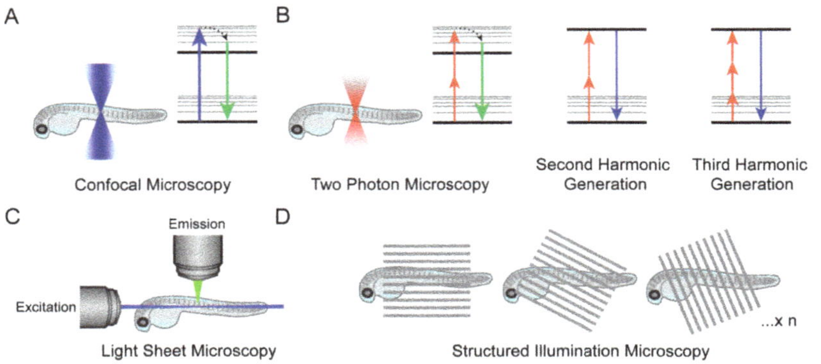

Figure 1. Schematic representation typical setups and state-of-the-art of intravital microscopy. The figure depicts cartoons of the zebrafish embryo imaged by (a) confocal microscopy, (b) multi-photon microscopy (diagrams: 2 photon fluorescence (left), SHG (middle) and THG (right)), (c) light sheet microscopy and (d) structured illumination microscopy.

measuring of forces on the object. Recently, Brillouin microscopy was used to measure the stiffness of the extracellular matrix of the zebrafish notochord [3]. Brillouin microscopy is a label- and contact-free, optical method that enables measuring the biomechanical properties of cells and tissues. Spectral shifts of the scattered light, in relation to the incident light, are measured. These shifts are generated by inherent acoustic vibrations in the sample, which are representative of their biomechanical properties [3].

3 Biomedical relevance

3.1 Application range and relevance

Intravital imaging has been instrumental for many key discoveries in the fields of cancer biology, neuroscience, developmental biology and immunology. Critical in these fields is the impact of the specific microenvironment and the organism as a whole on the event that is studied. This includes interactions with neighboring cells, biophysical and structural elements of the surrounding tissue, and the complex interplay of the different elements of the immune system. The complexity of the model-organism can still not be fully recapitulated by existing 2D or 3D *in vitro* models. In cancer biology, intravital imaging has shed light on all key processes of tumour growth and metastasis, collectively called the 'hallmarks of cancer' [6]. Examples include communication between tumour cells themselves or with their microenvironment [2], invasion and metastasis [7, 8] (figure 2), and vascular remodelling during colonisation and growth [8].

Neuroscience has pushed the development of intravital imaging, allowing visualisation and/or manipulation of neuronal activity *in vivo*. Calcium imaging, where fluorescent reporters are used to highlight changes to intracellular Ca^{2+} concentrations, enables monitoring neuronal activity or communication between astrocytes. When the animal is not under narcosis while in the microscope, 'awake

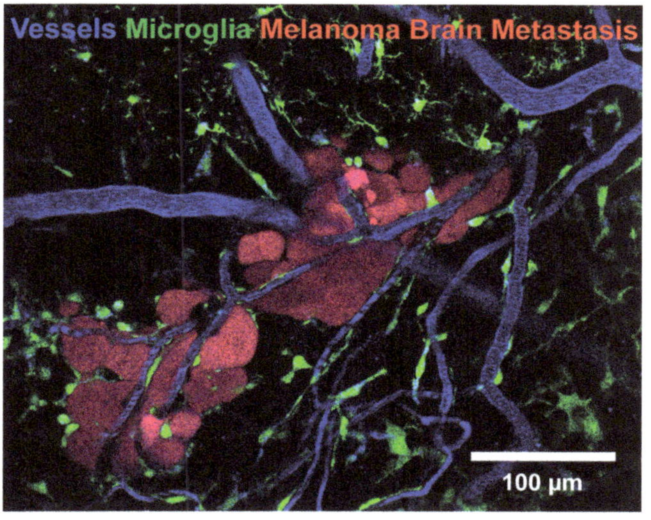

Figure 2. Intravital imaging of melanoma metastasis (red) in the brain of CX3CR1-GFP transgenic mice. The tdTomato-expressing melanoma cells were intracardially injected into cranial window-bearing mice and brain metastasis developed in the weeks following injection. The microglia in this mousestrain are expressing GFP (green), and the vasculature is visualised by i.v. injection of TRITC-dextran (blue). (Karreman, Winkler *et al*, unpublished data.)

imaging', enables one e.g. to study behavior or provide sensory input while monitoring the brain at high resolution.

Intravital microscopy is also shown to be critical to study developmental processes over time. Often small organisms are used, such as the drosophila larvae, the zebrafish embryo, and the chick embryo, which can be imaged for longer periods of time to track processes, e.g. gastrulation. To increase the imaging rate and reduce phototoxicity, light sheet imaging is a valuable tool for developmental biology (see section 2.2).

For immunologists, intravital imaging is of critical importance since no *in vitro* models sufficiently represent the complexity of the various components of the immune system, nor can they mimic the specific characteristics of the tissue compartment under study [4, 10]. The interaction of the adaptive and the innate and adaptive immune system, cytokines, platelets, vascular flow, and many other factors contribute to host-pathogen interactions or the inflammation response [10].

These fields mentioned above, and others, are heavily dependent on intravital microscopy for their scientific progress and are responsible for many scientific breakthroughs that enable the sophisticated intravital imaging methods we know today.

3.2 Sample preparation

Common animal models used for intravital imaging are the nematode (*C. elegans*, larvae/adult), the zebrafish larvae (*D. rerio*), the chick embryo or Chorioallantoic

Membrane (*G. gallus*), the fruitfly (*D. melanogaster*, embryo), and the mouse (*M. musculus*) [8]. Except for the mouse, the above-mentioned organisms are transparent during their embryonic stage, which facilitates intravital microscopy. Here, the organism can be embedded in an agarose matrix for stabilisation, and can be imaged over hours. To enable imaging of the internal organs of the mice, however, an endoscope can be used or an (chronic) optical window can be surgically engrafted. These include cranial windows to visualise neurological processes or brain tumours, a mammary window or dorsal-skin-flap window, e.g. to study breast tumours or skin tumours, respectively, or an abdominal imaging window to image the intestine, pancreas and/or liver [2, 6, 8, 9].

Although intravital imaging can be applied for label-free imaging (see section 2.2), in many applications fluorescent markers are used to highlight specific cells, cellular compartments, vessels or specific tissue components. Fluorescent dyes can be injected intravenously to highlight the vasculature, or specific cell-types [2]. Moreover, transgenic animal models that stably express fluorescent proteins in specific cell types are powerful models to visualise the microenvironment [1, 4, 6, 8]. Importantly, in both immunology and cancer biology, fluorescent cells can be introduced into the model system by injection and can be monitored long-term [1, 2, 5, 6, 8–10]. For preclinical studies, it can be necessary to inject human-derived cell lines (e.g. human cancer cells). In that case, immune-compromised or 'humanised' mice are required to circumvent an immune-response. In zebrafish, the adaptive immune system is not functional in the first 10 days, which allows for xenografts [8]. Aside from revealing morphological features of the tissue and dynamic processes, intravital imaging is currently increasingly employed for functional studies. Hereto, fluorescent reporters and dyes can be used to visualise critical processes such as cell signaling, protein expression, cell cycle status or tissue states (e.g. blood vessel permeability, hypoxia, changes in pH).

4 Parameters of image quality

4.1 Scattering

While imaging of living organisms, a challenge is the limited light penetration and image quality caused by light scattering in the tissue. Using transparent animals, this issue can be circumvented, however, for imaging in adult, non-transparent organisms, advanced imaging setups (see section 2.2) can be employed, such as multiphoton microscopy. Multiphoton microscopy generally uses longer wavelengths which enables deeper penetration into tissue (e.g. >1 mm in the mouse brain, <300 μm in mouse skin). Adaptive optics enables one to 'sculpt' the wavefront of the excitation light to compensate for the scattering properties of the tissue [4].

4.2 Spatial and temporal resolution

Depending on the research question, the possibility to perform high resolution, long-term imaging can be critical for an intravital imaging experiment. For cancer research, long-term monitoring of tumour growth or metastasis (i.e. over weeks/months) can be a powerful tool to uncover critical steps of these pathological processes [1, 2, 5, 6, 8]. High-speed imaging (i.e. movies or time-lapse imaging, for minutes/hours) can reveal cell–cell interactions or communication [2, 5]. Just like any other type of imaging, there is always a typical trade-off between imaging rate and spatial resolution: shorter pixel-dwell time or less averaging results in more noisy images. Repetitive imaging of the same area, however, can result in phototoxicity or bleaching, which could lead to damaging of the tissue, false results and/or may influence the event of interest.

4.3 Setup: movement artefacts and awake imaging

Image quality can suffer from movement artefacts, even those caused by respiration or heartbeat of an anesthetised animal. To circumvent this, a small animal can be embedded in an agarose matrix or, in e.g. a mouse, the imaging window/area can be stabilised in a specialised holder [8, 9]. Importantly, for some experiments in, for instance neuroscience or oncology [2], it can be critical that the animal is awake during imaging to circumvent inhibitory effects on functional processes in the brain. Here too, specialised stages were developed that allow free movement of awake mice while imaging through a cranial window. Moreover, triggered scanning to synchronise imaging with the respiration cycle can further reduce breathing artefacts.

5 Data processing

Depending of the imaging setup, time-span of the acquisition and required resolution, the size of the data sets obtained with intravital imaging can be in the range of multiple mega- or gigabytes. Moreover, in the case of light-sheet microscopy, the data sets could easily go up to terabytes. When embarking on an intravital imaging experiment, it is thus critical to consider storage, processing and analysis of the data.

Current image analysis software allows for (semi)-automated analysis of 3D and 4D data sets, including particle tracking, segmentation, alignment, image correction, classification, 3/4D visualisations and volume analysis. While excellent commercially available software is commonly used, such as Imaris, Amira and AIVIA, many scientists rely on the use of freely available software such as Fiji/ImageJ (open source), Ilastik, IMART, Voxx, ICY and others. The latter group are developed and kept by academics and are invaluable tools allowing for many data processing needs, ranging from basic image adjustments to high-end machine learning for segmentation and analysis. Furthermore, many tasks can be performed with the visualisation software of the microscope, such as LAS X/LASAF (Leica), Zen (Zeiss), and NIS-Elements (Nikon).

Image processing scripts can be generated and/or machine learning can be used to batch-process and analyse multiple data sets. Here too, advanced software such as Illastik, Imaris, AIVIA, ICY and Imaris can be used, for instance supported by Matlab. Due to ever-increasing need for such analyses, the number of suitable software and helpful online tutorials is steadily growing. However, a specific challenge for the (automated) analysis of intravital imaging is the heterogenous quality of the data sets, arising from more scattering deeper in the sample versus at its surface. Importantly, different model systems are used for different scientific questions, adding to the challenge of automating and normalising data analysis.

6 Conclusions

6.1 Strength and limitations

Intravital imaging enables unique insights into many key fields within life science, by revealing critical biological events at high spatial and temporal resolution, in real time, *in vivo*. Recapitulating developmental, pathological, and neurological processes demands a relevant model that enables studying the influence of and interactions with the microenvironment. Moreover, functional imaging (e.g. using fluorescent reporter-constructs) allows a 'live' readout of transcriptional, metabolic or structural changes to the object of interest, which can be correlated with specific behavior or treatment of the animal under study. For these and other reasons, many basic research questions but also biomedical studies greatly benefit from intravital imaging. However, intravital imaging is not easily accessible for all researchers. Importantly, some applications require high-end microscopes, which, for newer approaches, could even be homemade and thus unique to specific laboratoria. Moreover, not every lab has access to the required costly animal facilities and infrastructure. Many data sets generated while long-term imaging may require large amounts of storage-space and fully unsupervised automated analysis is still extremely rare. Of note, imaging software and data analysis require, or benefit from, additional training and often programming skills which are currently not part of basic training in life science.

Importantly, the ethical issues that arise with the use of living animals for research should always be taken into consideration. However, as long as even the most advanced *in vitro* models cannot fully recapitulate the complexity of a living organism, intravital imaging will remain the most powerful approach for dynamic 3D microscopy in life science.

6.2 Future developments

The increasing applications of intravital microscopy in life science is reflected in the technical developments taking place to further improve and support the technique. New developments focus on improving image quality by limiting the influence of scattering (such as adaptive optics) and fast super-resolution applications. Interesting to follow are the further developments that enable *in vivo* biomechanical analysis combined with or during intravital imaging. This, as well as lineage tracking or functional reporters could provide unique insights into local and chronological changes to cells or structures of interest in living tissue. Correlation with other imaging techniques, such as electron microscopy, enables extending the resolution of intravital microscopy to the nanometer scale and can reveal the full architecture of the microenvironment (see also chapter III.3, *Correlative multimodality imaging across scales*).

References and further reading

[1] Ritsma L, Ellenbroek S I J, Zomer A, Snippert H J, de Sauvage F J, Simons B D, Clevers H and van Rheenen J 2014 *Nature* **507** 362–5
[2] Osswald M *et al* 2015 *Nature* **528** 93–8
[3] Bevilacqua C, Sánchez-Iranzo H, Richter D, Diz-Muñoz A and Prevedel R 2019 *Biomed. Opt. Express* **10** 1420–31
[4] Turcotte R, Liang Y, Tanimoto M, Zhang Q, Li Z, Koyama M, Betzig E and Ji N 2019 *PNAS* **116** 9586–91
[5] Mulazzani M *et al* 2019 *PNAS* **116** 24275–84
[6] Ellenbroek S I J and van Rheenen J 2014 *Nat. Rev. Cancer* **14** 406–18
[7] Weigelin B, Bakker G-J and Friedl P 2016 *J. Cell Sci.* **129** 245–55
[8] Follain G, Mercier L, Osmani N, Harlepp S and Goetz J G 2017 *J. Cell Sci.* **130** 23–38
[9] Kitamura T, Pollard J W and Vendrell M 2016 *Trends Biotechnol.* **35** 5–8
[10] Handschuh J, Amore J and Müller A J 2020 *Cytometry* A **97** 458–70

Section 2

Optical coherence tomography

IOP Publishing

Imaging Modalities for Biological and Preclinical Research: A Compendium, Volume 2

Parts II–IV: *In vivo* preclinical imaging, multimodality imaging and outlook

Andreas Walter, Julia G Mannheim and Carmel J Caruana

Chapter II.2

Optical coherence tomography

Igor Meglinski and Alexander Bykov

Optical coherence tomography (OCT) is a highly effective non-invasive imaging modality, which is highly sensitive to the changes in the optical properties of biological tissues. Therefore, OCT is able to observe and quantify the internal changes within the biotissues, as a function of temporal variations of optical properties with very high in-depth and transversal resolutions (up to 1–10 μm) and high dynamic range (up to 120 dB). Over the last decade, OCT has been extensively used as an emerging non-invasive high-resolution imaging technique in various fields of biology and medicine. We have been actively involved in transferring OCT to the clinical and pre-clinical research. In this chapter, state-of-the-art, recent advances in OCT-based medical imaging, and future perspectives are highlighted and briefly discussed.

1 Introduction

Due to the recent intense development of high technologies and miniaturisation of solid-state optical/laser devices, a number of new revolutionary applications have arisen. This area is commonly known as 'photonics' since it has many parallels to electronics. A more recent trend has been the mapping of these new technologies to the life sciences, biology and medicine, hence the name 'bio-photonics' was coined. While the global photonics market has reached thousands of billions of Euros, we estimate that only 20% of the potential power and benefits of light technologies have been unlocked during the last decade. Understanding how light interacts with biological tissues gave rise to different optical modalities including OCT. OCT is an optical imaging modality that is extensively used in clinical and pre-clinical diagnostic applications [1]. One of the major advantages of OCT is its non-invasive or minimally invasive nature of the measurement due to the use of electromagnetic radiation (light) as a probe. OCT performs high resolution, cross-sectional

doi:10.1088/978-0-7503-3747-2ch2

tomographic imaging of the internal microstructure in materials and biological systems by measuring back-scattered or back-reflected light [2]. Image resolutions of 1–15 μm can be achieved, which is one to two orders of magnitude higher than with conventional ultrasound. Imaging can be performed *in situ*, non-invasively, and in real time. Conventional OCT synthesises cross-sectional morphological images of internal tissue structure from a series of laterally adjacent one-dimensional reflectivity profiles of the sample. The amplitudes of these profiles are proportional to the leap of the refractive index on the internal boundaries of the tissue. Information concerning tissue physiology/functionality may also be extracted by considering the wave properties and the polarisation state of sounding optical radiation. The development of such multifunctional OCT systems capable of delivering information from several OCT modalities simultaneously, e.g. Doppler OCT, has drawn the attention of many research groups worldwide [1–3].

OCT can be interfaced with a wide range of imaging delivery systems and imaging probes. OCT technology is based on fibre optics and uses components developed for the telecommunications industry. OCT systems can be engineered to be compact and low cost, suitable for applications in research, manufacturing, or the clinic. The imaging engine, intended as any optical detection device that performs high resolution and high sensitivity ranging and detection of back-reflected or back-scattered optical echoes, is the heart of the OCT system.

2 Principles and setups

2.1 Physical principles

Most of originally developed time-domain OCT (TD-OCT) systems have employed a reference delay scanning interferometer, such as a Michelson interferometer, in combination with a low coherence light source [1, 2]. The reference arm of the interferometer (see figure 1 (left)) contains an optical modulator (e.g. moving mirror). The studied object is placed in the sample arm. Light from a low coherent source is divided into two equal parts by a beam splitter. One part is guided into the reference arm, a separate part to the sample arm. In the reference arm, light is

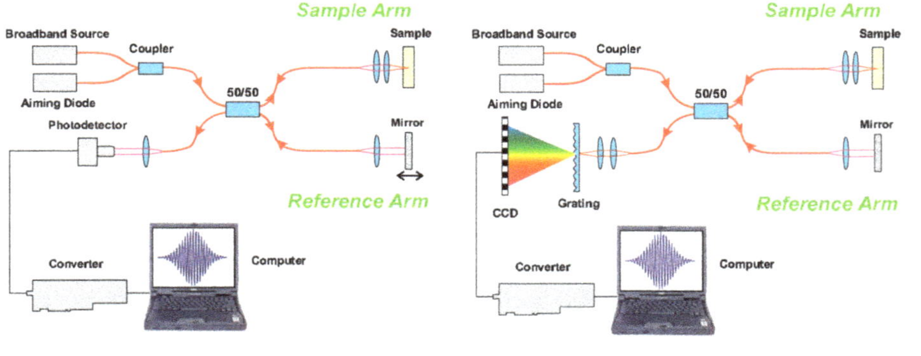

Figure 1. Schematic diagram of principles of TD-OCT (left) and SD-OCT (right) systems. Figures provided by Professor Kirill Larin, personal communication.

reflected by the mirror, in the sample arm—by the studied object. The reflected light from both arms is guided back to the beam splitter where the interference occurs. The interference signal is recorded by a photodetector. For the probing light with a long coherence length (e.g. laser source) the interference occurs practically at any optical path difference of the waves in the reference and sample arms of the interferometer. On the other hand, for the low coherence light of the broadband light source (e.g. superluminescent diode, supercontinuum laser, or even bulb), the interference occurs if the optical path difference is within the coherence length of the light source. This effect is called a coherence gating and determines the in-depth resolution of the system. In this case, for TD-OCT, the longitudinal scan of the object (A-scan) is performed by the motion of the reference mirror. The amplitude of the detected interference signal is proportional to the local value of the reflection coefficient of the medium at a certain depth determined by the position of the mirror in the reference arm. The local inhomogeneities with optical properties different from those of the surrounding medium cause a significant change in the amplitude of the detected signal. Thus, the time dependence of the amplitude of the detected signal during the longitudinal scan can be used to reconstruct the depth distribution of local values of reflection coefficient in a point of interest. In a transverse direction (B-scan) scanning is performed by a consecutive displacement of the probe beam position.

There are many different embodiments of the interferometer and imaging engine for specific applications such as polarisation diversity (insensitive) imaging, polarisation-sensitive imaging, and Doppler flow imaging. Polarisation-sensitive detection techniques have been demonstrated using a dual-channel interferometer. These techniques permit imaging of the birefringence properties of structures. Collagen and other tissues are strongly birefringent, and PS-OCT can be a sensitive indicator of the tissue microstructural organisation. Time encoded frequency domain OCT (TEFD-OCT), also known as swept source OCT (SS-OCT), combines some of the advantages of standard TD-OCT and Fourier domain-OCT (FD-OCT). By accommodation of a frequency scanning light source (i.e. frequency scanning laser), the system optical setup does not require any moving scanning element. The main advantage of this approach lies in the proven high SNR detection technology, while swept laser sources achieve very small instantaneous bandwidths (i.e. linewidth) at a very high scanning rate (in the order of 1MHz).

2.2 Typical setups and state-of-the-art

Currently, the OCT-based imaging technique may be divided into two major classes, TD-OCT and FD-OCT.

While originally developed TD-OCT methods were widely used in laboratory studies, the significant signal-to-noise advantage of FD-OCT over TD-OCT (up to 30 dB) opened a lot of new opportunities for the technique in biomedical applications [1–4]. Also, the simplicity of the interferometer construction, the absence of in-depth mechanical scanning, and the SNR advantage of FD-OCT are facilitating its further development for many applications. Moreover, the relaxed requirement for in-depth mechanical scanning in FD-OCT systems allows using a *common* beam path for the reference and sample arms of the interferometer with the reference plane located near the surface of the sample. Such a common-path interferometer is easier to align and is less sensitive to mechanical vibrations, and it allows for more-straightforward compensation for dispersion and polarisation differences between the reference and sample arms. FD-OCT may be further divided into two distinct methods of acquiring spectral information from an object under study: (i) Spectral-domain OCT (SD-OCT)—using a broadband laser source (typical FWHM > 50 nm) with a grating and a photodiode array in the detection arm (figure 1 (right)); and (ii) swept-source OCT (SS-OCT)—using a rapidly tunable, narrow-linewidth laser source over a broad optical bandwidth with a single photodiode in the detection arm. While SD-OCT and SS-OCT have similar sensitivity advantages over TD-OCT, the SS-OCT system can detect interference fringes over a substantially longer range of time delays between reflections from reference and sample interfaces. Phase-sensitive measurements of biological objects provide additional information on optical properties and play a key role in the development of functional OCT (fOCT) methods. TD-OCT with differential phase contrast shows great potential for imaging and classification of tissues as well as for fOCT. While conventional OCT is based on the detection and analysis of the intensity of back-scattered optical radiation, phase-sensitive OCT utilises the phase information obtained by the back-scattered light. Non-static objects will generate a non-zero phase difference, $\Delta\phi$, between two repeated measurements at one sample position. The phase-sensitive OCT technique (that could be both TD-OCT or SD-OCT) is capable of measuring nanometer scale path length changes in the time between two measurements at the same position (associated with the phase difference $\Delta p = \frac{\lambda}{4\pi}\Delta\phi$, λ is the wavelength of probing light) in clear and scattering media.

3 Biomedical relevance

3.1 Application range and relevance

Recent advances in medical imaging devices have gained interest also for endoscopic imaging of internal tissues. Arthroscopy is a widely used method in surgical operations of the knee joint. In combination with OCT it has been shown to be capable of detecting structural changes in the cartilage during surgery. However, due to the strong scattering, OCT imaging is only useful for superficial cartilage layers. Optical clearing significantly enhances the quality of OCT imaging of articular cartilage and subchondral bone (see figure 2(a)), including the cartilage–bone interface and enables direct screening of the subchondral bone [6].

Immunisation is one of the most efficient procedures for the prevention of diseases, but most vaccines have to be administered invasively. A new paradigm in medicine—transcutaneous delivery—has a number of advantages, such as the controlled release of the drug into the patient, it is user-friendly, convenient, painless, and offers the potential for multi-day dosing. OCT accompanied with cross-correlation analysis [7] was used for evaluation of drugs penetration through

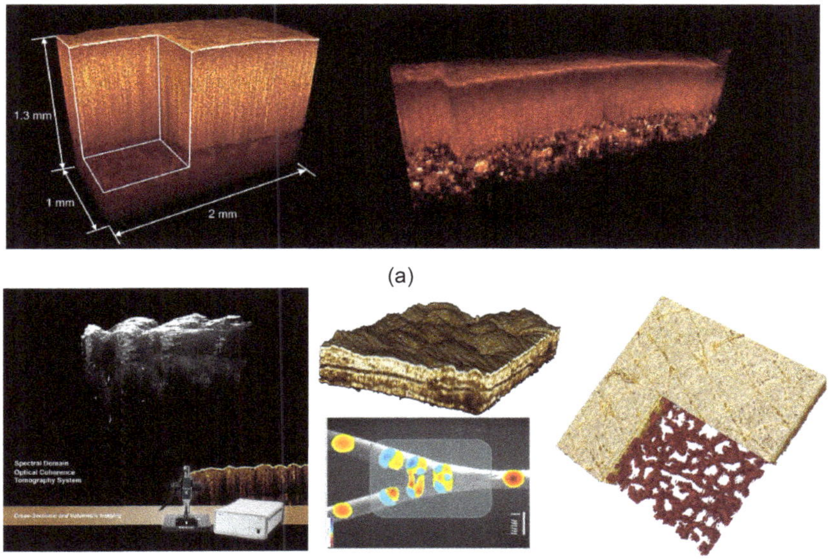

(a)

(b) (c) (d)

Figure 2. Examples of OCT applications. (a) OCT image of the cartilage after optical clearing by using iohexol (left) and propylene glycol (right) [6], copyright John Wiley & Sons. (b) Volumetric OCT image of human finger-pad (1.5 × 1.5 mm) showing epidermal, dermal layers, and multiple sweat channels. Image obtained/provided by Dr A Bykov. (c) 3D OCT image of skin during assessment of transcutaneous vaccine delivery [7] (top); Doppler OCT images of the velocity profiles of flow within complex vessel [8] (bottom); (d) 3D OCT image of spatial distribution of blood vessels in human skin observed at 3 × 3 × 1 mm volume [9], copyright John Wiley & Sons.

the skin *in vivo* (illustrated in figure 2(c) top). By using cross-correlation analysis of successive 2D OCT images, a front of molecular diffusion within the skin was clearly observed [7].

Doppler OCT is well suited for the quantitative mapping of flow velocity profiles within complex vessels (see figure 2(c) bottom) and the analysis of flow–vessel interactions [8]. The non-invasive imaging and quantitative analysis of blood flow in the complex-structured vascular bed are required in many biomedical applications, including those where the determination of mechanical properties of vessels or the knowledge of the mechanic interactions between the flow and the housing medium plays a key role. The change of micro-vessel wall elasticity could be a potential indicator of cardiovascular disease at the very early stage, while monitoring the blood flow dynamics and associated temporal and spatial variations in the vessel's wall shear stress could help to predict the possible rupture of atherosclerotic plaques.

The advantage of the Doppler OCT technique is a very high spatial resolution (up to 1–10 μm) of flow velocity profile measurements. Penetration depth in tissue-like turbid media is limited to around 1 mm due to the wavelengths range of the light sources typically employed in OCT systems (NIR and IR regions, 800–1300 nm). Furthermore, Doppler OCT allows the assessment of the retinal blood flow dynamics and seems a promising technique for diagnosis of diabetic retinopathy and glaucoma [4, 5]. The technique demonstrated also a high potential as needle-based Doppler OCT for real-time flow assessment including coronary flow measurements. The use of OCT for research on the flow-related changes in atherosclerotic plaque is promising, because of its demonstrated capacity to differentiate plaque constituents *ex vivo* and *in vivo*.

With the recent progress in data processing, performed on graphics processing units (GPUs) by using compute unified device architecture (CUDA) framework, and its further implementation to the OCT image, the spatial distribution of superficial blood vessels in human skin *in vivo* has been observed [9]. To remove background noise, reduce the artefacts associated with patient motions, and increase the overall quality of the experimental OCT images, an adaptive Wiener filtering technique has been employed. Fourier domain correlation has been subsequently applied to enhance the spatial resolution of images of the vascular network in human skin *in vivo*. This approach allows carrying out image processing in parallel significantly speeding up the computations. The presented results show that the double correlation method permits obtaining 2D/3D OCT images of subcutaneous microcirculation vascular network and its spatial distribution within the human skin *in vivo* (see figure 2(d)) with higher spatial resolution compared to the other OCT correlation-based techniques developed earlier. A similar approach is currently used to look at the blood vessels in the retina, known as OCT angiography (OCTA). With further developments, OCTA has excellent potential to be used in the diagnosis of diabetic retinopathy and other dangerous retina diseases.

Fiber-based OCT endoscopes allow direct visualisation of the gastrointestinal tract, as well as cartilage and tendon during arthroscopy. Polarisation sensitive OCT (PS-OCT) measures in depth the polarisation states of the light back-scattered from the sample. It is particularly well suited for musculoskeletal applications, as collagen alignment is critical to ensure appropriate mechanical responses to load and strain. The birefringent property of collagen molecules induces a variation of the polarisation state propagating in the tissue. Their mesoscopic alignment induces a cumulated delay between polarisation states which is quantifiable with PS-OCT. Arthroscopy is a widely used method in surgical operations of the knee joint. In combination with OCT, it has been shown to be capable of detecting structural changes in the cartilage during surgery [4, 5].

Finally, the unique features of OCT enable a broad range of research and clinical applications, and it is successfully applied in ophthalmology, for example, to obtain high-resolution images of the retina and glaucoma diagnosis (see figure 1(e)).

4 Conclusions

In conclusion, given the non-invasive nature of the OCT-based techniques, the high spatial resolution, reasonably simple hardware requirements, and relatively compact size of the system, OCT reveals itself as a potentially effective technique applicable in rheology, arthroscopy, ophthalmology, dermatology, skin care, and other areas in biomedicine, where non-contact, non-invasive, high resolution velocity mapping of moving constituents are of particular interest. OCT provides a unique tool for quantitative assessment of the dynamics of diffusion of drugs, target compounds, analytes, cosmetics, and various chemical agents in human skin and other biological tissues *in vivo*. Recent advances in OCT-based medical imaging have gained interest also for endoscopic imaging of internal tissues. Neuroimaging applications could also greatly benefit from Doppler OCT and OCTA to investigate neurovascular coupling mechanisms, an area of current intense research.

4.1 Strength and limitations

OCT is able to observe and quantify the internal changes within the biological tissues, as a function of spatial and temporal variations of optical properties with a very high in-depth and transversal resolution in the micrometer range and high dynamic range (up to 120 dB). FD-OCT has several significant advantages compared to the TD-OCT, including: (i) enhanced dynamic range; (ii) signal acquisition rate in megahertz range instead of kilohertz that allows volumetric real-time imaging *in vivo*. Drawbacks are the inability of beam focus control to be synchronised with the axial scanning, non-linearities in the wavelength, especially at high scanning frequencies, the broadening of the linewidth at high frequencies for SS-OCT and high sensitivity to movements of the scanning geometry or the sample.

For the application in ophthalmology, by placing neutral density filter in the reference of the interferometer, the actual power of incident light is attenuated to 0.5 mW that makes the technology much safer for routine diagnostic imaging.

4.2 Future developments

An extension of the Doppler OCT approach, optical coherence elastography (OCE) is an emerging functional imaging technique that quantifies the elasticity of biological tissue by analysing local tissue displacement as a function of the applied stress [5]. OCE has excellent potential and is expected to be extensively used in the biomedical diagnosis that requires precision phase measurement, e.g., eye pressure test in ophthalmology, blood flowmetry, tissue and cells biomechanics, and molecular imaging.

Besides, in the last decade, consistent and successful innovations have been achieved in the field of lasers and optics, collectively known as ‘photonics’, finding new applications in biomedicine. In fact, light can be much more ‘complex’—twisted, i.e., in addition to the conventional state of polarisation the light beams can be radially or azimuthally polarised and carry orbital angular momentum (OAM), and contain phase and/or polarisation singularities [10]. With the rapid development of quantum measurement technologies in the next five to seven years, we expect to see the application of OAM and quantum correlation for the visualisation of small tissue alterations at very high spatial resolution (beyond standard quantum limitations). A quantum version of OCT utilising OAM quantum entanglement photons is expected to be developed. Pilot studies suggest that a detection approach borrowed from quantum optics can be used to perform OCT with much lower light power than previously possible. This could greatly improve the imaging quality available from OCT used for medical imaging applications. Nevertheless, the effectiveness and future development of quantum OCT are strongly dependent on the ease with which the new approach can be transferred in clinical practice.

References and further reading

[1] Drexler W and Fujimoto J G 2015 *Optical Coherence Tomography—Technology and Applications* 2nd edn (Berlin: Springer)
[2] Bouma B E and Tearney G J 2002 *Handbook of Optical Coherence Tomography* (New York: Marcel Dekker Inc.)
[3] Tuchin V 2013 *Handbook of Coherent Domain Optical Methods: Biomedical Diagostics, Environment and Material Science* vol II (Berlin: Springer)
[4] Duker J S, Waheed N K and Goldman D 2014 *Handbook of Retinal OCT: Optical Coherence Tomography* (London: Elsevier)
[5] Akman A, Bayer A and Nouri-Mahdavi K 2018 *Optical Coherence Tomography in Glaucoma: A Practical Guide* (Berlin: Springer)
[6] Bykov A, Hautala T, Kinnunen M, Popov A, Karhula S, Saarakkala S, Nieminen M T, Tuchin V V and Meglinski I 2016 *J. Biophotonics* **9** 270–5
[7] Kamali T, Doronin A, Rattanapak T, Hook S and Meglinski I 2012 *Laser Phys. Lett.* **9** 607–10
[8] Bonesi M, Proskurin S G and Meglinski I 2010 *Laser Phys.* **20** 891–9
[9] Doronin A and Meglinski I 2013 *Laser Photonics Rev.* **7** 797–800
[10] Torres J P and Torner L 2011 *Twisted Photons* (Weinheim: Wiley)

Section 3

Photoacoustic imaging

IOP Publishing

Imaging Modalities for Biological and Preclinical Research: A Compendium, Volume 2

Parts II–IV: *In vivo* preclinical imaging, multimodality imaging and outlook

Andreas Walter, Julia G Mannheim and Carmel J Caruana

Chapter II.3

Photoacoustic imaging

Robert Nuster and Guenther Paltauf

Photoacoustic imaging (PAI) is a highly scalable, hybrid imaging technology that combines the favorable properties of pure optical and ultrasound imaging methods. Thereby gathered images reveal optical contrast based on endogenous (e.g. blood, fat, melanin, DNA) and/or exogenous (e.g. dyes, plasmonic nanoparticles) light absorbing chromophores contained in biological samples. It can be used to visualise anatomic structures such as the micro-vasculature and provides functional information in the form of blood oxygenation, blood flow and temperature. Major preclinical applications include imaging of angiogenesis, microcirculation, tumour micro-environments, drug response, brain functions, biomarkers and gene activities. As a rule of thumb, in PAI the ratio of imaging depth to spatial resolution is about 200 enabling investigation of organelles, cells, tissue and organs with one imaging modality.

1 Introduction

Photoacoustic or optoacoustic imaging (PAI) is an emerging imaging modality for various biomedical applications, providing anatomical, functional, and molecular information. Its hybrid nature combines the high contrast and spectroscopic specificity of optical imaging with the high spatial resolution of ultrasound imaging. Due to the optical absorption contrast, PAI offers the ability to detect biomolecules, such as hemoglobin, lipids, water, DNA, and other intrinsic and extrinsic light-absorbing chromophores by addressing their spectroscopic fingerprints via specifically selected photoacoustic excitation wavelengths. It can be used to visualise anatomic structures such as the micro-vasculature and provides functional information in the form of blood oxygenation, blood flow and temperature. Furthermore, depending on its technical implementation, PAI is highly scalable in terms of achievable imaging depth and spatial resolution. This opens the door for a wide

doi:10.1088/978-0-7503-3747-2ch3

range of *in vivo* applications investigating organelles, cells, tissue and organs with one imaging modality. Major preclinical applications include imaging of angiogenesis, microcirculation, tumour micro-environments, drug response, brain functions, biomarkers and gene activities [1, 2].

2 Principles and setups

2.1 Physical principles

The working principle is schematically shown in figure 1. An electromagnetic wave with temporally varying irradiance hits the tissue sample, causing a local elevation of temperature in optically absorbing structures. The heating gives rise to the generation of a distribution of locally enhanced pressure, from which ultrasound waves are launched. The photoacoustic (PA) image is produced by the information contained in the time-resolved ultrasound signals detected at different positions outside the sample. It uses the correlation of signal amplitude and its temporal sequence with the spatial distribution of the laser-induced pressure sources. These sources are in turn related to the heat produced by the deposited electromagnetic radiation energy. Typical values of local temperature and pressure rises are in the range of millikelvin to Kelvin and millibar to bar, respectively.

For maximum efficiency, the incident electromagnetic radiation is usually provided by laser pulses, which are fired onto the biological sample causing an impulsive heating. The requirement for an efficient excitation process is that the duration of the laser pulses is in the nanosecond range to fulfill the conditions of stress and thermal confinement. This means that the ranges of acoustic and thermal propagation within the laser pulse duration are small compared to the length scale of

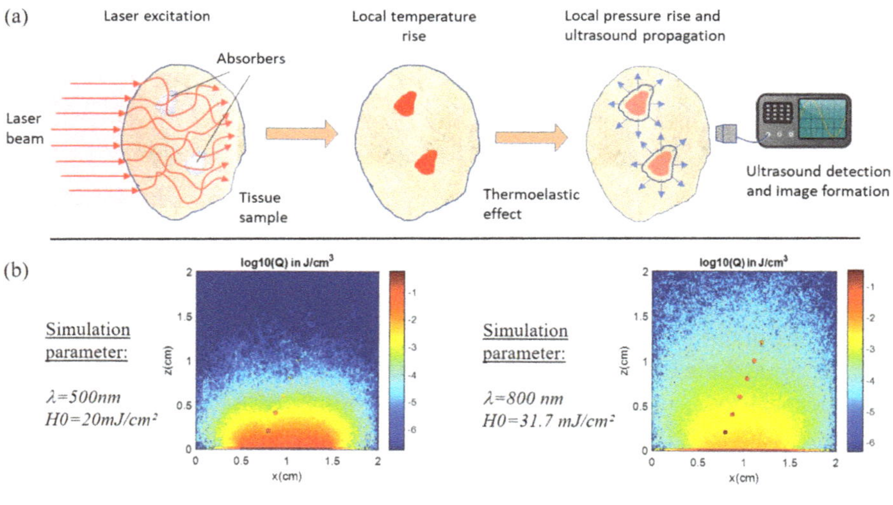

Figure 1. Schematic representation of the PAI working principle (a) and numerical simulation results of absorbed energy density distribution in a numerical phantom sample that mimics the physical properties of tissue (b). The simulation shows the dependence of optical penetration on the laser wavelength λ at the respective maximum permissible radiant exposure H_0.

the heated volume. Under these conditions, an initial pressure rise p_0 at time $t = 0$ right after the laser pulse is generated at a point $\boldsymbol{r}'$ within the heated volume, which is proportional to the absorbed optical energy $H(\boldsymbol{r}')$,

$$p_0(\boldsymbol{r}') = \Gamma H(\boldsymbol{r}'). \tag{1}$$

Γ is the dimensionless Grüneisen parameter describing the conversion efficiency of heat energy to pressure and is given by $\Gamma = \beta\, c^2/C_p$, where β is the thermal expansion coefficient, c the speed of sound and C_p the specific heat capacity at constant pressure. $H(\boldsymbol{r}')$ is the absorbed optical energy density distribution defined by the product of the absorption coefficient $\mu_A(\boldsymbol{r}')$ and the optical fluence $\phi(\boldsymbol{r}', \mu_A, \mu_S, g)$, where μ_S is the optical scattering coefficient and g is the scattering anisotropy factor. Substituting these quantities in equation (1) yields

$$p_0(\boldsymbol{r}') = \Gamma(\boldsymbol{r}')\mu_A(\boldsymbol{r}')\phi(\boldsymbol{r}', \mu_A, \mu_S, g). \tag{2}$$

This equation shows the dependency of the initial pressure distribution on a variety of mechanical, thermal and optical parameters. Since the mechanical and thermal properties are very similar in soft tissue, they are summarised in the constant, spatially invariant Grüneisen parameter $\Gamma(r') = const$. By contrast, a strong variation of optical absorption is observed between different tissue components, which becomes obvious by comparing the spectra of the light-absorbing chromophores in biological samples (figure 2(a)). Hence, the optical absorption parameters are dominating the contrast in PAI.

At this point it is worth mentioning that in general the extraction of quantitative spectroscopic information from PA images is challenging since p_0 depends on μ_A and the fluence ϕ, which is itself dependent on μ_A, with the result that p_0 is a nonlinear function of μ_A [1].

The pressure field $p(\boldsymbol{r}, t)$ emerging from p_0 is given by the solution of the PA wave equation for an acoustically uniform and non-attenuating medium with

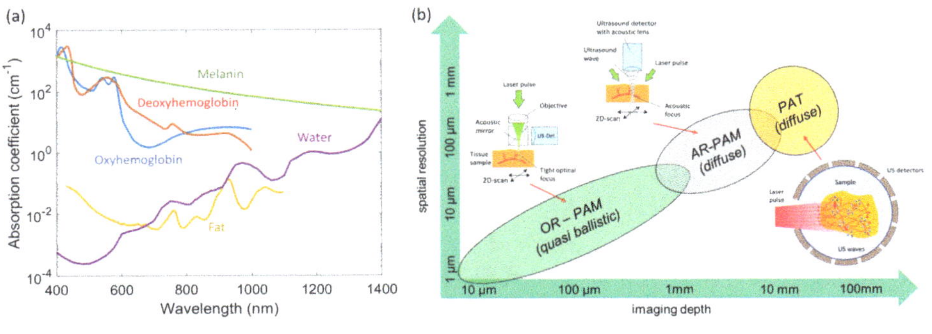

Figure 2. Absorption spectra of main endogenous tissue chromophores to which PAI contrast is based on (a) [4]. Relation between achievable spatial resolution and imaging depth depending on the specific PAI implementation (b).

$$p(\boldsymbol{r}, t) = \frac{1}{4\pi c}\frac{\partial}{\partial t}\left[\frac{1}{ct}\int_{S(r,\,t)} p_0(\boldsymbol{r}')d\boldsymbol{r}'\right], \tag{3}$$

where the integral is performed along a spherical surface with radius $|\boldsymbol{r}' - \boldsymbol{r}| = ct$ around the observation point $\boldsymbol{r}$. Commonly, this field is recorded with a single detector or an US detector array at various positions and the gathered information is used for image formation by solving the PA inverse problem. A recorded signal $S(\boldsymbol{r}_D, t)$ of a finite-size detector located at $\boldsymbol{r}_D$ is given by the convolution of the integral of the pressure field $p(\boldsymbol{r}, t)$ over the active sensor surface $A(\boldsymbol{r})$ with the electrical impulse response $EIR(t)$ of the ultrasound detector expressed by [3]

$$S(\boldsymbol{r}_D, t) = EIR(t) \times \int_{A(\boldsymbol{r})} p(\boldsymbol{r}, t)d\boldsymbol{r}. \tag{4}$$

Finally, the PA image formation is based on the physical principles described in the above equations and the specific implementation of the PA imaging modality as described in the following section.

2.2 Typical setups and state-of-the-art

A typical PAI setup consists of a wavelength-tunable pulsed light source, such as a nanosecond Q-switched Nd:YAG laser equipped with an optical parametric oscillator (OPO) or a dye-laser for efficient wideband PA excitation, a broadband US detection element (single or array), a data acquisition (DAQ) system for signal amplification and digitisation, a mechanical scanning stage for sample positioning and a computer for system control, data collection, and image formation. The bandwidth of the US detection element and the DAQ system should match the expected bandwidth of the PA signals that originate from small optical absorbers to obtain the best possible result in terms of signal-to-noise ratio (SNR), detection sensitivity and axial resolution [5].

In general, PAI can be divided in two different image-formation methods. The first method, commonly termed photoacoustic microscopy (PAM), is based on focused-scanning image formation, where a focused excitation light beam and a focused single element US detector are confocally aligned to each other (figure 3(a)) and are mechanically scanned across the sample surface to record the temporal signals [6]. In this case, the simplest way to form a 3D PAM image is to convert the signals into brightness values and to arrange them in the correct order according to the scanning process. The PAM methods are subdivided into optical-resolution (OR-) PAM, where the lateral resolution is given by the optical focus, and acoustical-resolution (AR-) PAM, where the acoustic focus determines the resolution. OR-PAM is limited to a depth of about 1 mm in biological tissue in which the light propagation is quasi-ballistic. AR-PAM can reach depths of several millimeters to centimeters, where light propagation is diffuse [8]. The second method is called photoacoustic tomography (PAT or PACT) and uses a wide-area light illumination with unfocused US detection at multiple locations surrounding the object (figure 4). The image is formed by applying a reconstruction algorithm (e.g. back-projection)

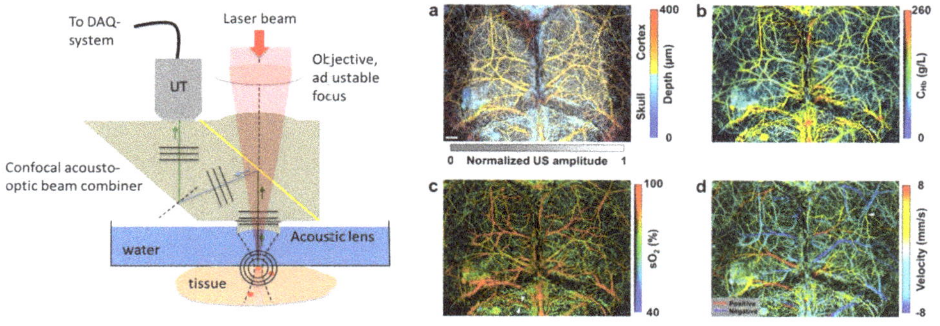

Figure 3. Schematic drawing of a commonly used reflection-mode OR-PAM setup (left) [6]. Representative PAM/SAM result of the brain of a mouse through the intact skull showing obtainable multiparameter information. (a) Depth-encoded skull vasculature and cortical vasculature separated by the SAM-determined skull (in grey). US: ultrasound. (b–d) Simultaneously acquired high-resolution maps of CHb, sO2, and CBF (speed and direction), respectively. Scale bar: 0.5 mm. Reproduced from [7] with permission of Springer (CC BY 4.0).

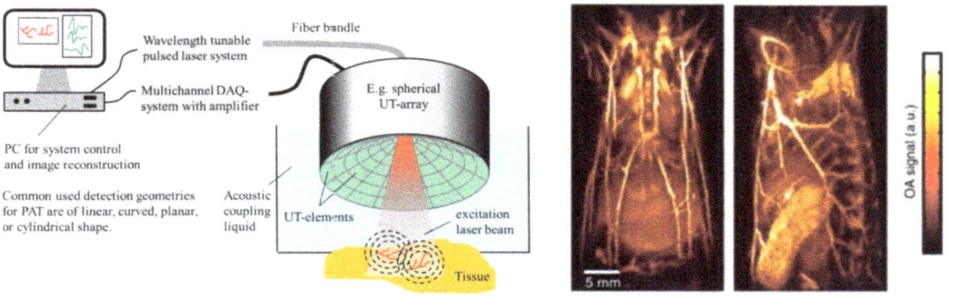

Figure 4. Schematic representation of PAT approach with spherical detection geometry (a). Whole-body mouse imaging result obtained by recording PA-data while scanning a spherical ultrasound detection array along a spiral scanning trajectory. A frame rate of 100Hz of a 1cm^3 volume is achievable, only limited by the pulse repetition rate of the excitation laser (b). Reproduced from [9] with permission of Springer (CC BY 4.0).

to the measured data. It is a particular strength of PAI devices that due to the combination of light supply and acoustic detection it can be easily combined with pure optical or pure ultrasound imaging modalities. In PAT, the use of ultrasound arrays has led to various systems combining it with ultrasound imaging. PAM can be combined with optical microscopy techniques, including optical coherence tomography (OCT). These additional imaging methods are able to depict tissue structure via back-scattered sound or light, in addition to the pure optical absorption contrast provided by PAI.

3 Biomedical relevance

3.1 Application range and relevance

The main *in vivo* imaging target in PAI is the vasculature with related functional parameters, such as oxygenation. This is made possible by the strong intrinsic contrast provided by hemoglobin and by the characteristic change of the absorption with the oxygenation level. Optically absorbing contrast agents (e.g. plasmonic nanoparticles) can be designed to target specific sites in tissue.

Applications:

- Tumour progression and treatment monitoring by imaging tumour-related neovascularisation;
- Pathological changes in vasculature;
- Wound healing.

3.2 Sample preparation

Samples typically investigated with PAI are small animals (mice, rats, zebrafish, chicken embryos, larvae), excised organs, tissue sections and cell cultures. Almost all PAI systems have in common that the US detectors need to be acoustically coupled to the sample to enable the interaction of PA waves propagating out of the sample with the detector sensing area. Hence, it is necessary to bridge the gap between the sample and the detector with some kind of liquid or gel (e.g. water, US-Gel) that is acoustically matched to tissue.

Whole-animal imaging with PAT systems requires the fixation of the animal in an animal-holder, the supply with anesthesia and contrast agents (if needed, e.g. ICG), and the positioning in a tempered water bath during imaging data acquisition. For such experiments, nude mice are used almost entirely, avoiding artefacts caused by strong signals from the hair absorption. In PAM it is recommended to carefully remove all hairs from the region of interest. High-resolution OR-PAM imaging requires that the sample surface is flat across the imaging area, since the detection sensitivity is highest close to the focal plane, which should maintain slightly beneath the sample surface. Sectioning of biological samples into thin layers and their fixation is an important preparation step for achieving high-resolution OR-PAM images over a volume that requires an imaging depth exceeding the ballistic photon propagation range.

4 Parameters of image quality

4.1 Excitation light source

The signal strength and depth to which structures can be detected scale with the incident radiant exposure (laser pulse energy per irradiated area). Laser safety considerations, however, set a limit to the applicable exposure. For instance, the maximum permissible exposure (MPE) of skin in the visible range is 20 mJ cm^{-2}. This value linearly increases in the near infrared up to 100 mJ cm^{-2}, as a result of the overall smaller optical absorption in this range and the corresponding deeper penetration. The different penetration depths at the MPE value can be seen in the simulation in figure 1(b).

4.2 Ultrasound detectors

Broadband detection increases the image quality: the low frequency content of signals is essential for accurate reconstruction of absorbing volumes, whereas the high frequency content provides the axial and lateral resolution. Another resolution limiting factor is the size of the detectors in PAT. Point-like detectors provide the best result in tomographic reconstruction. The combination of sensitive, broad-bandwidth detectors with small volume is almost impossible to achieve with piezoelectric technology. This has fostered the development of various optical sensors, which are optimised for PAT. Optical ultrasound sensors are based on modulations of a continuous light beam due to interaction with an ultrasound wave using effects like interferometry [10].

4.3 Reconstruction methods

Reconstruction in PAI aims at solving the inverse problem of sound generation and propagation. In PAT, an acoustical signal detected at a given time at the location of a (point-like) sensing element can have originated at any position on a sphere surrounding the point, with a radius equal to the time-of-flight multiplied by the speed of sound. Back projection reconstruction in PAT therefore uses a distribution of the information contained in the temporal signals onto spherical surfaces [11]. It is related to X-ray tomography, where the acquired transmission data are back projected onto straight lines by applying the inverse Radon transform. Non-uniformity of the speed of sound and acoustic attenuation cause the image quality to deteriorate. A refinement of the tomographic image formation uses iterative, model-based reconstruction. It is able to incorporate various shortcomings of the ultrasound detection and gives more accurate results than direct back projection at the expense of reconstruction time.

Quantitative reconstruction seeks the optical parameters, mainly the optical absorption, rather than the distribution of initial pressure. This requires solving the inverse optical problem by using a model for light propagation. Quantitative PAI is able to yield absolute concentrations of known chromophores, such as oxy- and deoxy-hemoglobin, after acquisition of spectrally resolved data.

4.4 Detection geometry

In PAT, the arrangement of detectors plays an important role. Similar to X-ray CT, information has to be sampled from a sufficiently dense coverage of the surface surrounding the sample by ultrasound detectors. Acoustic signals emanating from an absorber should at least be detected by sensors arranged over a 2π solid angle, otherwise the missing information leads to a loss in resolution. Furthermore, the back-projection reconstruction will produce streak artefacts, if the sensors are not densely enough distributed over the detection aperture.

5 Conclusions

5.1 Strength and limitations

Advantages:

- Scalability of imaging technology to image biological specimens of different sizes (nm–cm) ranging from organelles to small-animals and human organs.

- High resolution imaging beyond the depth of optical transport mean free path due to the hybrid nature providing optical contrast and ultrasound resolution.
- Can be used to address many preclinical research questions without the use of any exogenous contrast agents since a variety of endogenous chromophores (figure 2) already contained in biological samples provide sufficient absorption contrast.
- As in US imaging, depth information is contained in time-resolved US signals.
- Easily combinable with pure optical and ultrasound imaging modalities for hybrid multimodal imaging.
- Low cost compared to CT-, MRI-, PET or SEM/TEM imaging techniques.

Disadvantage:

- Necessity for acoustic coupling of the specimen to the US receiver. This may be problematic for samples where contamination is undesirable.
- Due to the limited penetration depth of non-ionising radiation into tissue and the large acoustic impedance differences between tissue and bones, cross-sectional whole-body human imaging is impossible.

5.2 Future developments

PA quantitative imaging methods will benefit from refinements of reconstruction approaches and the use of machine learning techniques. Implementations of large, dense ultrasound arrays for fast, three-dimensional PAT are envisaged, partly by using optical sensing techniques. Developments of multimodal techniques, combining PAI with pure optical and ultrasound imaging will continue, e.g. towards ultrasound transmission tomography or pure optical technology, using laser-generated ultrasound.

References and further reading

[1] Beard P 2011 *Interface Focus* **1** 602–31
[2] Wang L V and Hu S 2012 *Science* **335** 1458–62
[3] Deán-Ben X L and Razansky D 2019 *Phys. Med. Biol.* **64** *18TR01*
[4] Jacques S L 2013 *Phys. Med. Biol.* **58** R37–61
[5] Wang L V and Yao J 2016 *Nat. Methods* **13** 627–38
[6] Hu S and Wang L V 2013 *Biophys. J.* **105** 841–7
[7] Ning B, Sun N and Cao R *et al* 2015 *Sci. Rep.* **5** 18775
[8] Jeon S, Kim J, Lee D, Baik J W and Kim C 2019 *Photoacoustics* **15** 100141
[9] Deán-Ben X L, Fehm T F, Ford S J, Gottschalk S and Razansky D 2017 *Light: Sci. Appl.* **6** e16247
[10] Laufer J, Johnson P and Zhang E *et al* 2012 *J. Biomed. Opt.* **17** 56016
[11] Xu M and Wang L V 2005 *Phys. Rev. E Stat. Nonlin. Soft Matter Phys.* **71** 16706

Section 4

Optical imaging

Chapter II.4.a

Fluorescence

Felix Gremse and Stefanie Rosenhain

Fluorescence imaging is a noninvasive imaging technique that enables the visualisation of fluorescent proteins or dyes to acquire information about molecular processes and structures in cells, tissues, or animals. Fluorescence signals in specific regions or even in the whole murine body can be assessed longitudinally in two or three dimensions with high sensitivity. Thus, it plays an indispensable role in preclinical research offering versatile applications, for example, in cancer, neurology, surgery, and pharmacokinetic research. From the basic two-dimensional (2D) reflectance imaging to three-dimensional (3D) fluorescence tomography, new promising fluorescence techniques such as intravital microscopy and intraoperative imaging have been developed and are evolving quickly. However, fluorescence imaging was practically restricted to superficial structures for a very long time, because heterogeneous absorption and scattering coefficients of biological chromophores, such as hemoglobin, hamper a correct fluorescence signal reconstruction in deep regions. This obstacle has been overcome by either (i) combining fluorescence imaging with an additional anatomical imaging modality, e.g. computed tomography (CT) or magnetic resonance imaging (MRI), that promotes more accurate organ segmentations, (ii) getting closer to the tissue of interest by using endoscopes or optical chambers, or (iii) using higher excitation wavelengths in near infrared or short-wavelength infrared region.

1 Introduction

Fluorescence imaging is based on the interaction of light with tissue, generating a response which is measured by optical imaging devices. Both anatomical, functional, and molecular parameters can be assessed allowing various applications. Thus, fluorescence imaging is characterised as one of the most universal visualisation technique in basic research [1]. Fluorescence *per se* is the ability of certain molecules

doi:10.1088/978-0-7503-3747-2ch4

to absorb photons of a specific wavelength and emit photons of another, usually longer, wavelength. Optical filters block the incident light and allow capturing images of the emitted light, which often is orders of magnitude less intense than the excitation light. Since the discovery of the first fluorescent protein (green, GFP) in 1962, live cell fluorescent microscopy has been widely used in biological research, but macroscopic optical imaging modalities have emerged and evolved rapidly within the last decades [2]. This chapter will concentrate on the latter, mainly 2D fluorescence reflectance imaging (FRI) and 3D fluorescence-mediated tomography (FMT, also called fluorescence molecular tomography or fluorescence tomography). 3D FMT makes use of multiple excitation patterns, such as a grid of laser injection points, to reconstruct a 3D fluorescence distribution image. Additionally, more invasive procedures such as intraoperative imaging, endoscopic, and intravital imaging, where a transparent chamber is implanted into the animal, exist. With all these fluorescence techniques, multispectral imaging can be performed to assess multiple fluorochromes in one imaging session. The biomedical applications of fluorescence imaging in preclinical research are extensive since several highly specific molecular targets can be visualised and a broad spectrum of fluorescent proteins and small molecules exist.

2 Principles and setups

2.1 Physical principles

Fluorescence imaging involves the interaction of photons with fluorochromes, with the surrounding tissue but also with optical filters and finally a detector to form an image. A photon is characterised as an elementary particle or light quantum with a defined energy and photons with adequate energy can excite fluorophores. Photons are typically generated and applied by a monochromatic laser or a polychromatic lamp with an optical filter. Each photon carries a quantum amount of energy ($h\nu_{ex}$), inversely proportional to its wavelength (λ_{ex}). When hitting a fluorochrome, the photon transfers its energy to the fluorochrome (the photon is absorbed), inducing a temporary state of higher energy (called S_2, figure 1(a)). In many cases, the fluorochrome transfers this excess energy as heat to the environment, e.g. the surrounding tissue and returns to its ground state (S_0). In some cases, the fluorochrome converts only part of the energy into heat (internal conversion, IC), reaches a state of lower energy (S_1), emits another photon carrying the rest of the energy ($h\nu_{em}$), and returns to its ground state. The chance of absorption (and excitation) depends on the molar attenuation coefficient, which describes how strongly a substance attenuates light. The Beer–Lambert law describes how the attenuation of a light beam through a homogeneous material is determined by the molar attenuation, the substance concentration, and the path length of the light beam. Not all absorbed photons cause emission of a photon, and the convergence ratio is called quantum yield. Furthermore, the absorption of a substance is a function of the wavelength which is called the absorption spectrum of the substance. The emitted light (λ_{em}) is also called fluorescent light and has a spectrum. The difference in peak excitation and emission wavelengths (λ_{ex}-λ_{em}) is called Stokes shift, see figure 1(b). Photons of excitation and fluorescent

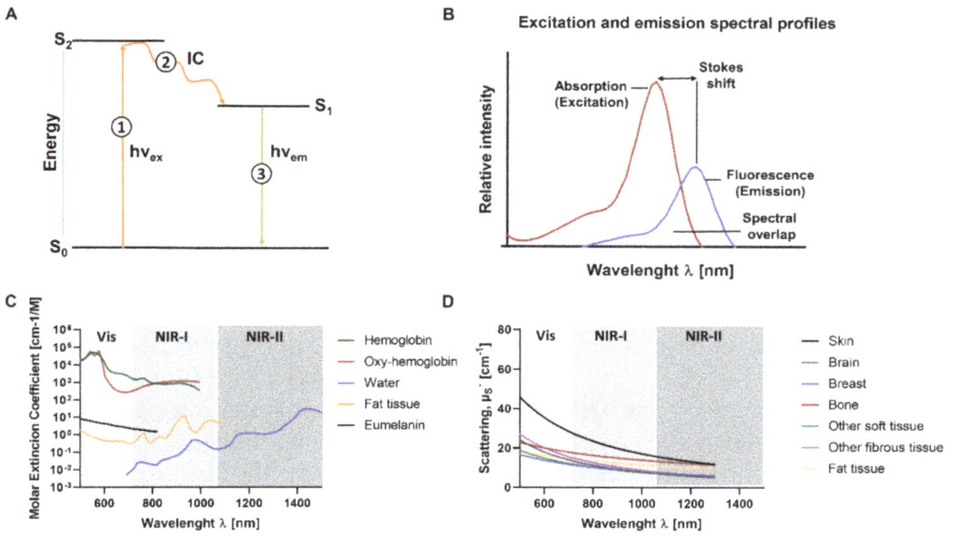

Figure 1. (a) Jablonski diagram showing the energy states during excitation of a fluorochrome. (b) Excitation and emission spectra of a typical fluorochrome. (c) Absorption coefficients of oxy- and hemoglobin, water, lipids, and melanin as function of wavelength. (d) Optical scattering decreases with higher wavelengths for different tissue types.

light travel diffusely through the tissue because photons are frequently scattered, roughly every 100 μm in biological tissues. Each scattering event slightly changes the direction of the photon leading to a random path through the tissue, until the photon is eventually absorbed or leaves the object again [3]. Instead of reasoning about individual photons, the statistical distribution of light in the tissue is considered. Photons of a thin laser beam, for example, cause a diffuse light distribution much larger than the beam diameter, which can be observed when illuminating a finger with a red laser pointer. A green laser pointer, however, hardly penetrates the finger due to the much higher absorption at that wavelength.

Furthermore, each photon can be absorbed, i.e. converted into heat, when encountering an absorbing molecule. Hemoglobin, (eu)melanin, and lipids are the main absorbers of visible and near-infrared light, respectively, but have their lowest absorption coefficient in a region of approximately 650–900 nm [4]. Hence, near infrared fluorescence (NIRF) imaging within the first optical window (NIR-I), see figure 1(c), is the preferred technique for *in vivo* fluorescence imaging. The second optical window (NIR-II), also known as short-wavelength infrared (SWIR), features lower scattering, potentially enabling sharper imaging in deeper tissues.

2.2 Typical setups

Fluorescence imaging requires the following infrastructure: an external light source, either monochromatic (laser) or polychromatic (filtered polychromatic light), for the excitation of the fluorophore. To detect the emitted light leaving the object, a camera is needed, hence a cooled CCD (charge-coupled device) detector is often used to achieve high sensitivity. Different optical bandpass filters located in front of the lens restrict detection to a selected optical spectrum, e.g. to block the excitation light. Semi-conductors in the detector convert the detected photons into electrons and an electric charge is formed. An output register and amplifier as well as a camera controller connected to a computer system generate the fluorescence image. A single illumination configuration does not provide sufficient information to reconstruct the 3D fluorescence distribution. Therefore, recording at multiple configurations where the anaesthetised mouse is illuminated at different scanning points, either by a moveable light source or mouse bed is necessary [3]. The typical setup for epi-illumination and transillumination imaging is illustrated in figure 2.

For preclinical *in vivo* fluorescence imaging, an appropriate anesthesia unit connected to the modality is needed. Several fluorescence imaging devices are commercially available. The PerkinElmer FMT performs fluorescence epi-illumination imaging and is also equipped with a movable laser below the animal for transillumination imaging, enabling reconstruction of the 3D fluorescence distribution. It is available with up to four lasers (635 nm, 675 nm, 745 nm, 790 nm). It is not meant for bioluminescence imaging. The IVIS Spectrum can also perform 2D and 3D fluorescence imaging but works with a filtered light source for transillumination which is weaker than the laser of the PerkinElmer FMT, potentially causing longer scanning durations. It is equipped with a highly sensitive CCD camera and is also suitable for bioluminescence imaging. If multispectral bioluminescence images are acquired, a 3D reconstruction can be performed. Additionally, the IVIS Spectrum can be upgraded with a CT to provide anatomical information for improved optical reconstruction and multimodal analysis. The MILabs VECTor^6 is a modular system for CT, optical, PET, and SPECT imaging. The optical module contains a

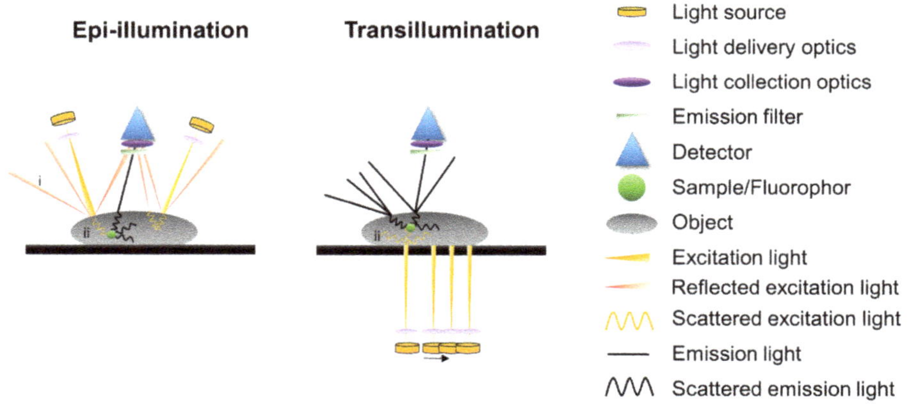

Figure 2. Schematic set-up to explain the principles of epi-illumination and transillumination imaging. When the excitation light encounters the probe, it is either (i) reflected (not shown in the transillumination scheme) or (ii) absorbed and scattered. The photons travel diffusely through the tissue, hit the fluorophore, and fluorescent light is emitted. Only a fraction of the photons of this light eventually reach the tissue surface from where it continues in a straight path in the air. Some photons then hit the light collection optics and are guided into a detector, e.g. a highly sensitive CCD camera, to form an image. An optical filter is mounted in front of the lens to filter out the excitation light.

sensitive CCD camera suitable for fluorescence and bioluminescence imaging. A laser is mounted below the animal for transillumination fluorescence imaging, enabling 3D fluorescence reconstruction. If available, the CT data is used for improved 3D fluorescence and bioluminescence reconstruction and multimodal analysis. The IR VIVO from Photon etc is the newest of these systems and is equipped with a CMOS (complementary metal–oxide–semiconductor) camera which is sensitive to higher wavelengths, i.e. in the NIR-2 range. It is not meant for bioluminescence imaging.

3 Biomedical relevance

3.1 Application range

Fluorescence imaging offers versatile applications in biomedical research, because of the wide variety of fluorescent proteins, reporter genes, fluorescence dyes, and probes, many of which are commercially available. Imaging of GFP is suitable for intravital and intraoperative imaging, because only a low tissue penetration is needed. However, GFP can hardly be used for *in vivo* whole body imaging due to the high absorption of visible light. Hence, the following applications focus on NIRF imaging, because it enables imaging of deeply located structures. Imaging of specific molecular biomarkers or using activatable probes, which are sensitive to cathepsins or matrix-metalloproteases, can detect and monitor tumours and metastases in early and late stages. In activatable probes, the fluorescence is quenched until the complex reaches the target tissue, where it is cleaved by proteases to release fluorescence signals only locally in the target tissue. Fluorescence quenching refers to processes that result in a decrease in the fluorescence intensity of a fluorophore without destroying the fluorophore. There are various effects that can lead to fluorescence quenching, for example: complex formation, internal conversion, or energy transfer to other molecules, so-called quenchers. In summary, activatable probes can assess tumour treatment response, recurrence, tumour angiogenesis, remodelling tumour microenvironment, cancer (stem) cell movement, tumour cell signaling, progression of cancer cell cycle, and enzymatic activity. Additionally, apoptosis in tumours or inflamed tissue can be visualised using annexin-probes or caspases [1, 5–7]. Biodistribution analysis using fluorescence labelling of newly synthesised nano-medicine speeds up drug development, because it offers an early screening of promising potentially active substances for toxicity, efficacy, and side effects. Fluorescence imaging is frequently applied in theranostic research to investigate different drug delivery systems and can be used for pathway investigation [8, 9].

3.2 Sample preparation

Fluorescence can be introduced into the tissue by (i) genetically encoded material such as fluorescent proteins (GFP and its derivatives), and (ii) exogenous material such as fluorescent dyes to synthesise an optical imaging probe or commercially available probes. In detail, (i) fluorophores can be inserted in cells by transfection with fluorescent protein or by fusion of cDNA to the gene encoding for a specific target protein. Even completely fluorescent organs or animals can be generated.

For (ii), the dye typically needs to be linked to or incorporated into a molecule such as a nanoparticle, micelle, liposome, antibody, etc. Furthermore, many probes including targeted and activatable ones are commercially available. Beside the potential cytotoxicity of some imaging agents such as cadmium-based quantum dots, there is a great variety of different biocompatible, non-toxic NIRF fluorochromes [3, 6]. Fluorescent dyes can have specific interaction with cells and labelling can alter the body biodistribution even for large molecules such as nanoparticles or antibodies. Moreover, the intensity of a fluorochrome can depend on the surrounding media or tissue type which should be considered during experiment planning and analysis. Organ-dependent calibration could be required, for example, for *ex vivo* fluorescence imaging of excised organs.

Prior to *in vivo* fluorescence imaging, it is recommended to reduce the autofluorescence especially in the gastrointestinal system by feeding chlorophyll-free chow. Since the hair and particularly dark fur will absorb a main part of the excitation and emission light, careful hair removal of the scanned area by shaving or using depilatory cream is necessary. Whereas the hair of white mice can be removed and fluorescence imaging on their clean skin can be easily performed, black mice often reveal remaining black patches on their skin, posing additional challenges for optical imaging. Since this removal might cause skin irritations, nude mice are more suitable for optical *in vivo* imaging studies. To ensure that the mouse will not move during imaging, it needs to be anaesthetised and anesthesia must be maintained during the whole scanning procedure. A multimodal mouse bed for the scanning procedure is useful to acquire fluorescence images and anatomical images in the same position of the mouse. Furthermore, slight squeesing of the mouse body in an appropriate mouse bed reduces the thickness of the mouse and may increase the accuracy of the result [10].

4 Parameters of image quality

4.1 Spatial and temporal resolution

Fluorescence imaging offers great spatial resolution of superficial structures but decreases with higher penetration depth. While a 2D epi-illumination whole body image may resolve superficial structures of ~100 μm, a 3D FMT reconstruction has a spatial resolution of ~2 mm also in deep regions. 2D reflectance imaging can be acquired at high temporal resolution (multiple images per second) but tomographic imaging requires minutes, because many images need to be acquired.

4.2 Tissue penetration depth

Due to the high scattering and absorption coefficients of biological tissue in the visible or near-infrared wavelength, fluorescence measurements are highly surface-weighted. Most photons are absorbed on their way from the surface to the fluorochrome or on the way back and this effect is stronger if the fluorochrome is in deep or optically dark regions. While the sensitivity can be improved by increasing the measurement time, there are practical limitations, i.e. the imaging duration should not be substantially longer than 20 min. Tomographic reconstruction techniques try to resolve this issue using a numerical model but are also limited by the amount of the detected light. Therefore, fluorescence detection is practically limited to a few centimeters tissue thickness which is sufficient for imaging small animals, particularly mice, but not for imaging of whole human bodies, at least with the currently available detector technology [7].

4.3 Bleed-through and crosstalk

The emission filter in front of the camera should be chosen to block as much of the excitation light as possible and as little of the emission light as possible. Typically, a high-pass or band-pass filter is used, whereby different filters are optimal for different fluorophores. Unfortunately, realistic filters cannot completely remove the excitation light and a tiny fraction may pass through the filter and appear as fluorescence light in the detected fluorescence images, called **bleed-through**, limiting the effective sensitivity of the system. Furthermore, many optical filters are angle-dependent in their performance and positioning the filter close to the animal can lead to more incident light that is not perpendicular, increasing the bleed-through. Fluorophores typically have a broad emission spectrum. Therefore, an emission filter selected for one fluorochrome will also capture an inevitable amount of another fluorochrome, called **crosstalk**, causing an incorrect contribution to the reconstructed fluorescence distribution. Special combinations of fluorophores, filters, and light sources can be chosen and a postprocessing technique called **spectral unmixing** can be used to reduce crosstalk.

5 Data processing and visualisation

Fluorescence imaging is highly specific to the fluorochromes, particularly at higher wavelengths where the autofluorescence is negligible. This means that another source of information is needed for orientation, i.e. to provide an anatomical reference. For 2D imaging, an image without optical filter is usually acquired and used as 'underlay' below the fluorescence 'overlay'. The underlay is shown in grey scales while the overlay is visualised using a colour table. For 3D imaging, an anatomical modality such as CT is often used as 3D underlay. Using 2D reflectance imaging, the resulting fluorescence imaging is highly surface-weighted, which means that fluorescent sources close to the surface, such as subcutaneous tumours, appear brighter in the resulting image than deep sources such as the liver. Therefore, it is practically not possible to derive quantitative numbers describing organ uptake *in vivo*. Quantification could be performed *ex vivo* for excised and homogenised organs. In that case, the depth is not discriminating the results, but the different tissue composition should be taken into consideration. However, this issue of fluorescence signal quantification can be overcome partially with 3D imaging when using transillumination acquisition and tomographic reconstruction. Furthermore, optically dark organs (those with high relative blood volume) such as liver and spleen result in lower intensities which can be improved using an absorption map for 3D reconstruction. During image acquisition, the exposure duration and illumination strength are adjusted, e.g. increased in case of low

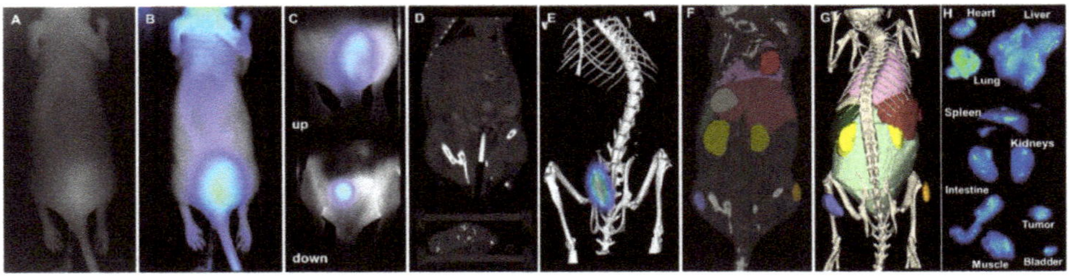

Figure 3. (a) 2D image without optical filter ('underlay'). (b) Fluorescence image ('overlay'). (c) For 3D fluorescence imaging, many transillumination images can be acquired, from different angles (only two are shown). (d) CT scan provides anatomical information. (e) CT and fluorescence distribution can be visualised in 3D, here showing fluorescence in a rectal insertion. (f, g) Organ segmentation based on CT data. (h) Excised organs with fluorescent uptake (2D fluorescence reflection imaging).

amounts of fluorescence to achieve sufficient photon counts for a sharp image. This needs to be considered during analysis of 2D images, i.e. by normalising the actual intensities by the product of the exposure time and illumination strength.

For quantitative analysis of 2D or 3D images, regions of interest are defined, e.g. by delineating the tumour or the liver (figure 3). This is performed using special software which then assigns the fluorescence to the regions. In the case of multi-modal 3D imaging, anatomical CT data can be used to define organs using manual delineation or even automated segmentation algorithms.

6 Conclusion

6.1 Strengths and limitations

In vivo fluorescence imaging offers several advantages in terms of spatiotemporal resolution, specificity, and its application possibilities. Beside a broad range of available fluorescence dyes and probes, reporter genes enable the generation of fluorescent cells or animals. Owing to of the high sensitivity of fluorescence imaging, even small amounts under 1 pmol of fluorescence dyes can be detected. The absence of radiation damage due to the usage of low energy photons enables **longitudinal imaging**, e.g. to investigate tumour growth and disease progression over time. The fast acquisition time of 2D fluorescence imaging allows high throughput screening of animals and even multiple mice can be scanned together. Another advantage of fluorescence imaging is the long time period in which the fluorescence signals can be detected in the animals. While radiotracers are often limited by a short half-life and magnet resonance contrast agents are often toxic in long time exposure, fluorophores can be observed for several weeks. Furthermore, multiple probes can be imaged simultaneously in the same animal as long as their wavelengths can be differentiated spectrally, which is called multispectral imaging [6]. The equipment for fluorescence imaging modalities is normally easy and inexpensive to maintain and operate. Fluorescence tomography resolves some problems of 2D imaging such as surface-weighted images and enables potentially quantitative analysis in 3D, e.g. to determine biodistribution and elimination of fluorescent probes, but should be combined with an anatomical modality such as CT or MRI for accurate reconstruction and analysis. Fluorescence imaging has high sensitivity and specificity for superficial tissue regions, which enables clinical translation for imaging of limbs, with endoscopes, or in image-guided surgery.

6.2 Future developments

Even though fluorescence imaging has evolved fast within the last years, its potential has not yet been fully exploited. Trends are moving to improve optical imaging probes, fluorescent proteins and fluorescent dyes, as well as detection technologies, and tomographic reconstruction of fluorescence signals. The combined advantages of multi-modal imaging suggest the development of dual-functionalised probes and hybrid devices such as FMT-PET or FMT-MRI. Multispectral imaging requires research and selection of suitable combinations of probes, lasers, and filters as well as spectral unmixing techniques. Tomographic imaging supports new applications of fluorescence imaging in deep tissue regions, for example the detection and therapy monitoring of inflammatory bowel diseases. Short-wave infrared (SWIR) imaging uses wavelengths in the range of 1000–2000 nm, whereby photon scattering is reduced in comparison to the NIR-I window, enabling sharper imaging of deeper tissue regions. This technique gained a lot of attention in the last years and its future applications in preclinical research still need to be examined.

References and further reading

[1] Shah K and Weissleder R 2005 Molecular optical imaging: applications leading to the development of present day therapeutics *NeuroRX* **2** 215–25

[2] Ntziachristos V 2010 Going deeper than microscopy: the optical imaging frontier in biology *Nat. Methods* **7** 603–14

[3] James M L and Gambhir S S 2012 A molecular imaging primer: modalities, imaging agents, and applications *Physiol. Rev.* **92** 897–965

[4] Karasev M M, Stepanenko O V and Rumyantsev K A *et al* 2019 Near-infrared fluorescent proteins and their applications *Biochem. (Moscow)* **8** 32–50

[5] Imamura T, Saitou T and Kawakami R 2018 *In vivo* optical imaging of cancer cell function and tumor microenvironment *Cancer Sci.* **109** 912–8

[6] Weissleder R and Pittet M J 2008 Imaging in the era of molecular oncology *Nature* **452** 580–9

[7] Li Q, Zeng J, Miao Q and Gao M 2019 Self-illuminating agents for deep-tissue optical imaging *Front. Bioeng. Biotechnol.* **7** 326

[8] Handgraaf H J M, Boonstra M C and Prevoo H A J M *et al* 2017 Real-time near-infrared fluorescence imaging using cRGD-ZW800-1 for intraoperative visualization of multiple cancer types *Oncotarget* **8** 21054–66

[9] Etrych T, Janoušková O and Chytil P 2019 Fluorescence imaging as a tool in preclinical evaluation of polymer-based nano-DDS systems intended for cancer treatment *Pharmaceutics* **11** 471

[10] Gremse F, Doleschel D, Zafarnia S, Babler A, Jahnen-Dechent W, Lammers T, Lederle W and Kiessling F 2015 Hybrid μCT-FMT imaging and image analysis *J. Vis. Exp.* **2015** e52770

IOP Publishing

Imaging Modalities for Biological and Preclinical Research: A Compendium, Volume 2

Parts II–IV: *In vivo* preclinical imaging, multimodality imaging and outlook

Andreas Walter, Julia G Mannheim and Carmel J Caruana

Chapter II.4.b

Bioluminescence

Laura Mezzanotte and Natasa Gaspar

Bioluminescence (BL) imaging relies on the detection of photons emitted by luciferin when reacting with a luciferase enzyme. The detection of photons is achieved using a CCD camera usually mounted on top of a light-tight box that serves as the imaging chamber. Animals are placed in the imaging chamber under anesthesia and acquisition of signals takes seconds or minutes. The technique has become routinely used in preclinical research to follow the fate of injected cells or to monitor gene expression. In fact, luciferases act as reporter genes and can be encoded and expressed in many organisms such as bacteria, parasites, plants, mammalian cells and whole animals. Although spatial resolution is lower than in other imaging techniques, BL imaging offers the advantage of high sensitivity and versatility, is easy to use and fast at comparatively low cost. Recent developments in the field greatly expanded the available tools for 2D and 3D BL imaging, promising a new era of more sensitive and deeper molecular imaging.

1 Introduction

Bioluminescence, is a kind of chemiluminescence occurring in a number of living organisms. It is based on the enzymatic reaction of a luciferase with a substrate called luciferin in the presence of O_2 and other cofactors, such as ATP and a magnesium ion. The terms 'luciferin' and 'luciferase' were first used by Raphael Dubois (1849–1929) who discovered that mixing two aqueous extracts from the light organs (one of them treated with hot water) of a luminous Pyrophorus generated luminescence. The reaction yields an excited-state oxidation product that decays to produce a photon of emitted light that can be sensitively detected and accurately measured. There are many luciferase/luciferin systems derived from terrestrial organisms like firefly (*Photinus pyralis*), green or red click beetle (*Pyrophorus plagiophthalamus*) and other luminous insects. Other systems are derived from

doi:10.1088/978-0-7503-3747-2ch5

marine organisms like the coelenterate sea pansy (*Renilla reniformis*), the crustacean *Gaussia princeps*, and the ostracod Cypridina species. Luciferase enzymes from terrestrial organisms generally use D-luciferin as substrate and luciferase from marine organism coelenterazine. The bacterial luciferases use a different reaction to produce light which involves fatty acid and reduced flavin mononucleotide ($FMNH_2$) as essential factors for the BL reaction. These BL systems have been used for non-invasive *in vivo* imaging of biological processes due to the possibility of genetic engineering of cells to express luciferase as reporter gene and the ability to detect light by instrumentation equipped with CCD cameras. The firefly luciferase (Fluc)/D-luciferin pair is the most exploited BL reporter system due to its relatively high quantum yield (0.44) and red-shifted light emission that penetrates better through tissue (530–640 nm). Compared to firefly luciferases, marine and bacterial luciferases are less commonly used for *in vivo* studies because the light emitted from these systems falls in the blue/green region (430–490 nm) that is known for its poor tissue penetration. To date many different BL systems have been developed by discovery and cloning of novel luciferase genes, creation of mutants or synthetic enzymes. Moreover, intense research led to synthesis of luciferin analogues able to enhance brightness of BL systems and alter the spectral emission. Some of them are reported in table 1.

Table 1. List of common luciferase reporter genes that are currently used for *in vivo* imaging applications.

Luciferase name	Origin and modifications	Emission peak *in vivo*
Firefly luciferase (LUC2)	*Photynus pyralis*	610 nM with D-luciferin
Firefly luciferase red (PPyRE8; PpyRE9)	*Photynus pyralis mutants*	618 nM with D-luciferin
Firefly luciferase green (PPyGR9)	*Photynus pyralis mutant*	547 nM with D-luciferin
Click beetle luciferase red (CBred)	*Pyrophorus plagiophthalamus mutant*	620 nM with D-luciferin
Click beetle luciferase red 2 (CBred2)	*Pyrophorus plagiophthalamus mutant*	720 nM with Naphthyl luciferin
Click beetle luciferase green (CBG99)	*Pyrophorus plagiophthalamus mutant*	542 nM with D-luciferin
Gaussia luciferase (Gluc)	*Gaussia princeps*	480 nM with Coelenterazine
Renilla luciferase	*Renilla reniformis*	480 nM with Coelenterazine
NanoLuc luciferase	*Ophlophorus gracillocistris mutant*	460 nM with Furimazine
Bacterial luciferase	*Vibrio fisheri*	480 nM with $FMNH_2$
Akaluc luciferase	*Synthetic luciferase derived from Photinus pyralis*	650 nM with Akalumine-HCl

2 Principles and setups

2.1 Chemical and physical principles

The catalytic reaction between D-luciferin and Fluc begins with the adenylation of D-luciferin (D-LH_2) to yield D-LH_2-AMP, which is then oxygenated. The luciferin dioxetanone anion (Dx^-) is the light emitting structure. It consists of a four-membered strained ring and it is formed upon AMP release. Subsequently, a light-emitting intermediate, the excited singlet state of oxyluciferin, $^1(OL^-)^*$, is generated, which relaxes to the ground state and produces luminescent light with a peak intensity of around 560 nM upon (OLH; figure 1).

The BL of coelenterazines, instead is attributed to their imidazopyrazinone structure. To initiate a BL reaction, an oxygen binds to position 2 of the imidazopyrazinone of coelenterazine to form a peroxide. This leads to the formation of the dioxetanone ring, which then decomposes to produce CO_2 and the amide anion of coelenteramide in its excited state. The return to ground state emits blue light (450–470 nm). Coelenterazine spontaneously emits a low level of chemiluminescence in aqueous buffer in absence of luciferase, which generates a background emission for bioimaging applications. Moreover, coelenterazines are poorly soluble in neutral aqueous buffers and unstable in air, though they are soluble in methanol and stable after the addition of trace HCL. In bacterial cells, a long-chain aldehyde is produced by the reduction of a corresponding long-chain acid through a recycling process. $FMNH_2$ is also produced by the NADH-mediated reduction of FMN, catalyzed by FMN reductase also through a recycling process. To use the bacterial BL system for light generation in other organisms a cassette encoding all the genes required for the expression of enzymes involved in the generation and recycling of luciferin and luciferase has been developed. The emission peaks of various luminescent bacterial species range from 472 nM to over 500 nM. This autologous BL system has been used mostly to image bacteria in small animals.

The light generated by different luciferase/luciferin pairs is able to travel diffusely through the tissue. Photons are scattered in biological tissues approximately every 100 μm and are absorbed mainly by hemoglobin and melanin. The lowest absorption occurs in the region of 650–900 nM which is the optimal optical window for *in vivo* imaging. Firefly, Click beetle red and Akaluc luciferases emit ample amount of photons in this region.

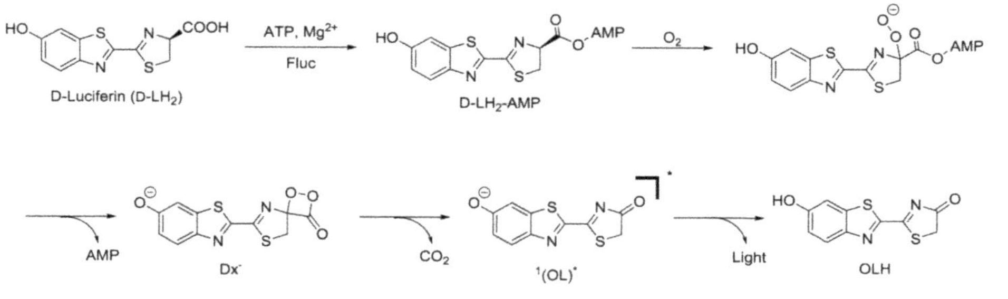

Figure 1. Oxidation of D-luciferin by firefly luciferase (Fluc).

2.2 Typical setups

In order to detect the low light generated from the BL reactions occurring *in vivo*, sensitive charge couple device (CCD) cameras are needed. A CCD camera is able to convert the photons that hit the image sensor (CCD chip) into electrons. A CCD chip consists of an array of photosites (squares). The electron charges for every photosite are assigned a pixel value and then displayed to generate the raw images. The CCD cameras for *in vivo* imaging are cooled to –90 °C to –100 °C using a thermoelectric cooling module (also called Peltier) in order to reduce noise and increase sensitivity of detection.

The camera is usually mounted in a light-tight box where external illumination cannot penetrate. In many instruments, optical filters are positioned in front of the camera which can be band pass (generally 20–40 nM) or cut off to allow multi-spectral acquisitions. For animal imaging the imager is coupled to an anesthesia module. In general, the animal is anesthetised after the injection of the BL substrate and then placed inside the light-tight box under anesthesia. Acquisition time ranges from seconds to minutes depending on the intensity of the light emission. However, certain types of instruments permit real time imaging in freely moving animals: these instruments are equipped with ultrasensitive iCCD or EMCCD cameras which detect high numbers of photons within millisecond acquisitions (figure 2).

3 Biomedical relevance

3.1 Application range

BL imaging is nowadays a standard technique for cell tracking in small animals and for studying gene expression.

Since luciferase genes can be engineered in the genome of many organisms from virus to bacteria, fungi, mammalian or plant cells and transgenic animals, BL has been applied to all fields of life sciences. The first applications 20 years ago of Firefly and Renilla luciferase were in the field of oncology and immunology. Researchers were able to visualise tumour growth or T cell trafficking in nude mice by injection of cells expressing the reporter gene. At the same time a bacterial luciferase system

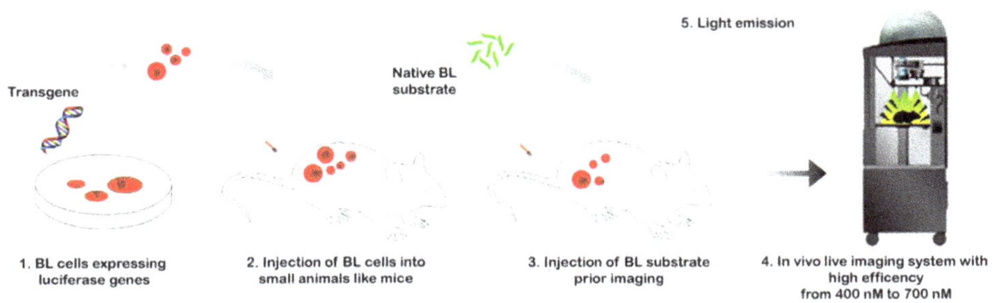

Figure 2. Workflow of a BLI experiment. After establishment of genetically engineered cells for the expression of luciferase enzyme (1), a BL substrate is administered to animals that received an injection of bioluminescent cells (2). The animal is placed under anesthesia in the imaging system and images acquired.

was applied to track bacterial infections. Nowadays BL imaging is commonly used for preclinical research in oncology, neuroscience, microbiology, immunity and developmental biology.

3.2 Sample preparation

As a gene reporter, luciferase needs to be introduced in the organism of interest for imaging purposes. Expression of the luciferase can be transient or stable. Generally, the latter is chosen for longitudinal experiments where repeated imaging session are performed at different times. For example, stable expression in mammalian cells is achieved by transducing them with lentiviruses which possess the luciferase sequence in their genome. Lentiviruses are able to integrate into the genome of the host. On the other hand, the luciferase gene can be cloned into species specific plasmids, that are used to transform bacteria and parasites. To study gene expression, the luciferase gene is cloned under the control of specific regulatory elements or fused to specific endogenous genes. Such constructs are employed to generate transgenic animals by transfection of oocytes. Another option for the generation of luciferase expressing animals is to employ recently developed knock in strategies using CRISPR/CAS systems.

4 Parameters of image quality

4.1 Spatial and temporal resolution

BL imaging offers a spatial resolution of millimeters in superficial tissue (0–1 mm) which decreases for light sources located in greater depths. 2D images can be acquired at a temporal resolution ranging from 1 s to 5 min depending on the level of needed sensitivity. For tomographic imaging longer acquisition times are needed because many images need to be acquired and enough photons should be detected using band pass filters to allow 3D reconstruction of the signal.

4.2 Tissue penetration depth

Scattering and absorption of light through biological tissue in the visible or near-infrared wavelength limits the applicability of BL imaging in the range of 1–10 millimeters. Both sensitivity and spatial resolution are affected by depth. The sensitivity can be improved by increasing the acquisition time, or by employing a brighter BL system which generates a substantial number of near infrared photons. Tomographic reconstruction techniques may improve resolution. Therefore, BL emission is sufficient for imaging small animals, particularly mice, but lacks translational potential to clinical applications also due to the use of GMOs the potential toxicity of some BL substrates in human cells (e.g. coelenterazine).

4.3 Background signal

BL has been described many times as glow in the dark because the light is generated from a chemical reaction and it is not dependent on another source of light (e.g. the excitation light needed for generation of the fluorescent signal). For example, in the case of firefly luciferase/D-luciferin the signal-to-noise ratio is of two orders of magnitude and the background emission of animals is negligible. However, bioluminescent substrates may be unstable and luminescent even in the absence of the enzyme. This is the case of coelenterazine which undergoes oxidation and production of background emission in live animals.

5 Data processing and visualisation

A 2D bioluminescent image is usually overlaid on the photographic image of the animals to identify the source of emission. In standard experiments 2D images are acquired with open filter. Regions of Interest (ROIs) are defined in the acquired image and quantified using specific software.

Since some instruments allow multispectral acquisitions, it is possible to perform dual colour BL imaging by applying a spectral unmixing algorithm to the series of images obtained using multiple filters. The algorithm decomposes a mixed pixel into a collection of constituent spectra and calculates the contribution of each spectrum to the signal in that pixel.

As for fluorescence, BL imaging is highly surface weighted so the light signals coming from a source in superficial tissue appear brighter than the ones that would come from the same source located in deep tissue. A method to quantitatively measure light coming from locations at different depths in the animal is 3D BL tomography. Tomography can partially reconstruct the location of the light source and the amount of light coming from different depths since the coefficient of absorption of different tissues is known. In the case of tomography, subsequent anatomical CT data can be used to locate organs mostly using automated segmentation algorithms.

6 Conclusion

6.1 Strengths and limitations

BL imaging offers many advantages that makes it one of the most widely used and versatile imaging modalities in preclinical research. These are: ease of use, high sensitivity (nowadays few numbers of neurons can be visualised in marmoset brain, for example), relatively low cost. The main limitation of the technique remains the low spatial resolution compared to techniques like MRI and the lack of translation to clinical settings.

6.2 Future developments

Progress in the development of novel bioluminescent systems with different colours and brighter emission, engineering of new luciferase mutants and modification of luciferins to generate analogues or caged substrates are further expanding the range of applications of BL.

References and further reading

[1] Shimomura O 2019 *Bioluminescence: Chemical Principles and Methods* 3rd edn (Singapore: World Scientific Publishing)

[2] Contag C H and Bachmann M H 2002 *Annu. Rev. Biomed. Eng.* **4** 235–60

[3] Yao Z, Zhang B S and Prescher J A 2018 *Curr. Opin. Chem. Biol.* **45** 148–56

[4] Mezzanotte L, van 't Root M and Karatas H *et al* 2017 *Trends Biotechnol.* **35** 640–65

[5] Darne C, Lu Y and Sevick-Muraca E M 2014 *Phys. Med. Biol.* **6** 59 R1–64

[6] Hall M P, Woodroofe C C and Wood M G *et al* 2018 *Nat. Commun.* **9** 132

[7] Stowe C L, Burley T A and Allan H *et al* 2018 *Elife* **15** 8

[8] Iwano S, Sugiyama M and Hama H *et al* 2018 *Science* **359** 935–9

[9] Kotlobay A A, Sarkisyan K S and Mokrushina Y A *et al* 2018 *Proc. Natl. Acad. Sci. USA* **115** 12728–32

[10] Yeh H W, Karmach O and Ji A *et al* 2017 *Nat. Methods* **14** 971–4

IOP Publishing

Imaging Modalities for Biological and Preclinical Research: A Compendium, Volume 2

Parts II–IV: *In vivo* preclinical imaging, multimodality imaging and outlook

Andreas Walter, Julia G Mannheim and Carmel J Caruana

Chapter II.4.c

Cerenkov luminescence imaging

Magdalena Skubal, Benedict McLarney and Jan Grimm

Cerenkov luminescence imaging (CLI) is a powerful alternative imaging technique to positron emission tomography (PET) and single photon emission computed tomography (SPECT). Cerenkov luminescence (CL) is produced by radioisotopes emitting charged particles that can travel through medium faster than the speed of light in that medium. These suprarelativistic particles emit light in a continuous spectrum that ranges from ultraviolet (UV) to visible, predominantly blue (400–500 nm) wavelengths of light. In this chapter, we will firstly outline the discovery of CL and its physical principles, followed by a discussion of the current state of the art along with best practices. The current range of preclinical and clinical implementations and developments will be highlighted, including image-guided tumour resection, dosimetry monitoring and agent development. Future developments in this field, such as approaches to shift the blue weighted emission of CL into less scattered and absorbed red light will be discussed. It is likely that CLI will be established as a common preclinical molecular imaging modality. It is also possible that CLI will have some impact in clinical imaging with reduced cost, similar sensitivity, but added quantitative features compared to other optical imaging (OI) entities.

1 Introduction to Cerenkov luminescence

Cerenkov luminescence (CL) was characterised in the first third of the twentieth century by Russian physicist Pavel Cerenkov whilst CL imaging (CLI) has only emerged in biomedical imaging as recently as 2009 [1]. CL is produced when a charged particle, such as a positively charged positron or negatively charged electron from e.g. radionuclides such as ^{18}F-FDG, travels through a dielectric medium (e.g. water or tissue) with a velocity faster than the velocity of light in that medium. The suprarelativistic particle transfers its kinetic energy to the bipolar molecules in the surrounding medium, which in biological tissues is predominantly water. As the particle travels it causes the electrons of the atoms in the surrounding field to be polarised. They remain in this polarised state until the particle has passed, at which time the electrons of the surrounding atoms return to their ground state and discharge the transferred energy in the form of blue light, i.e. CL (figure 1(a)). When the medium relaxes the emitted optical photons travel in the direction of movement of the charged particle (figure 1(b)) [2].

In a preclinical setting, the implementation of CLI has led to an attractive alternative to expensive positron emission tomography (PET) scans. CLI has significant advantages over conventional PET scanning including shorter imaging times, comparable sensitivity and superior resolution, albeit surface weighted. CLI is currently being adapted for clinical imaging where it could be used to ensure cost efficient tumour imaging and staging, complete tumour resection using commonly administered radionuclides and for dosimetry monitoring in therapy. The development of highly sensitive charge coupled device (CCD) and electron multiplying CCD (EMCCD) cameras has propelled the field. Newer methods use the emitted CL to excite fluorescent proteins (FPs) or an array of quantum dots (QDs) and nanoparticles (NPs) that emit light at longer (red-shifted) wavelengths. In comparison to the blue CL, absorption and scattering is reduced for these longer wavelengths and therefore can aid in signal detection from deeper seated tissue.

2 Principles and setups

2.1 Physical principles

As outlined in the introduction, when molecules in a dielectric medium are polarised by a passing charged particle, they reach a higher energetic state. The departure of the particle causes the atoms to return to their ground state and the acquired energy is released and emitted as luminescence. The energy state achieved by the molecules is directly related to the velocity of the charged particle. There are certain physical

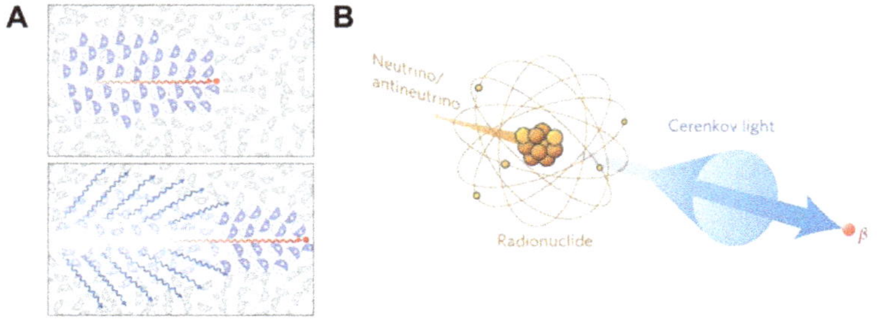

Figure 1. Principles of Cerenkov luminescence. (a) Cerenkov luminescence is emitted by a charged particle (red dot) traveling through a dielectric medium faster than the speed of light in that medium. In doing so the particle polarises the medium (top image). Relaxation of the molecules in the medium produces visible blue weighted light (Cerenkov luminescence—blue wavy lines) emitted in a forward direction (bottom image). (b) Cerenkov luminescence is produced by the surrounding medium in which a charged particle is traveling. Radionuclides that are emitting β particles with energies greater than the Cerenkov threshold (261 keV (0.261 MeV) in water) result in the generation of Cerenkov luminescence. Reprinted with permission from [5] copyright 2017, Springer Nature.

conditions and factors required in order for CL to be achieved. Fortunately, the energy threshold required to reach this state is readily met by the majority of beta (β) (electron or positron)-emitters that are used in nuclear medicine. In turn, these isotopes and external beam radiation therapy (performed using a linear accelerator—LINAC) can easily produce CL in tissues. The velocity of a β particle can be calculated with equation (1) below, where c is the speed of light, E is the energy of particle and E_0 is the mass of the β particle at rest [3]. In addition, many medically used α-particles from radioactive decay and those generated during external beam radiation do produce CL emission via their decay daughters whilst most secondary electrons (emitted via the photoelectric effect from gamma-emitting isotopes) do not [5].

$$v = c\left\{1 - \frac{E_0^2}{(E + E_0)^2}\right\}^{1/2} \tag{1}$$

The refractive index (RI) of the medium in which the emitter is located is another determining factor for the efficiency of CL emission. The higher the RI, the easier it is for CL to be generated [4]. Therefore, the emitted photon intensity of CL for CLI is related to both the refractive index of the tissue and the particle energy. Accordingly, differing tissues produce varied levels of CL and the dependence upon the RI to produce CL reduces as particle energy increases. The Frank–Tamm relationship proved that by increasing the particle energy, more CL is produced. The Frank–Tamm equation (2) allows the precise quantification of the number of photons produced over a distance (dN/dx), where α is the fine structure constant (1/137), β is the velocity of the particle relative to the velocity of light in a vacuum, n is the RI of the medium, and λ is the wavelength of interest [4].

$$\frac{dN}{dx} = 2\pi\alpha\left(1 - \frac{1}{\beta^2 n^2}\right)\int_{\lambda_1}^{\lambda_2} \frac{1}{\lambda_2} d\lambda \tag{2}$$

2.2 Typical setups and state-of-the-art

Since 2009, CLI has undergone a rapid development and implementation, especially in the preclinical field. Numerous in-house systems currently exist to carry out CLI along with commercial systems like the IVIS® Spectrum imaging system (Perkin Elmer). The most common systems are based on the implementation of a highly sensitive CCD camera, often an EMCCD, equipped to appropriate optics. Considering the high sensitivity of these cameras and that CL is a weak source of light, CLI of radioisotopes is commonly performed within a light excluding enclosure. This prevents the detection of photons from non-CL sources, which can affect accuracy and produce false positive signal detection. Custom made and commercial CCD cameras have begun to enable the translation of CLI into the clinical applications, e.g. the LightPath CLI system (Lightpoint Medical). The LightPath CLI system has been used for the CLI of the back-table perioperative assessment of tumour resection margins in cancer surgery [6]. The optics are tuned to the size of the region of interest (ROI) to ensure maximum coverage and sensitivity, and can readily deliver 2D (planar) CL images [3, 6].

Tomography is a commonplace method in medical imaging and has been adapted to CLI. CL tomography (CLT) improves the spatial localisation of highly scattered and weak CL. Two methods are commonly applied for CLT, the first is to capture CL images from multiple views by rotating the object or detection system in line with similar tomographic approaches such as X-ray CT. Finite element models in combination with *a priori* knowledge have been shown to provide 3D models of radiotracer distribution *in vivo* [7]. The second method takes advantage of varying photon scatter and absorption in tissues across wavelengths. Multispectral and accurate optical models outlining the difference between tissue and tumours has enabled CLT imaging to localise radiotracers from planar 2D images. This method can be achieved without rotation and uses filters to select specific wavelength bands of CL to determine the depth and original source of light as detected at the sample surface [8]. Direct correlation has been shown between PET and SPECT with CLT [1, 9]. The predominant mode of reconstruction in these methods is back projection. However, additional factors and newer reconstruction algorithms have improved the resolution and quality of reconstructed images [9].

More advanced setups have shown that endoscopy can be one of the most applicable methods to CLI. The inherent lack of light within the body reduces the detection of non-CL photons. These systems combine the anatomical information of endoscopy with the functional information of CLI, such as organ or tumour uptake and metabolism. This is achieved by coupling a CCD camera to optical fibres [2, 10]. Studies using CL endoscopy (CLE) have shown how CLI can be used to increase the resolution and information from SPECT and PET [10]. However, the advantages of CLE are to be balanced against the inherent loss of signals seen with endoscopes due to their low aperture number. Additionally, special consideration should be made to reduce potential scintillation effects (bright spots during imaging from radionuclides interacting with the fibre) seen in CLE.

One area where CLI has successfully been applied clinically is the monitoring of dosimetry, where the dose is delivered using a clinical LINAC. This radiation produces far more CL than seen with radionuclides (dosimetry commonly uses energies from 4 to 25 MeV whilst radionuclides range from 0.4 to 2.4 MeV). When imaging LINAC produced CL, the acquisition of the camera is synced with the pulses of the LINAC. This aids in mitigating against the detection of non-CL photons whilst also enabling imaging under ambient light conditions. Serial acquisition of images along with rotation of the detector along an axis has achieved highly resolved 3D models of dose delivery using back projection reconstruction [11]. Imaging in this way with suitable agents can be used to determine the oxygen saturation of the tissue as it undergoes dosimetry [12].

As is common with many OI methods in tissues, the absorption and scattering of light poses a large obstacle, particularly in the visible regime of light (400–700 nm). CL is predominantly blue weighted and hence the ability to resolve and precisely determine radionuclide location is greatly reduced [13]. However, a vast array of fluorochromes absorb blue light and emit at longer wavelengths. Previous attempts have applied this to excite fluorescein [14] while the latest attempts have used a cascade of fluorescent dyes to transfer blue light to the NIR spectrum (700 nm and above) where the absorption and scattering of light by tissues is greatly reduced.

3 Biomedical relevance

CLI is undergoing rapid development on both preclinical and clinical fronts. In this section these developments will be outlined based on their main application.

3.1 Preclinical application range and relevance

Radioactive decay is the process by which high energy α, β, and γ radionuclides achieve similar energy levels with their surrounding medium. The characteristics of this radioactive decay of a radionuclide determines the applicable imaging method (e.g. if a radiotracer will produce CL and be suitable for CLI) and the degree of cellular damage due to the radiation dose (figure 2(a)) [5]. This has implications for both pre- and clinical imaging environments. CLI has been accomplished preclinically (cells to model organisms) with a range of β emitter radionuclides including the β^+ emitters: ^{64}Cu, ^{18}F, ^{89}Zr, ^{86}Y, ^{68}Ga, ^{72}As and β^- emitters: ^{177}Lu, ^{32}P ^{90}Y, ^{131}I (figure 2(b–d)) [4, 15]. Here, we will focus on β-emitting particles from the most frequently used medical tracers and further discuss selected applications of CLI in preclinical research.

The first CLI preclinical study (in 2009) centered around the detection of CL emitted by the frequently used PET imaging agent, ^{18}F-FDG [1]. ^{18}F-FDG is a marker of cellular glucose consumption in metabolically active cells with increased uptake in tumours versus healthy tissue due to their high metabolic rate and glucose intake (Warburg effect). Accordingly, ^{18}F-FDG was used to localise malignant prostate lesions with CLI [1]. Similar studies confirmed ^{18}F-FDG activity in mice bearing C6-FLuc glioma tumours. CLI and PET images were acquired sequentially using the commercial IVIS® Spectrum system and a microPET. ^{18}F-FDG was shown to be retained in C6-FLuc tumours with detected CLI signals correlating with microPET images (figure 2(e–f)) [16]. Having served as a proof of concept, the potential of CLI was then explored with a wider range of radionuclides [15, 16]. ^{89}Zirconium (^{89}Zr)-radiolabelled monoclonal antibodies (^{89}Zr-desferrioxamine B-DFO-J591) against the prostate-specific membrane antigen (PSMA) were used to show the first example of *in vivo* dual optical imaging of CLI and immune-PET of prostate tumours [15]. ^{89}Zr-DFO-trastuzumab targeted to HER2/neu (human epidermal growth factor receptor 2) positive tumour cells was used to show the intraoperative application of CLI in breast cancer surgery. The distribution of the probe was evaluated by CLI and PET, enabling the differentiation between HER2/neu positive (BT-474) and HER2/neu negative (MDA-MB-468) tumours. HER2/neu positive tumours were resected using CLI, and CLI was used to confirm complete removal of the targeted cells after surgery [9]. Further applications have used ^{90}Y labelled peptides, RGD (an Arginine-Glycine-Aspartic acid motif targeting $\alpha v\beta 3$ integrin) coupled with bombesin (targeting the gastrin-releasing peptide

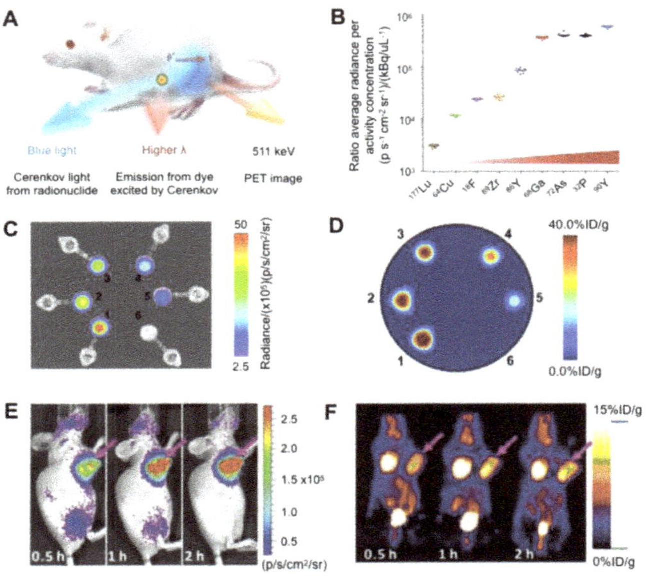

Figure 2. Preclinical phantom and animal studies using CLI. (**a**) Methods for the detection of CLI: blue CL natively generated, the red-shifted fluorescence achieved from a dye excited by blue (CL) light and the most commonly used PET signal. Reprinted with permission from [5], copyright 2018, Springer Nature. (**b**) The utility of different radionuclides for use in CLI was assessed by the relative light output from each of the available radionuclides. This is a plot of the ratio of the average radiance of select nuclides to the activity concentration ([p/s/cm^2/sr]/[kBq/μL]) of the radionuclide. Modified and reprinted with permission from [4], copyright 2018, Springer Nature. Phantom images of ^{89}Zr activity in water using (**c**) CLI and (**d**) PET. Eppendorf tubes (1–6) correspond to activity concentrations of 40.3, 32.6, 27.4, 20.4, 13.3 and 0.00 kBq/μL. Reprinted with permission from [15], copyright 2010, SNMI. *In vivo* imaging of ^{18}F activity with examples of (**e**) CLI and (**f**) PET imaging of nude mice bearing C6-FLuc tumour injected via tail vein with ^{18}F-FDG at 0.5, 1, 2 h post-injection. Reprinted with permission from [16], copyright 2010 Liu *et al*, used under Creative Commons Licence.

receptor) to selectively visualise PC3 (human prostate cancer cell line) tumours in mice with CLI [16].

3.2 Clinical application range and relevance

The combination of nuclear and OI techniques has great potential in the clinical setting to guide tumour resection during surgery [2]. This can be achieved using the multitude of targeted nuclear agents that specifically accumulate in diseased tissue, e.g. radiolabelled antibodies. Currently, there are far more clinically approved and disease targeted radionuclides in comparison to a selected few OI agents. Therefore, the intraoperative imaging of clinically approved radiopharmaceuticals with CLI is a logical step providing a low risk procedure. Current clinical CLI applications take advantage of the naturally light-free anatomical compartments of the human body. The lack of interference from external light (non-CL photons) has enabled the application of CLI in endoscopies [2].

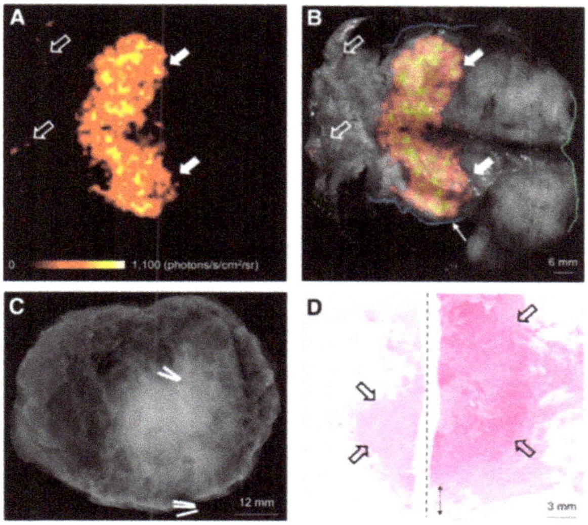

Figure 3. Clinical intraoperative assessment of tumour margins using CLI. (a) CL image and (b) grey-scale photographic image overlaid with CL image of patient with grade 3, estrogen receptor–negative/human epidermal growth factor receptor 2-negative, NST carcinoma. White arrows indicate increased signal from the tumour (mean radiance 871 ± 131 photons/s/cm^2/sr, mean tumour to background ratio (TBR) is 3.22). The posterior margin (outlined in blue, indicated by a small arrow) was measured as 2 mm. The medial margin (outlined in green) was measured as 5 mm. (c) Radiography image of specimen. (d) Histopathology image from adjacent slides on which the posterior margin is visible at the bottom of the image and open arrows indicate a part of the primary tumour. Reprinted with permission from [6] copyright 2017, SNMMI.

Due to the number of approved and tumour targeting radionuclides, CLI is positioned as a particularly interesting and promising technique for clinical tumour margin assessment [6, 9]. Complete tumour resection during surgery is an essential part of cancer treatment. Thorough tumour resection remains a challenge as currently surgeons depend on palpation and the visual appearance of tissues in order to distinguish malignant from healthy tissue. Incomplete removal of the tumour with positive tumour margins is directly linked with a high risk of local recurrence and reduced prognosis. Thus, the entire removal of the lesion with a surrounding margin of healthy tissue is essential and these surgeries would greatly benefit from intraoperative image guidance [2].

The first in-human CLI clinical para-operative imaging trial was carried out using ^{18}F-FDG and a CLI camera box system, which was used to image excised specimens. The employed system and protocol enabled the assessment of tumour margins in breast cancer. The malignant tissue was excised and imaged using CLI. CLI delivered high resolution images which enabled successful evaluation of the status of tumour margins. These images were well correlated with gold standard histopathologic analyses (figure 3). ^{68}Ga-PSMA, a PSMA specific probe, was used for primary staging in high risk patients by PET and CLI [2]. In this case, the exact tumour size and location was determined pre-surgery via PET, followed by definition of exact tumour margins intraoperatively using CLI. Finally, the efficacy of the operation was determined using CLI before closure and post-surgery PET [9].

Real-time imaging during the surgical procedure and not solely excised tissue (not currently possible), would allow a surgeon to use PET and radiotherapy agents that are already in clinical use for real time guided CLI [9]. Commonly used tracers, such as ^{18}F-FDG, or radiolabelled monoclonal high specificity antibodies, such as ^{89}Zr-DFO-trastuzumab, could be used for surgery of several solid cancers, including breast cancer, lung cancer, colorectal cancer, melanoma, head and neck cancer [2, 9]. It should be noted that the number of tracer injections required for CLI guided surgery would need to be factored into surgical planning. This will depend on the half-life of the radionuclide and as a result, pre-, intra- and post- operative imaging could be achieved using single or multiple tracer injections.

CLI of PET radionuclides can also be used to facilitate surgery planning. CLI enables the recognition of the same structures as seen in PET during staging, prior to surgery. Simultaneously, this has provided a comparison between the two modalities where, in some cases, CLI could provide a cheaper and faster alternative to PET [6]. Due to the OI component of CLI, white-light images can be captured immediately prior to/after CLI image acquisition. Combining the CLI image with this white light image, can readily provide functional and anatomical information on the nature, location, and extent of tissue uptake [2]. In 2013, CLI was used to visualise the thyroid gland of a patient treated for hyperthyroidism with ^{131}Iodine [17], which was followed soon by the first imaging of ^{18}F-FDG in patients with CLI in a small pilot trial [18]. This clinical application of CLI indicates that imaging the uptake of β-emitting radiopharmaceuticals could provide a rapid and inexpensive alternative for monitoring radiation doses given to superficial organs [17].

3.3 CL activated agents

Thus far this chapter has focused on the technical implementation and examples of CLI; attention will now be turned to the vast array of CL activatable and interacting agents. These will be briefly discussed here with examples, with most of the agents currently having a preclinical application. CL, whilst weak in comparison to fluorescence from laser excitation, has many benefits over external excitation. First, the light source comes from within the body, overcoming the difficulty faced with photoactivatable therapy or monitoring when illuminating deep seated tissues. Second, the short wavelength of CL can excite many common and approved fluorochromes (FCs). Using CL to excite fluorochromes that emit in the NIR can overcome the attenuation of UV and blue light in tissues by shifting the emission to red wavelengths, aiding in detection.

Using CL in this way is referred to as Cerenkov Radiation Energy Transfer (CRET) or Secondary Cerenkov Emission Fluorescence Imaging (SCIFI). This is possible due to the spectral overlap of CL emission and FC absorption, effectively eliminating the need for an external excitation source and merging fluorescence and CLI [19]. Combining these factors with the targeting capability of many radionuclides and the rapid absorption of CL by tissues can enable highly localised and selective CL in targets of interest, with limited off-target effects. The CL produced by ^{18}F-FDG has been capable of uncaging caged ethyl ester-luciferin (cleavage is driven by UV to blue light). Tumours were implanted in mice and transfected to express luciferase, which in turn catalyzed the bioluminescence process of uncaged ethyl ester-luciferin [20]. The bioluminescence (i.e. tumour location) could then be monitored and this successfully showed how CL delivered to the tumour by ^{18}F-FDG could release a caged fluorophore. Additionally, CL has been shown capable of exciting FDA approved fluorescent agents with the potential for direct clinical translation. One such example enabled triple-modality imaging (PET, CRET and guided confocal laser endomicroscopy (CLE)) for tumour identification with fluorescein sodium excited by radiopharmaceuticals such as ^{11}C-CHO and ^{18}F-FDG, all of which are clinically approved. CRET extended imaging depths from 1 to 8 mm compared to CLI with increases in SNR and optical intensity of 3.7 and 4.7 at tumour sites, respectively. Having identified the tumour location, CLE was used for surgical navigation. CRET and CLI were then used to validate entire tumour removal [21].

CRET and SCIFI have been applied to excite a wide range of quantum dots (QDs) and nanoparticles (NPs) both *in vitro* and *in vivo*. QDs and NPs have significant advantages in comparison to fluorescent dyes (fluorophores) as they continuously absorb light up to their emission profile, are less susceptible to photobleaching (high photostability) and can be functionalised to target tissues or sense chemical processes using suitable molecules. Due to their ability to emit with varying stokes shifts, QDs and NPs have enabled multiplexed imaging with similar absorption but varied emission profiles and the appropriate emission filters [22]. This characteristic of CL has significant advantages over PET imaging, which is only sensitive to 511 keV generated photons and is currently unable to differentiate between different radionuclide sources. One such example is the ability to perform short wave infra-red (SWIR) imaging (>1000 nm) with CL from a LINAC exciting PdSe based QDs [23]. SWIR has significant advantages over the visible and NIR spectrum as light scattering and absorption are greatly reduced.

CL is not limited to merely exciting FCs, QDs and NPs for Stokes shift purposes. Further smart agent development has resulted in an activatable imaging agent where over time, signal conversion is reliant upon enzyme expression in a tumour [19]. In this case, a gold (Au) NP was synthesised with FAM (fluorescein) bound to the surface of the NP via a peptide. The gold of the NP acted as a quencher preventing fluorescence when in proximity. When exposed to the matrix metallopeptidase-2 (MMP-2) enzyme (commonly found in tumours) FAM was no longer quenched by the NP and could undergo SCIFI based fluorescence [3]. This highlighted not only the location and extent of the tumour but also the extent to which MMP-2 activity was occurring within the tumour microenvironment.

Tumours are commonly treated using radiation therapy from a LINAC. However, hypoxic tumours (deprived of O_2) are less responsive to external beam radiation therapy and determining the oxygenation of the tumour microenvironment is difficult to study in a non-invasive manner. Accordingly, determining tumour oxygenation during external beam radiation is especially important. Recently, a phosphorescent probe; Oxyphor PtG4 was used to determine the oxygen partial pressure (pO_2) value of tumours during LINAC therapy. The phosphorescent lifetime of the probe (activated by CL and imaged using CLI) was directly related to tumour oxygenation based on the absorption of blood (hemoglobin) in tissue. Tumours in mice were found to have higher phosphorescence levels for longer periods of time post LINAC treatment (less absorption by blood). This was used to determine the consistently lower levels of pO_2 in the tumour compared to normal tissue [24].

4 Parameters of image quality

As CLI is an optical based method, image quality is determined by the same physical limitations OI is subject to in terms of light scattering and absorption. Previously mentioned fluorochromes can aid in overcoming these limits. The main influences on image quality in CLI revolve around the sensitivity of the camera to the photons released by the radionuclide, the dark noise of the camera, the number of photons of light released by the radionuclide, removal and exclusion of non-CL sources of light, the quantum efficiency of suitable fluorochromes and shielding the camera from radiation that can cause scintillation within lenses, fibres or the camera itself. In turn, the choice of radionuclide greatly affects the signal-to-noise ratio of the system and its resolution as CL production is directly related to the radionuclide. Even though the scattering and absorption of light by tissue is strongest in the blue spectrum the resolution of CLI is often submillimeter and far superior to that of PET and SPECT (on the order of millimeters). It should be noted that the higher resolutions are only achievable with superficial organs like the thyroid as the detection of photons after a few mm of tissue is challenging. The acquisition time for CLI of radionuclides tends to be on the order of minutes where motion artefacts can play a role in image quality. In comparison, dosimetry monitoring from external irradiation sources produce far more CL reducing acquisition times (μs). External light sources (room lights) require time-gating of the acquisition but do enable imaging without darkness.

5 Data processing

The data processing and visualisation of CLI is in line with OI methods. A major improvement in image quality is the subtraction of any dark current noise in the system from the resulting images. The majority of publications overlay CLI images on white light background images using a range of colourmaps. These tend to be either 'hot' or 'jet' colourmaps. Visualisation is carried out in 2D aside from tomographic imaging which is inherently 3D.

6 Conclusions

6.1 Strength and limitations

CLI provides the opportunity to perform OI with the multitude of clinically used radiotracers, allowing for a new imaging paradigm. Preclinically, CLI is a cost-effective and faster alternative to nuclear imaging instruments like PET and CT.

CLI provides a robust method to test the preclinical development of new and targeted radiotracers. The ability to image dosimetry in real-time can have direct positive implications to patient outcome and quality of life. CLI has shown promise in tumour resection and provides a real-time guide for ongoing surgical procedures. CL in general has already shown strong capabilities in activating a range of fluorescent agents, enabling the monitoring of enzymatic activity, and pO_2 with CLI. A major limitation of CLI resides in the blue light spectrum of emission and its low intensity; however, it is likely this will be overcome with the development of large Stokes shift dyes converting blue to NIR or SWIR light and improved cameras. The other limitation is the current need to remove all external light sources, e.g. room lights and equipment lights, with a completely dark enclosure for CLI, which is slowing current clinical translation. However, new developments are striving to overcome these issues.

6.2 Future developments

In the near future, CLI will likely become a widely adopted method for radionuclide monitoring, dosimetry, tumour imaging for superficially located tumours and guided tumour resection. The continuous development of sensitive cameras and their reduction in price will further highlight CLI's advantages over PET and SPECT with superior resolution and cheaper implementation. CLI's true potential will be realised when a clinically approved dye capable of converting CL to NIR or SWIR light is developed.

Acknowledgements

This work was in part funded by the NCI grant R01CA183953 to JG.

References and further reading

[1] Robertson R, Germanos M S, Li C, Mitchell G S, Cherry S R and Silva M D 2009 *Phys. Med. Biol.* **54** N355–65

[2] Grootendorst M R, Cariati M, Kothari A and Tuch D S 2016 *Purushotham Clin. Transl. Imaging* **4** 353–66

[3] Tamura R, Pratt E C and Grimm J 2018 *Semin. Nucl. Med.* **48** 359–66

[4] Pratt E C, Shaffer T M, Zhang Q Z, Drain C M and Grimm J 2018 *Nat. Nanotechnol.* **13** 418

[5] Shaffer T M, Pratt E C and Grimm J 2017 *Nat. Nanotechnol.* **12** 106–17

[6] Grootendorst M R, Cariati M, Pinder S E, Kothari A, Douek M and Kovacs T *et al* 2017 *J. Nuc. Med.* **58** 891–8

[7] Hu Z, Liang J, Yang W, Fan W, Li C and Ma X *et al* 2010 *Opt. Express* **18** 24441–50

[8] Guo H, He X, Liu M, Zhang Z, Hu Z and Tian 2017 *IEEE Trans. Med. Imaging* **36** 1337–46

[9] Holland J P, Normand G, Ruggiero A, Lewis J S and Grimm J 2011 *Mol. Imaging* **10** 177–86 1–3

[10] Liu H G, Carpenter C M, Jiang H, Pratx G, Sun C and Buchin M P *et al* 2012 *J. Nucl. Med.* **53** 1579–84

[11] Glaser A K, Voigt W H A, Davis S C, Zhang R X, Gladstone D J and Pogue B W 2013 *Opt. Lett.* **38** 634–6

[12] Zhang R X, Davis S C, Demers J L H, Glaser A K, Gladstone D J and Esipova T V *et al* 2013 *J. Biomed. Opt.* **18** 050503
[13] Jacques S L 2013 *Phys. Med. Biol.* **58** 5007–8
[14] Ma X, Wang J and Cheng Z 2014 *Front. Phys.* **2** 4
[15] Ruggiero A, Holland J P, Lewis J S and Grimm J 2010 *J. Nucl. Med.* **51** 1123–30
[16] Liu H, Ren G, Miao Z, Zhang X, Tang X and Han P *et al* 2010 *PLoS One* **5** e9470
[17] Spinelli A E, Ferdeghini M, Cavedon C, Zivelonghi E, Calandrino R and Fenzi A *et al* 2013 *J. Biomed. Opt.* **18** 020502
[18] Thorek D L, Riedl C C and Grimm J 2014 *J. Nucl. Med.* **55** 95–8
[19] Thorek D L, Ogirala A, Beattie B J and Grimm J 2013 *Nat. Med.* **19** 1345
[20] Ran C, Zhang Z, Hooker J and Moore A 2012 *Mol. Imag. Biol.* **14** 156–62
[21] Zheng S, Zhang Z, Qu Y, Zhang X, Guo H and Shi X *et al* 2019 *Adv. Sci.* **6** 1900159
[22] Liu H G, Zhang X F, Xing B G, Han P Z, Gambhir S S and Cheng Z 2010 *Small* **6** 1087–91
[23] Shell J R, LaRochelle E P, Bruza P, Gunn J R, Jarvis L A and Gladstone D J *et al* 2019 *J. Biomed. Opt.* **24** 036001
[24] Cao X, Allu S R, Jiang S, Jia M, Gunn J R and Yao C *et al* 2020 *Nat. Comm.* **11** 1–9

Section 5

Confocal and multiphoton endomicroscopy

IOP Publishing

Imaging Modalities for Biological and Preclinical Research:
A Compendium, Volume 2
Parts II–IV: *In vivo* preclinical imaging, multimodality imaging and outlook
Andreas Walter, Julia G Mannheim and Carmel J Caruana

Chapter II.5

Confocal and multiphoton endomicroscopy

Claire Lefort

Endoscopic examination is a well-established medical practice and helps hospital practitioners to orientate diagnostic and treatments with a painless protocol which also contributes to improving the care of patients. But the endoscopic images generated are restricted to surface imaging and image quality is limited to a millimetric resolution. The precision of the diagnostic can be impacted by these restrictions and some significant alteration of the structure, metabolism or physiology of organs can be missed. Today, many technical developments are in progress for the development of endomicroscopes. It is therefore necessary, for example, to resort to the intermediary of an optical waveguide whose internal structure imposes many technical and computational constraints, as discussed in the present chapter. The miniaturisation of distal optics or scanning devices is still in progress for elaboration simultaneously with an optimal computational strategy for erasing image defaults linked to the necessity of extremal miniaturisation.

1 Introduction

Endoscopic examination is a routine practice striving to guide medical diagnosis, without resorting necessarily to a surgical biopsy. This medical procedure is mainly indicated for the visual exploration of the interior of organs accessible by natural paths, directly inside the patient, *in vivo* and *in situ*. Laryngoscopy for larynx exploration, bronchoscopy for lung exploration and colonoscopy for colon exploration are a few examples of routine endoscopic practices. The main interest of an endoscopic exploration rests on its ability to substitute for invasive surgical biopsies.

The resolution associated with standard endoscopes is millimetric, an insufficient range in some cases. In that context, technical researches are now focused on the development of endomicroscopic systems having a micrometer and sub-cellular resolution [1–4]. The distinction between an endoscopic and an endomicroscopic

doi:10.1088/978-0-7503-3747-2ch7

device mainly concerns ranges of resolution, field of view and process of image generation. Today, Mauna Kea Technologies[1] is the international company leader in the field of confocal endomicroscopy. The final goal leading to the technological development of endomicroscopic systems is to deliver a tool generating *in vivo* and *in situ* endomicroscopic images having a quality similar to those generated by bulk optical microscopes. Thus, imaging *in vivo* structures with a microscopic resolution becomes possible in a minimally invasive way. The emergence of optical guides such as optical fibres, miniaturised lenses and scanning devices is associated with the highest progress for the development of endomicroscopes. Nevertheless, numerous technical challenges are not yet sufficiently mature and are in fact still in the spotlight in order to obtain a final tool able to generalise the routine practice of endomicroscopy in hospitals.

2 Principles and setups

2.1 General presentation of an endoscopic exploration

Three kinds of endoscopic systems are discussed in the present chapter, i.e., standard, confocal and multiphoton endoscopic systems. The standard endoscope is considered as a reference; its practice at the hospital is now well-established and well-mastered by hospital practitioners. Its principle rests on the miniaturisation of a video camera associated with a white light. An image of the surface of the interior of organs is generated with a millimetric resolution. Figure 1 illustrates schematically the endoscopic exploration of the lung. Such an endoscopic system is composed of an optical tube dedicated to enter a human cavity for its examination—here lungs—a

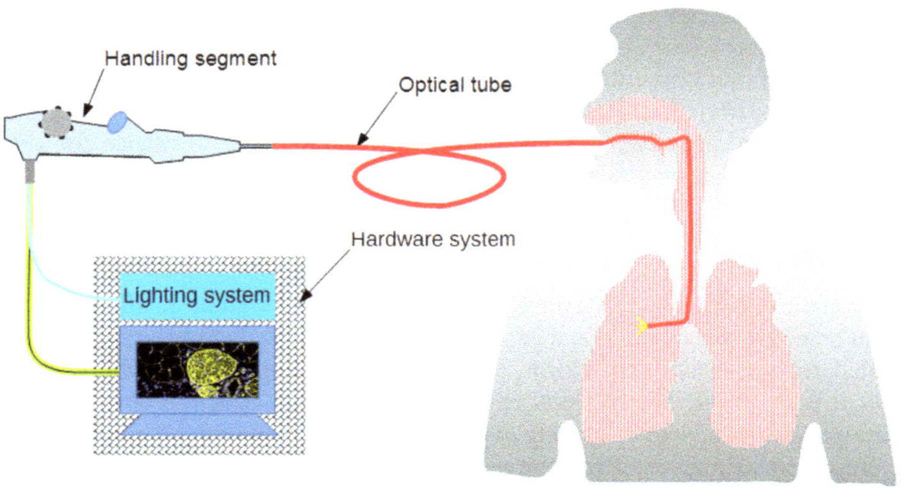

Figure 1. Scheme illustrating an endoscopic system.

[1] https://www.maunakeatech.com/int/

handling segment to manually guide the optical tube and a hardware system containing the lighting and the imaging parts. The region of the endoscope the closest to the target imaged is named 'distal end'. Reversely, the optical tube input is the proximal section of the endoscope.

The images generated by an endoscopic system are restricted to a surface image with a millimetric resolution. The precision of the diagnostic can be impacted by these restrictions and some significant alterations of the structure, metabolism or physiology of organs can be missed. Thus, a clinical need to have access to an endomicroscopic system is meaningful.

2.2 Physical principles of an endomicroscope

From the viewpoint of clinicians, the overall handling of an endomicroscope is identical to the handling of an endoscope. Only technical differences exist but do not impact the tool handling. Currently, confocal endomicroscopes are as yet marginally implemented in hospitals and no commercial multiphoton endomicroscopic system exists yet; the technical specificities of confocal and multiphoton endomicroscopes are detailed in the following sections. Their physical principles rest mainly on the generation of fluorescence signals. Thus, identical properties of the excitation light (coherent light, wavelength ranges, average powers…) and emission signals are involved between endomicroscopes and bulk microscopes. We can notice here that harmonic generation is an alternative imaging modality that can be generated in multiphoton endomicroscopy. More details about signal generation can be found in chapter I.1.b *Fluorescence and confocal microscopy* and in chapter I.2.c *Multiphoton microscopy*. The specificity of endomicroscopes compared to bulk microscopes rests on the miniaturisation and the transfer of the excitation beam and the detection strategy *in vivo* and *in situ*, the closest to the imaged target.

2.3 Technical principles, typical setup and state-of-the-art of endomicroscopy

The optical waveguide contained in the optical tube (figure 1), delivers the excitation beam and collects the emission signals. This element constitutes the key point of an endomicroscope. Two kinds of optical waveguides can be implemented in an endomicroscopic system: a double clad fibre (DCF) with a single core and a fibre-optic bundle composed of a few thousands of optical cores [6, 7]. Figure 2 presents technically the structure of the endomicroscope and the differences induced by each of the two strategies. The choice of the optical waveguide impacts the time delay for generating an image, the scanning strategy, the level of miniaturisation of the distal end and the complexity of the computational strategy required for the optimisation of image quality. The optical waveguide is also an element potentially responsible for self-fluorescence generating background noise. A solution of pure silica structure should be preferred in that case [8, 9]. Furthermore, whatever the optical waveguide chosen, a focusing element must be positioned at the endoscopic distal end.

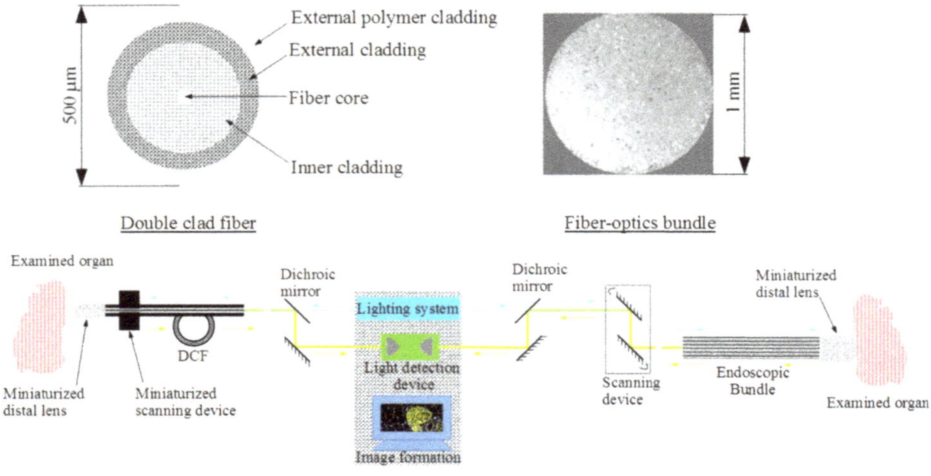

Figure 2. Presentation of the two architectures of an endomicroscope depending on the optical waveguide chosen. A double clad fibre (DCF, left) and a fibre-optics bundle (right) are the two technical strategies that can be encountered in an endomicroscopic system.

Focusing element at endoscopic distal end

Regardless of the optical waveguide chosen, reaching microscopic performances of an endomicroscopic system requires the use of a miniaturised focusing element positioned at the endoscopic distal end. This optical microsystem governs the optical resolution, the image field of view and working distance. The design of the dimensions and optical properties of the focusing element can be modelled numerically by computational software, the best known being Zemax[2].

Commercially, two main strategies for building a passive and miniaturised focusing element exist: construct a GRIN lens[3] or assemble miniaturised achromatic doublets or triplets[4]. Numerical aperture, chromatic and geometric aberrations and their impact on image properties must be carefully considered during the modelling step with regard to the dimensions of the optical waveguide.

Optical waveguide and scanning device

A double clad fibre with a miniaturised scanning device at endoscopic distal end

A double clad fibre (DCF) is an optical fibre composed of a single core which guides the excitation beam and two claddings, as illustrated in figure 2 (left). The inner cladding contributes to guide the excitation beam into the fibre core. The collection of the emission light is led by the inner cladding. Photonics crystal fibres (PCFs) are microstructured fibres often chosen for their optimised guiding properties for both excitation delivering and emission collection [11]. With a single core fibre and in

[2] https://www.zemax.com/
[3] https://www.grintech.de/en/
[4] https://www.bernoptics.com/ or https://www.edmundoptics.com/

order to generate an image, a distal scanning device must be associated with the focusing element positioned at the endoscopic distal end. This can be realised thanks to (i) a piezoelectric device [7] or (ii) a micro electromechanical *mirror* system (MEMS) [4, 7]. This last point, and the specificities associated with its miniaturisation, constitutes the main technical difficulty of this strategy.

A fibre-optics bundle with a proximal galva-resonant scanning device
A fibre-optics bundle is an optical waveguide composed of a few thousands of fibre cores with a millimetric external diameter [2, 6]. For an optimal light guiding and avoiding optical crosstalk, each core must be independent from its neighbor. That imposes optical properties of each core drastically different to each other and results in a significant difference in optical index between two neighboring cores. In that condition, each core of the bundle constitutes a pixel of the image. On an indicative basis, a usual image weakly resolved and generated with a bulk optical microscope is constituted by a square of 256 pixels × 256 pixels which contains about 65 000 pixels. An image generated with an endomicroscope based on a fibre-optics bundle is thus more weakly resolved than usual microscopic images. Moreover, parameters such as the distance between two cores, the core diameter and their relative homogeneity are impacting image quality: resolution and collection efficiency. This last point constitutes the main technical difficulty of this strategy. Thus, generating an exploitable image with a fibre-optics bundle is closely linked to a computational strategy erasing the impact of core inhomogeneities [5].

The main interest of a fibre-optics bundle concerns the scanning device. Indeed, with each fibre core constituting the pixel of the image, the image is generated thanks to the sequential injection of the excitation beam into each core. Thus, the scanning device is positioned at the proximal input of the optical tube and does not require to be miniaturised. Many solutions of scanning devices are thus offered and do not impose any miniaturisation: two galvanometric mirrors or a single galvanometric mirror combined with a resonant mirror, for example.

Difference between a confocal and a multiphoton endomicroscopic system
Similar physical principles are involved between confocal (or multiphoton) microscopy and confocal (or multiphoton) endomicroscopy. The main difference between microscopy and endomicroscopy concerns the delivering of the excitation laser light at the optical waveguide distal end. For confocal microscopy, no temporal constraint governs the excitation beam. In that situation, no significant optical adjustments are required for confocal endomicroscopy except the use of an optical waveguide.

However, an essential adaptation of the optical excitation strategy must be developed in the case of multiphoton endomicroscopy. Indeed, considering that a temporal confinement of the excitation beam optimises drastically the generation of multiphoton signals, the shortest pulse duration is favorable for efficient generation of multiphoton processes. Thus, near infrared femtosecond pulses with a MHz repetition rate give a good solution for generating multiphoton processes. But when injected into an optical waveguide, linear and nonlinear effects appear and result in spectro-temporal distortions of pulse profile. Delivering femtosecond pulses at fibre output imposes a prechirp unit positioned at the proximal input, before the endoscopic fibre [7, 14]. Prism compressor, grating stretcher, grisms line or chirp mirrors are a few examples of passive optical elements currently used to compensate for linear and nonlinear effects occurring in the core of the optical fibre. Many active solutions, involving active systems such as spatial light modulators are also sometimes used. Such an optical mask is the exclusive solution adapted with a fibre-optics bundle due to the drastically different dispersion properties between two neighboring cores.

3 Biomedical relevance

In vivo and *in situ* imaging of the interior of organs allow hospital practitioners to gather information about anatomical structure and its relation to pathology. Structure or the metabolic state of organs become accessible *in vivo* and are of first interest for helping to orientate a medical diagnostic. An early diagnostic of cancers (lung or larynx for example) is associated with a procedure less painful when no surgical biopsy is required. Moreover, in the context of preclinical researches, for spleen studies or neuroscience, the use of small animals and their ineluctable sacrifice is not a satisfying practice ethically. Furthermore, much information about metabolic states and evolutions becomes complex to identify *in vivo*, especially after the animal's death. Resorting to endomicroscopy is an alternative solution which opens new opportunities in preclinical researches.

4 Parameters of image quality

4.1 Label-free imaging of biological constituents

Beyond the technical complexities specific to endomicroscopy, the endomicroscopic examination is also governed by the ability of the biological constituents to emit endogenously an optical signal. Several substances are known for their label-free fluorescence or for their ability to generate a second harmonic signal [8]. In the context of label-free imaging of biological substances, many alternative solutions of optical imaging are also currently in the spotlight. Fluorescence life time imaging (FLIM) and coherent anti-Stokes Raman scattering (CARS) are two examples of alternative solutions for generating an endogenous signal. However, their adaptation to endoscopy is even more complex technically than confocal or multiphoton fluorescence where the main limitation concerns the low level of emission flux difficult to separate from image noise. More information can be found about these two alternative techniques in chapter I.8.a *Förster resonance energy transfer—fluorescence lifetime imaging microscopy* and in chapter I.9.a *Raman imaging* and chapter I.9.b *Coherent Raman imaging.* Endogenous fluorescence is also studied with alternative laser systems based on optical parametric oscillators [15] or ultrawide band laser systems [16]. Today, we could consider that the biological substances endogenously fluorescent have not yet all been characterised due to the lack of an adapted laser system.

4.2 Movements and endomicroscopic examination

In vivo endomicroscopic imaging of living guinea pig is necessarily associated with body movements, at least due to breathing movements, and unavoidable in the case of endomicroscopic imaging of a small animal in living conditions. In that situation, the question of the optical waveguide fixation becomes of great importance and micrometric movements of the optical waveguide cannot be avoidable. Moreover, movements of the optical waveguide can also be associated with modifications of optical properties which impact the image quality and especially contrast and intensity levels. These external constraints are affecting precision and sensitivity of measurement. In the case of double contrast agent, the geometric alignment and the fuse of an image pair becomes complex and imaging depth a high challenge. A computational strategy of image registration becomes indispensable [17]. This computational step estimates the geometric transformation and corrects the spatial superposition of the anatomical or functional structures imaged.

5 Data processing

Data processing is key point for several steps of the image formation through an endomicroscopic system. First, depending on the optical waveguide chosen, the spiral oscillations in the case of DCF or the pixel merging in the case of a fibre-optic bundle entails the aid of a specific computational strategy for image reconstruction. Then, in the case of movements of the optical waveguide, the image registration is indispensable and can also benefit from the strategies already developed for alternative medical imaging such as magnetic resonance imaging, positron emission tomography (PET) or optical coherence tomography (OCT).

6 Conclusions

6.1 Strength and limitations

The micrometric imaging of the interior of organs and imaging in live its metabolic state are of first interest for helping clinicians for an early diagnostic of diseases such as cancers or for the discrimination of healthy or cancerous tissues. Currently, the maturity of endomicroscopic systems may not be enough and they are probably not yet well implanted into hospitals. Many reasons could explain this fact, starting from the habits of hospital practitioners. Technical constraints are also limiting image formation with an endomicroscopic system. For example, image field of view is limited to a square of 200 μm on each side. Moreover, only a few biological substances can be imaged without labelling and the question of image registration stays significant. Considering these points, the use of a surgical biopsy, associated with external labelling and imaging with standard and bulk optical microscopes remains in the forefront for guiding the research of a reliable diagnosis.

6.2 Future developments

The future developments of endomicroscopy concern in fact many scientific and technological key points. Developing and endomicroscope presenting (1) at least a millimetric field of view, (2) with a sub cellular resolution, (3) in millimetric depths, (4) with many endogenous signals, could be realised in many years thanks to great progress in: (i) miniaturisation of optics with reduced optical aberrations; (ii) optimised optical waveguide; (iii) a new system for axial scanning; (iv) the characterisation of endogenous signals from new substances of interest. More generally, the future developments of endomicroscopic systems rest on general progress of technologic and scientific knowledge and discoveries, covering at least the fields of physics, chemistry, biology, informatics and mathematics.

References and further reading

[1] Rousse A R and Gmitro A F 2000 *Opt. Letters* **25** 1708–10
[2] Juskaitis R, Wilson T and Watson T F 1997 *Scanning* **19** 15–9
[3] Fu L, Gan X and Gu M 2005 *Opt. Express* **13** 5528–34
[4] Myaing M T, MacDonald D J and Li X 2006 *Opt. Lett.* **31** 1076–8
[5] Garofalakis A, Kruglik S G, Mansuryan T, Gillibert A, Thiberville L, Louradour F, Vever-Bizet C and Bourg-Heckly G 2019 *J. Biomed. Opt.* **24** 106004
[6] Gu M, Bao H and Kang H 2014 *J. Microsc.* **254** 13–8
[7] Ducourthial G, Lefort C, Peyrot D A, Mansuryan T, Kruglik S, Vever-Bizet C, Thiberville L, Lacombe F, Bourg-Heckly G and Louradour F 2013 *Proc. SPIE* **8575** 85750H
[8] Liang W X, Hall G, Messerschmidt B, Li M-J and Li X 2017 *Light* **6** e17082
[9] Lefort C, Hamzeh H, Louradour F, Pain F and Haidar D A 2014 *J. Biomed. Opt.* **19** 076005
[10] Andreana M, Le T, Drexler W and Unterhuber A 2019 *Opt. Lett.* **44** 1588–91
[11] Lakowicz J R 2006 *Principle of Fluorescence Spectroscopy* 3rd edn (New York: Springer)
[12] Fu L, Jain A, Xie H, Cranfield C and Gu M 2006 *Opt. Express* **14** 1027–32
[13] Fu L and Gu M 2007 *J. Microsc.* **226** 195–206
[14] Lefort C, Kalashyan M, Ducourthial G, Mansuryan T, O'Connor R P and Louradour F 2014 *JOSA* B **31** 2317–24
[15] Tragardh J, Robb G, Amor R, Amos W B, Dempster J and McConnel G 2015 *J. Microsc.* **259** 210–8
[16] Poudel C and Kaminski C 2019 *JOSA* B **36** A139–53
[17] Rueckert D and Schnabel J A 2014 *Registration and Segmentation in Medical Imaging* (Berlin: Springer)

Section 6

Ultrasound

IOP Publishing

Imaging Modalities for Biological and Preclinical Research: A Compendium, Volume 2

Parts II–IV: *In vivo* preclinical imaging, multimodality imaging and outlook

Andreas Walter, Julia G Mannheim and Carmel J Caruana

Chapter II.6

Ultrasound

Carmel M Moran and Adrian J W Thomson

The real-time nature of preclinical ultrasound imaging, the small footprint of the ultrasound scanner and the inexpensive nature of the technique, have made ultrasound scanning one of the most versatile imaging technologies in the preclinical laboratory. Ultrasound imaging is now routinely used to assess the anatomy and physiology of small animal models and to guide organ-specific injections. It is primarily used in studies which (1) require imaging at multiple time-points such as that undertaken in longitudinal studies studying recovery of cardiac function after myocardial infarction, embryonic growth, tumour growth etc, and (2) real-time imaging to guide injections into specific organs. The physical processes underlying formation of preclinical ultrasound images are the same as those that form clinical ultrasound images, however, the small body weight (1 g (zebrafish), 25 g (mouse), 350 g (rat)) and faster heart-rates (600 bpm (mouse), 350 bpm (rat), 90 bpm (zebrafish)) of preclinical models compared to a human (70 kg, 60 bpm), necessitate much higher spatial and temporal resolution to enable useful diagnostic measurements to be undertaken. In this chapter, the basic physics underlying ultrasound imaging and Doppler blood flow techniques will be introduced and an overview provided of the range of applications to which ultrasound can provide useful diagnostic information. The use of contrast agents in ultrasound imaging, both targeted and untargeted and the use of newer techniques such as elastography to measure the stiffness of tissues will be introduced. The role of artificial intelligence applied to ultrasound diagnosis is increasingly being used with clinical scanners and is likely to have increased impact in functional analysis in preclinical applications and may be of particular benefit for the non-specialist ultrasound user as it is largely operator-independent.

doi:10.1088/978-0-7503-3747-2ch8

1 Introduction

Ultrasound imaging is a real-time imaging technique which can be used to form grey-scale images of structures within the body enabling identification and quantitative measurements of soft tissue anatomy. Doppler ultrasound can also be used to quantify blood flow both within the heart and in blood-vessels within the body. Preclinical applications for ultrasound include, but are not limited to, assessment of cardiac function; embryonic and placental development, abnormal growth/cancer diagnosis and assessment of blood flow both within large vessels (aorta and carotid) and within smaller vessels within larger organs such as the kidney, liver and heart [1]. Additionally, it is useful for visualising and guiding real-time injections into specific organs where a high level of accuracy and precision is required. The use of ultrasonic contrast agents which are confined to the vascular space, enables indicators of vascular perfusion to be measured. Additionally, the use of ultrasonic contrast agents targeted to specific biological markers expressed on the surface of vessels enables assessment of the presence and quantity of these markers.

2 Principles and setups

2.1 Physical principles

Ultrasound images are formed from the ultrasound echoes backscattered from structures and boundaries within the path of a transmitted ultrasound beam. Ultrasound waves of megahertz frequency are transmitted into the body using a transducer. Most transducers are formed of piezoelectric material which converts the electrical signal into an ultrasound pulse on transmit and the converse for the returning ultrasound signal. As the ultrasound travels through the body, it is reflected from interfaces and scattered by small structures within tissues. The magnitude of these echoes is dictated by the change of acoustic properties between adjacent structures in the path of the beam and the size of the structures relative to the wavelength of the ultrasound beam. Those echoes, which are scattered back to the transducer (back-scattered), are converted to electrical signals and processed to form a two-dimensional cross-sectional image that is displayed on the scanner. Structures of greater backscatter intensity are displayed as brighter structures and those with reduced backscatter as darker echoes on the image. Location of structures within the image is determined by the time the scattered ultrasound echoes take to return to the transducer and rely on the assumption that ultrasound travels within soft tissue at a speed of sound equal to 1540 m s^{-1}, hence echoes from deeper organs will take longer to return to the transducer than echoes from shallower organs [2]. The high speed of sound within soft tissue coupled with the limited penetration depth required to image small animals means that ultrasound image data can be acquired very rapidly and displayed in real-time. Thus, despite the rapid heart-rate of the mouse, sufficient frames can be acquired within a single cardiac cycle to study both diastolic and systolic function. Lateral resolution of ultrasound images is dictated by the frequency of the ultrasound beam with higher frequencies giving better spatial resolution than lower frequencies. However, there is a limitation to the maximum frequency that can be used, as attenuation of the ultrasound beam within soft tissue is frequency dependent. Hence, clinical ultrasound imaging is performed at frequencies between 2–20 MHz, while frequencies between 20–80 MHz are used to image small animal models, with higher frequencies reserved for superficial structures and lower frequencies for larger animal models such as rats.

Blood flow within preclinical models can be measured using Doppler ultrasound. There are two principal display modes of blood flow on ultrasound scanners: colour Doppler, which overlays the B-mode image and displays mean blood flow within a pre-selected region-of-interest, and spectral Doppler, which displays the Doppler velocity as a function of time displayed as a continuous scroll at the bottom of the screen. Both rely on the Doppler principle, i.e. the change in observed frequency (pitch) due to the relative motion between a transmitting source and observer. For Doppler ultrasound, the blood is the moving target and the transducer is assumed stationary, hence the frequency of the ultrasound wave scattered from moving red blood cells (f_s) relative to the frequency of the transmitted wave (f_t) is proportional to the speed of the blood cells (v) the cosine of the angle between the direction of blood flow and the transmitted wave (φ) and inversely proportional to the speed of sound in soft tissue (c) (equation 1).

$$f_s - f_t = \frac{2 f_t v \cos \varphi}{c} \tag{1}$$

In preclinical imaging, colour Doppler is of most use for locating small vessels and determining the direction of flow within them and spectral Doppler for measuring the magnitude of flow within specific vessels or organs.

Preclinical ultrasound imaging is particularly useful for many of the applications for which ultrasound is utilised clinically, e.g. cardiac imaging, imaging of developing embryos and ultrasound-guided injections. However, unlike clinical subjects, for almost all preclinical studies, the mouse will require anaesthesia which will impact on the whole animal homeostasis, respiratory and thermoregulation and must be considered as a potential confounding experimental effect [3].

2.2 Typical setups and state-of-the-art

The preclinical ultrasound imaging market is composed principally by companies which have spun-out from academic ultrasound research laboratories where the initial design, and fabrication of high frequency transducers and electronics were developed in-house to image preclinical models. Currently, the European preclinical ultrasound imaging market is dominated by Visualsonics (www.visualsonics.com) who provide a series of preclinical, array-based ultrasound scanners. SonoVol (www.Sonovol.com) manufactures a scanner (Vega) which acquires wide-field 3D ultrasound images. Iconeus (www.iconeus.com) has developed a functional neuro-imaging system for brain imaging utilising plane-wave technology. S-Sharp Corporation (www.s-sharp.com) uses single element transducer technology to acquire real-time 2D images. In addition, the S-Sharp scanner (Prospect T1) measures the elastic properties of soft tissues utilising a technique known as shear-wave imaging to determine the Young's modulus (stiffness) of the tissue of interest. A small footprint transducer provides the 'push' pulse that generates the shear waves within the tissue. The speed (c (ms^{-1})) at which the shear waves propagate can be directly related to the Young's modulus (E (kPa)) of an organ under two assumptions: (1) the tissue is incompressible (density (ρ) does not change) and (2) the tissue has uniform elastic properties (equation 2). If these assumptions are valid, the Young's modulus of the tissue can be estimated by measurement of the shear wave velocity. Clinically the stiffness of tissue has been shown to be related to pathology. For preclinical models, this technique is predominantly used to assess mouse liver fibrosis.

$$E = 3\rho c^2 \tag{2}$$

Other research laboratories have utilised clinical ultrasound scanners and transducers that have been developed for musculo-skeletal or intraoperative applications, operating at frequencies up to 20 MHz for imaging larger animal models. Commercial preclinical scanning systems for real-time imaging generally include a heated scanning table with physiological monitoring capabilities embedded: electrodes for ECG monitoring and for measurement of respiration and a heat temperature sensor to monitor the temperature of the animal. In addition, a versatile transducer mount to hold the ultrasound transducer requiring minimal user interaction is also essential as at high spatial resolutions, even small movements generated by the operator can impact significantly on the quality of the images obtained (figures 1 and 2). For preclinical imaging, transducer mounts or the heated table are controlled by

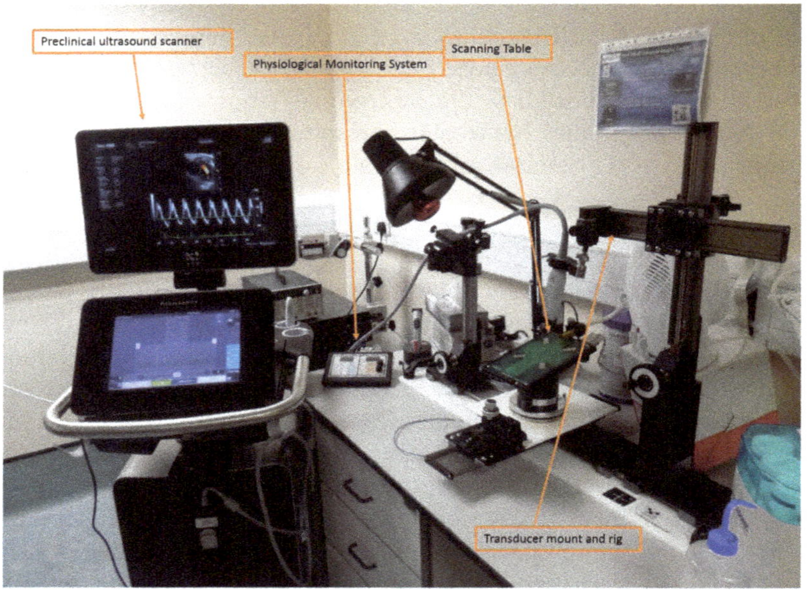

Figure 1. Typical preclinical ultrasound imaging laboratory setup with the scanner, physiological monitoring system, scanning table and transducer mount highlighted.

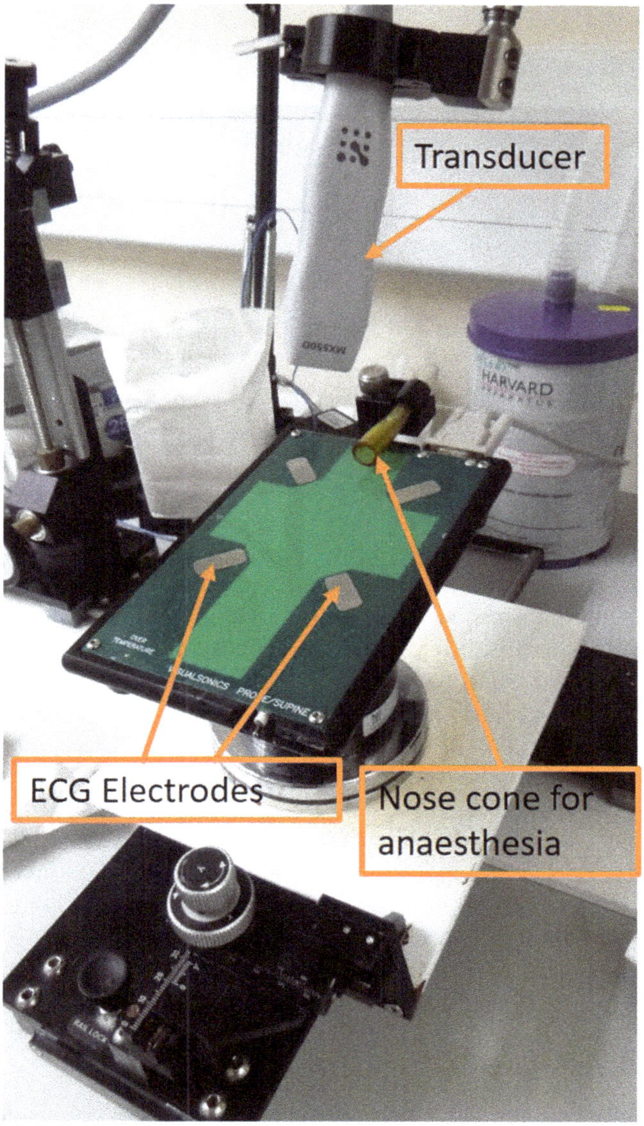

Figure 2. Scanning table showing anaesthetic nose-cone and ECG electrodes for monitoring of heart-rate and single element transducer in mount.

micro-manipulators to enable discrete finite steps of the transducer relative to the mouse position to be undertaken in any direction. Most companies offer the option of a combined small animal anaesthetic unit and a range of transducers, either single element or array dependent on the scanner hardware. The frequency and number of transducers required is also dependent on the range of animal models and organs of interest.

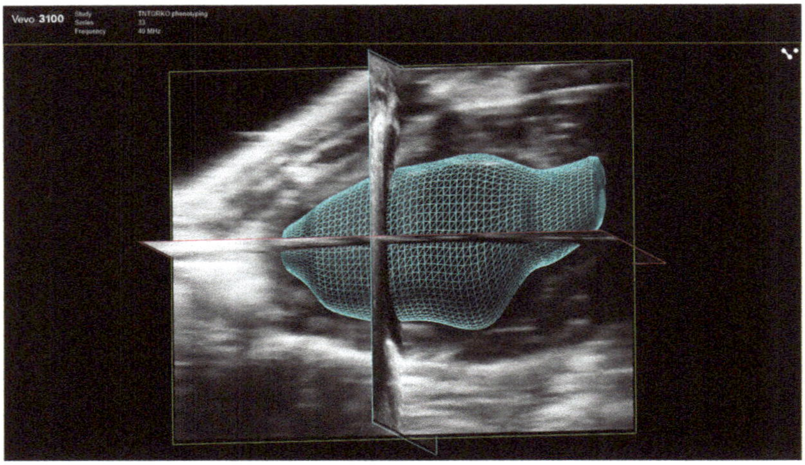

Figure 3. 4D reconstruction of murine heart over one cardiac cycle.

The Vevo 3100 (Visualsonics, Fujifilm, Toronto, Canada) is the fifth-generation preclinical scanner manufactured by the company. The first three generations utilised single element transducer technology which gave very high resolution but only over a limited depth-of-field. More recent scanners utilise array technology enabling the ultrasound beam to be electronically focussed and steered. This results in a broader depth of field over which the ultrasound beam can be focussed and the ability to use colour Doppler to visualise blood flow within user-selected regions of interest. Dependent on application, and manufacturer, supplementary features can be purchased. These include for cardiology applications, strain and strain-rate imaging techniques which are gaining traction within the cardiology community as early indicators of systolic and diastolic dysfunction or regional myocardial impairment. A second feature is a module to acquire and reconstruct 4D movies of cardiac motion (figure 3). For soft tissue organs, contrast-specific imaging techniques are available with associated analysis software to enable the acquisition and quantitative analysis of dynamic enhanced images of contrast within vessels and organs

3 Biomedical relevance

In a similar manner to clinical ultrasound imaging, preclinical ultrasound imaging can be used to provide diagnostic information on a large range of soft tissue organs including, but not limited to, the heart, vascular structures, placenta, liver, kidney, testes and soft tissue tumours. In the sections below, reference will be made predominantly to the images acquired from mouse models, although similar images can be obtained from rat models using a lower frequency transducer.

Unlike humans, rodents are generally anaesthetised prior to scanning using gaseous isofluorane mixed with 100% oxygen or medical air. Induction of anaesthesia is undertaken in an induction box using 3%–4% isofluorane. Once the animal is anaesthetised it is transferred to the heated scanning table where anaesthesia is maintained as required but generally at 1%–2% isofluorane using a nose cone.

For anaesthesia of rats, a larger nose cone is required, and a higher anaesthetic maintenance level of around 3%. For longer scan sessions, eye gel or ointment is applied to avoid corneal drying. If additional heating is required to maintain the body temperature a heated fan can be utilised. Infrared light can be used at a safe distance.

3.1 Application range

3.1.1 Cardiac scanning

Due to its rapid motion, the heart is one of the most technologically challenging organs to image [4, 5]. This is especially true in the mouse, where unanaesthetised heart-rates can be as high as 700 bpm. For cardiac scanning, once the mouse is anaesthetised, it is placed supine on the heated scanning table and each leg is gently extended so that the paw can be attached (micropore tape) to the fixed electrodes on the heated table enabling respiration and ECG data to be acquired and displayed on the scanner. The hair on the chest of the mouse is removed with a depilatory cream which should be washed out with care to avoid skin irritation. Warmed ultrasound coupling gel is then placed on the chest. The mounted-transducer is lowered towards the animal until the scanning-surface is embedded in the gel and close to the surface of the skin. Unlike clinical ultrasound scanning, in preclinical ultrasound scanning, the transducer does not touch the animal to ensure that the pressure of the transducer does not affect the cardiac dynamics of the mouse (figure 4).

Parasternal long- and short-axis views and apical views of the heart can be obtained in real-time. Colour Doppler can be used to identify vessels and to determine the direction of flow with red colour indicating movement of blood towards the transducer and blue indicating movement away from the transducer. Spectral Doppler can be utilised to measure the magnitude and direction of blood flow (figure 7) throughout the cardiac cycle. In cardiac studies a common site to

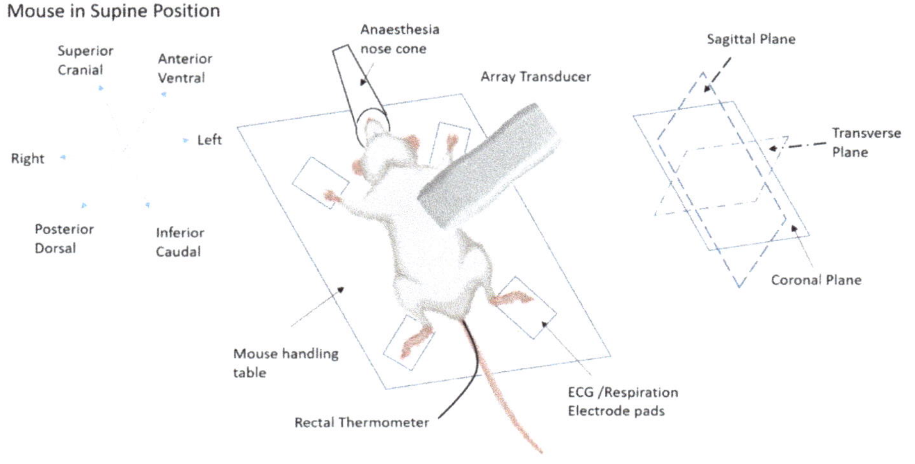

Figure 4. Typical experimental setup for cardiac imaging of mouse. Note that the anaethetised mouse is lying supine on a heated table with legs extended and attached to ECG/respiration pads (Created with BioRender.com).

acquire measurement of blood flow is from the apical view enabling a four-chamber view of the heart to be obtained. In this view, the inflow of blood into the left ventricle from the left atrium through the mitral valve can be measured in addition to flow from the left ventricle into the aortic outflow. Typically in mice, the inflow Doppler blood flow signal displays a larger early (E) peak followed by a late (A—atrial) peak followed by an outflow signal. From this signal, it is possible to measure ejection time and isovolumic contraction and relaxation times.

Myocardial wall motion can also be studied and displacement, velocity, strain and strain-rate can be measured regionally within the myocardium or globally using a technique known as speckle tracking which tracks the movement of the wall using the unique pattern (speckle) generated from the coherent addition of scattering from structures too small to resolve. Strain and strain-rate imaging are of particular interest in studies of cardiomyopathic animals and in studies of myocardial remodelling after induced myocardial infarction (figure 5).

Myocardial function can also be studied in neonates from as early as first day post-natal. In such studies, care must be taken to ensure that the pup is appropriately handled, monitored during recovery and reintroduced appropriately back to the mother using standardised praxis. Additionally, the nose cone for inducing anaesthesia requires modification to ensure that sufficient anaesthesia is being delivered to the pup and a smaller rectal probe is required to ensure continuous monitoring and control of the pup's body temperature. The small size of the pup tends to require extension of the ECG electrodes on the scanning table.

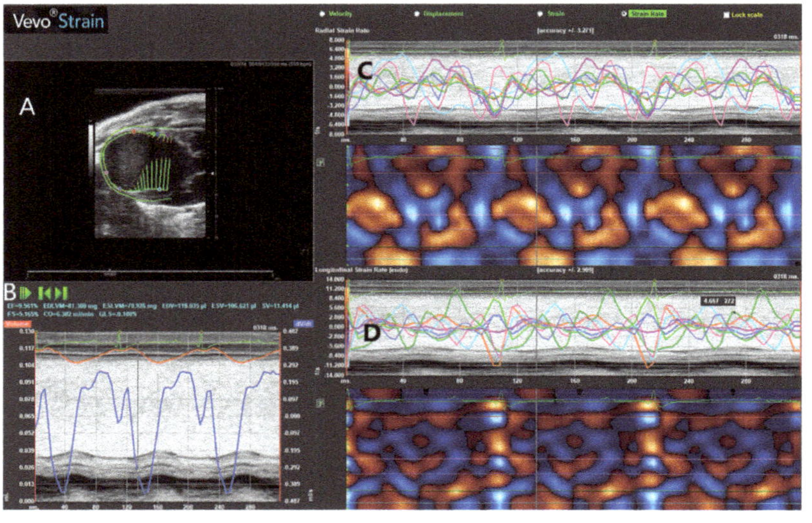

Figure 5. Data obtained from aC57 black mouse. (a) Parasternal long-axis view of the heart with arrows indicating direction and magnitude of the movement of the endocardial border. (b) Left ventricular volume and rate-of-change of left-ventricular volume over consecutive cardiac cycles with ECG and respiration displayed. (c) Radial strain-rate curves from five points selected by the user on the endocardial border on B-mode image. (d) Longitudinal strain-rate curves from the same five points. Notice the reproducibility of the strain-rate over consecutive cardiac cycles.

In pregnant dams, by mapping the position or each embryo within the horns, the cardiovascular function of individual embryos can be monitored over sequential days [6]. Indeed, it becomes increasingly easier to follow individual embryos as they grow larger within the maternal horns as they are less able to move position or be reabsorbed. The duration of a cardiovascular scan of embryos is limited to 60 min as beyond this, the effects of the anaesthesia can significantly affect the cardiovascular function of the embryos. Additionally, since it is only the heart-rate of the mother which is monitored, embryonic heart-rates can be calculated from Doppler waveforms or from a determination of myocardial wall motion. Also noteworthy is that in embryos, the blood within the heart gives an enhanced ultrasound signal, likely due to nucleation of the red blood cells, and hence differentiation between the myocardial walls and blood within the chambers can be more challenging [7]. This enhancement disappears within 2 to 3 days post-natally.

3.1.2 Soft tissue scanning

Dependent on the organs to be scanned, the mouse can be placed supine or prone on the heated scanning table. To scan the *liver*, the mouse is placed in the supine position and a probe of insonating frequency of 30–40 MHz may be used. The liver is made up of four lobes (median, left, right and caudate). The hepatic artery and portal vein may be visualised, tumours may be sized and staged and fibrosis may be determined using shear wave imaging techniques (figure 6) (section 2.2.) or the hepatic to renal echogenicity ratio. The pancreas and spleen lie adjacent to the liver with the spleen having a 'more 'holey' appearance than the liver.

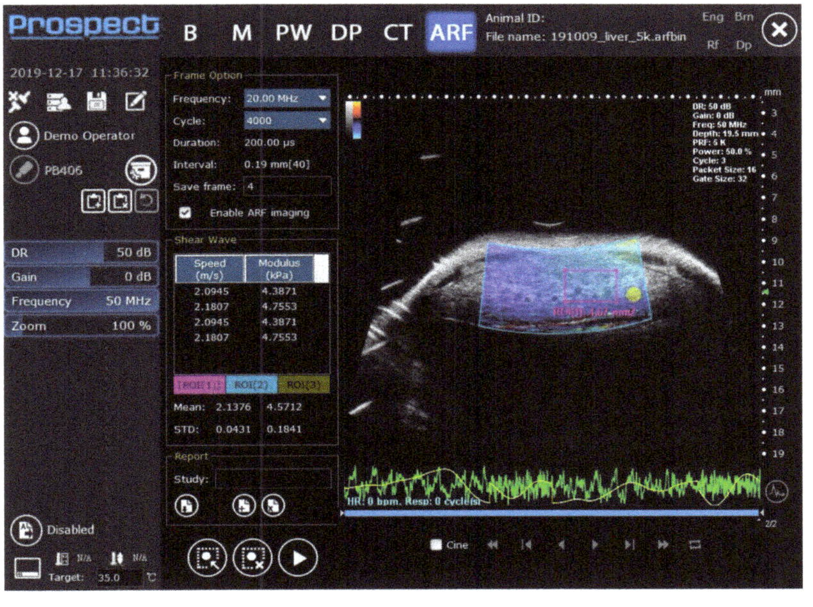

Figure 6. Image showing mouse liver with shear wave imaging window superimposed on grey-scale B-mode image. Values for both shear speed (m s^{-1}) and Young's modulus (kPa) are generated. Four measurements are made and the mean and standard deviation are generated. In this case the mean shear wave speed is 2.1 m s^{-1} and the mean Young's modulus is 4.5 kPa.

For scanning of *kidneys*, the mouse may either be placed in the supine or prone position—the left kidney is positioned slightly more posterior in the abdomen than the right kidney. Both the cortex and medulla can be identified within the kidney capsule: the medulla within the centre of the kidney tends to be darker than the surrounding cortex. Both the renal arteries and veins can be identified in colour Doppler mode and the flow within these vessels can be quantified using spectral Doppler. Longitudinal studies can be undertaken using B-mode imaging for locating anatomical landmarks to ensure measurements are taken from the same position and both colour and spectral Doppler enabling the measurement of quantitative changes in blood flow including the resistive and pulsatility indices. Such changes can be used as indicators of changes in renal pathology [8]. The adrenal glands can also be visualised superior to the kidney.

Both the female and male *reproductive systems* may be visualised with ultrasound of centre frequencies between 40–50 MHz. The ovaries are embedded within fat pads and tend to be easier to identify if ovulation has occurred recently. Scanning the male mouse testes using power Doppler, or 3D power Doppler enables the blood vessels to be identified and percentage vascularity to be calculated. Although the prostate gland can be visualised in B-mode, blood velocities within prostate tend to be low and contrast agents can be used to visualise changes in blood flow due to pathology.

3.1.3 Embryonic scanning

In addition to embryonic cardiac scanning highlighted in section 3.1, blood flow within umbilical arteries, veins and placentas for each embryo can be measured. Colour Doppler is of specific value in this application in identification of the vessels (figure 7). Changes in resistive index, a figure which indicates the resistance of an organ to perfusion, can be calculated from the peak systolic velocity and end diastolic velocity and are useful as an indicator of changes in placental pathology. As the embryo develops, the embryonic brain can also be visualised due to the lack of mineralisation of the skull, with the circle of Willis being easily identified by E12. Abdominal organs may also be identified as the embryo develops with kidneys, liver and stomach and vessels clearly identifiable post E16.

3.1.4 Ultrasound-guided injections

Both organ and embryonic-guided injections can be undertaken using ultrasound to guide localisation and placement of the needle. For embryonic injections, if the embryos are externalised a pulled glass capillary is used to enable small volumes to be injected under sterile aspetic practices. For injection into non-externalised organs and embryos a Hamilton syringe with a hypodermic G30 needle attached can be used with good effect [9].

3.2 Sample preparation

Almost all ultrasound scanning of small animals is undertaken with the animal anaesthetised using either gaseous isoflurane with medical air or 100% oxygen.

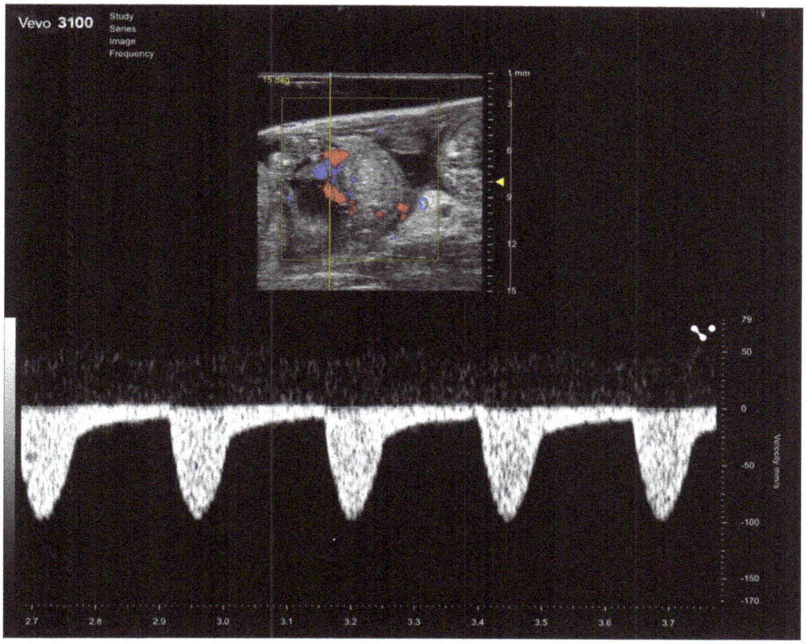

Figure 7. B-mode, colour Doppler and spectral Doppler image of umbilical artery in E18.5 mouse embryo. Note peak velocities of approximately 100 mm s^{-1}.

Induction of anaesthesia is performed in an induction box using 3%–4% isofluorane and anaesthesia is maintained at isofluorane levels between 1%–2%. Once the animal is anaesthetised it is transferred to the heated scanning table where anaesthesia is maintained using a nose cone. An active charcoal scavenger is used to protect the operator from inhaling any excess isoflurane. If monitoring of the ECG or respiration is required, all four limbs of the animals are gently stretched and attached to the embedded electrodes within the table utilising electrolyte gel. While under anaesthesia, heart-rate should be maintained between 400–600 bpm in the normal mouse while in the normal rat between 300–400 bpm. Hair-removing cream is applied to the animal dependent on the area to be scanned and removed gently using warmed water. If the hair is dense, an electric clipper may be used to remove the hair initially but it should be followed up with hair removing cream to ensure no stubble remains which would may affect the quality of the ultrasound image. The temperature of the animal should be monitored and maintained at all times using a heated table and if necessary an electric fan. This is especially important if a large amount of body hair is removed from the animal for scanning purposes. Body temperature for a mouse is 37 °C and for a rat 37.5 °C. At the end of an experiment, moisturising cream should be applied to the area from which hair has been removed to reduce skin irritation.

If an ultrasonic contrast study is to be undertaken, a tail vein injection is generally required. Currently there is only one commercially available contrast agent marketed for preclinical applications manufactured by Bracco Research SA (Geneva, Switzerland). However many research groups manufacture microbubbles in-house for preclinical applications [10, 11]. Ultrasonic contrast agents are gas-filled microbubbles encapsulated by a lipid-shell. The gases tend to be fluorocarbons and in some instances a combination of nitrogen and perfluorcarbon. The contrast agents can be untargeted where they are used to visualise and quantify blood-flow or targeted to specific biological markers expressed within the vasculature for molecular imaging applications. The contrast agents are optimally visualised in the B-mode image using contrast-specific imaging techniques. These techniques rely upon the non-linear scattering properties of the microbubbles to multiple transmitted ultrasound signals of varying amplitude and phase compared to the predominantly linear scattering from soft tissues exposed to the same pulses [12]. On reception of the scattered signals, software enables the non-linear components to be isolated without significant background tissue structure being evident within the image. Selection of regions-of-interest within organs or vessels enables time-intensity curves of backscattered intensity from specific regions to be plotted as a function of time. Dependent on whether the contrast is administered as a bolus or infusion (typical maximum volume 50 μL in murine model) will influence the shape of the curves but typical parameters such as peak intensity, time-to-peak intensity, area under the curves graphs and mean transit time can all be measured. Either in-house or commercial software can then enable such parameters to be displayed graphically or in colour-coded parametric maps. This can be used to great effect in studying tissue perfusion in the kidney or in tumours [13].

Preclinical targeted contrast agents can be manufactured using either a commercial preparation conjugated with a biotinylated antibody or using in-house microbubble preparations attaching targeting ligands in a wide variety of ways [14]. After injection of the targeted microbubbles, the determination of the number of microbubbles which have bound to their target-sites relies upon the differentiation between freely circulating microbubbles and those attached to their targets. This can be achieved in a number of ways. The most straightforward way is to acquire a low-amplitude ultrasound image of both attached and freely circulating microbubbles, then utilise a large amplitude ultrasound pulse to destroy both circulating and attached microbubbles within the field of view, and then rapidly return to low amplitude mode and acquire images of freely circulating microbubbles. Subtraction of pre- and post-destruction frames will enable the magnitude of signal due to attached microbubbles to be quantified. Alternative methods to assess the quantity of microbubbles attached include analysis of dwell-time of microbubbles and interframe correlation of echo-signals [14].

4 Parameters of image quality

4.1 Resolution of ultrasound scanners

The spatial resolution of an ultrasound imaging system is determined by the smallest distance between two objects which can be resolved. In ultrasound imaging, spatial resolution is defined in the axial, lateral and azimuthal (slice thickness) planes. In the axial plane, the resolution is determined by the length of the transmit pulse of the transducer. This transmit pulse is typically several cycles in duration and axial resolution is generally of the order of half the transmit pulse length. The lateral resolution of the ultrasound beam is defined by the transmit beam width with optimal lateral resolution occurring at the focus of the beam and of the order of half the beam-width. With the use of array technology, multiple focal positions are possible in one image so optimal lateral resolution can be maintained over a much greater depth—however, this results in a slower frame-rate which may have implications for high speed imaging such as that required for preclinical cardiac imaging. The resolution in the third plane—the azimuthal plane—is also determined by the width of the beam in this plane. In array transducers, a lens on the surface of the transducer produces a fixed focus and optimal resolution in the azimuthal plane is obtained at this depth. Outwith this depth, slice thickness will be larger. For preclinical ultrasound scanners, spatial resolutions of 40 microns are achievable with ultrasound transmit frequencies up to 55 MHz.

4.2 Artefacts in preclinical ultrasound imaging.

Formation of a B-mode image relies upon several assumptions. Firstly, as described in section 2, ultrasound scanners assume a mean speed of sound in soft tissue of 1540 m s^{-1}. Actual speed of sound values in soft tissue can vary considerably—with lower values in fat tissue (1430 m s^{-1}) and higher values in liver tissue (1578 m s^{-1}). This variation may result in range errors where structures displayed within the image are misregistered with respect to their true anatomical position. It can also result in defocussing of the ultrasound beam. Secondly, it is assumed that attenuation is constant within soft tissue and that time-gain compensation can be applied which will compensate for this attenuation. However, for structures which attenuate strongly, a strong shadow may appear distal to the structure such as that observed with calcifications with vessels. In such instances, the attenuation due to the calcification is larger than that observed for soft tissue and the grey-scale values are not compensated to the level to compensate for this attenuation. Conversely with fluid-filled structures, such as cysts, which do not attenuate to the same extent as soft tissue, marked enhancement can be observed distal to the cyst structure. Thirdly the scanner assumes that the transmitted pulse travels in straight lines to a target and the scattered echo is transmitted back directly to the transducer. However due to the irregularity of tissue surfaces and the angle of incidence of the ultrasound beam to these surfaces, reflection artefacts can occur which may result in reduced echo backscattered amplitude, duplication of structures if there is a strongly reflecting interface or multiple reverberations due to a reflecting surface perpendicular to the transmitted beam.

5 Data processing

Ultrasound is a real-time imaging technique and, as such, much of the analysis and measurement of structures is performed during the scanning procedure. However, off-line analysis is generally performed for ultrasound studies using company-specific software where there is significant processing required or extensive analysis.

This is particularly the case for 4D cardiac imaging, strain and strain-rate analysis and contrast-enhancement studies.

6 Conclusions

Over the past 20 years, since the launch of the first preclinical ultrasound scanner, preclinical ultrasound imaging has gained increasing traction within preclinical laboratories as a versatile quantitative imaging modality providing real-time, high-resolution, diagnostic information within small animal models. The relative inexpense of the hardware compared to MRI, CT and PET scanners has enabled many research communities to use the technique for a broad range of organ scanning. The increasing use of artificial intelligence for analysis of images and detection of abnormalities will further enhance and support the non-specialist user in their use of preclinical ultrasound imaging.

6.1 Strength and limitations

The strengths of preclinical ultrasound scanning rest in the ability to rapidly acquire and view anatomical and blood flow information with spatial resolutions better than 50 microns and blood velocities ranging from 15 mm s^{-1} to 4 m s^{-1}. This enables the impact of real-time physiological events to be monitored and makes it an ideal technique for guiding injections.

The potential disadvantage of ultrasound is that when imaging at increasing depth, a lower frequency transducer is required and thus spatial resolution and frame-rate of the scanner will be reduced. In addition, artefacts as described in section 4.2 can occur, so it is essential that training is undertaken to ensure that the user is sufficiently skilled to acquire and interpret the real-time images in a meaningful way. Ultrasound is strongly reflected by gas structures (e.g. lungs and bowel) so imaging of the bowel and organs distal to the bowel are difficult unless cleared of gas. Imaging of the brain is possible in embryos but increasingly difficult in adult mice due to the strong attenuation of the ultrasound by the overlying skull. However, using Doppler techniques and nonlinear contrast imaging modes the vascular structure of the brain can be mapped.

6.2 Future developments

The increasing use of artificial intelligence and machine learning are likely to have increasingly high impact on all imaging techniques—both clinical and preclinical [15]. Within preclinical ultrasound imaging, this is likely to result in a more supportive landscape for acquisition and analysis of ultrasound data for biological users. In addition, over the past five years we have seen increased use of multispectral optoacoustic tomography (photoacoustic imaging) where a pulsed laser is used to excite tissues [16]. The tissues undergo thermoelastic expansion with the emission of ultrasound waves which can be detected using an ultrasound transducer. Use of different laser wavelengths allows detection of different photoacoustic molecules and has been used to good effect to measure haemoglobin concentration and changes in oxygenation levels within tissues. The rapid development of novel light-absorbing nanoparticles and organic dyes for specific applications is likely to result in rapid expansion in this field.

Secondly, the relatively new field of functional ultrasound utilising multi- plane-wave ultrasound imaging to enable imaging of neuronal activity has immense potential in the future for study of neurological diseases and neuroscience [17].

References and further reading

[1] Foster F S, Pavlin C J, Harasiewicz K A, Christopher D A and Turnbull D H 2000 *Ultrasound Med. Biol.* **26** 1–27

[2] Hoskins P R, Martin K and Thrush A 2019 *Diagnostic Ultrasound: Physics and Equipment* 3rd edn (Boca Raton, FL: CRC Press)

[3] Tremoleda J L, Kerton A and Gsell W 2012 *EJNMMI Res.* **2** 44

[4] Lindsey M I, Kassiri Z, Virag J A I, de Costra Bras L E and Scherrer-Crosbie M 2018 *Am. J. Physiol. Heart Circ. Physiol.* **314** H733–52

[5] Zacchigna S, Paldino A, Falcao-Pires I, Daskalopoulos E P and Dal Ferro *et al* 2021 *Cardiovasc. Res.* **117** 43–59

[6] Corrigan N, Brazil D P and McAuliffe F M 2010 *Reprod. Sci.* **17** 147–57

[7] Le Floc'h J, Cherin E, Zhang M Y, Akirav C, Lee Adamson S, Vray D and Foster F S 2004 *Ultrasound Med. Biol.* **30** 1303–19

[8] Jones N K *et al* 2020 *Hypertension* **75** 1213–22

[9] Rinaldi S F, Makieva S, Frew L, Wade J, Thomson A, Moran C M, Norman J E and Stock S J 2015 *Am. J. Pathol.* **185** 1201–6

[10] Peyman S A, McLaughlan J R, Abou-Saleh R H, Marston G, Johnson B R G, Freear S, Coletta P L, Markham A F and Evans S D 2016 *Lab Chip* **16** 679

[11] Beguin E, Bau L, Shrivastava S and Stride E 2019 *ACS Appl. Mater. Interfaces* **11** 1829–40

[12] Sidhu P S, Cantisani V, Dietrich C F, Gilja O H and Saftoiu A *et al* 2018 *Ultraschall Med.* **39** e2–44

[13] Shirinifard A, Thiagarajan S, Johnson M D, Calabrese C and Sablauer A 2017 *Ultrasound Med. Biol.* **43** 1628–38

[14] Wang S, Hossack J A and Klibanov A L 2018 *J. Drug Target.* **26** 420–34

[15] Liu S, Wang Y, Yang X, Lei B, Liu L, Xiang S, Ni D and Wang T 2019 *Engineering* **5** 261–75

[16] Costa M M, Shah A, Rivens I, Box C, O'Shea T, Papaevangelou E, Bamber J and ter Haar G 2019 *Photoacoustics* **13** 53–65

[17] Deffieux T, Demene C, Pernot M and Tanter M 2018 *Curr. Opin. Neurobiol.* **18** 128–35

Section 7

Magnetic resonance imaging

IOP Publishing

Imaging Modalities for Biological and Preclinical Research: A Compendium, Volume 2
Parts II–IV: *In vivo* preclinical imaging, multimodality imaging and outlook
Andreas Walter, Julia G Mannheim and Carmel J Caruana

Chapter II.7.a

Nuclear magnetic resonance spectroscopy and imaging

Teodora-Adriana Perles-Barbacaru, Angele Viola, Thomas Helbich and Viktoria Ehret

Magnetic resonance spectroscopy (MRS) and imaging (MRI) are a non-invasive means to obtain metabolic, physiological and functional information at the system, organ, tissue and cellular level in the living organism. Nuclear magnetic resonance (NMR) is based on the interaction of non-ionising electromagnetic waves with nucleons of atoms having net nuclear magnetic moment. Most clinical and research MRI studies acquire the signal originating from the hydrogen nuclei of water. For morphological MRI, several endogenous contrast mechanisms exist that can be weighted or enhanced by exogenous contrast agents to optimise tissue contrast. Dynamic or repeated studies are feasible with virtually no side effects to monitor spatial and temporal aspects of physiological and pathophysiological processes such as development, ageing, disease progression and treatment effects.

Multiparametric MRI consists in combining distinct MR methods to collect multiple relevant and often quantitative physiological parameters within one or several imaging sessions. In preclinical research, NMR techniques therefore contribute to the refinement of the experimental design and the reduction of the number of animals used. Drawbacks of NMR are its relatively low sensitivity, in particular, if NMR signals from atomic nuclei other than the water hydrogen are acquired, the need for complex physical models to obtain quantitative parameters, the compromise between acquisition duration, image quality and spatial resolution, and the high cost of the equipment and its maintenance.

1 Introduction

MRI and MRS are noninvasive techniques using non-ionising electromagnetic (EM) waves to probe tissue structure, composition and metabolism as well as many

doi:10.1088/978-0-7503-3747-2ch9

Table 1. MRI and MRS accessible parameters.

Measurable parameters	Examples
Structure of organs	Shape, borders, macroscopic components
Motion and deformation of organs	Speech articulation, muscle contractility, cardiac cycle, volumetry
Physical properties of tissues	Stiffness, viscosity, magnetic susceptibility
Tissue microarchitecture	Diffusivity, anisotropy
Vascular morphology, hemodynamics	Branching, stenosis, perfusion
Structural brain connectivity	Fibre tracts, network
Tissue activity, functional brain connectivity	Perfusion, oxygen consumption
Chemical composition, tissue distribution of compounds, metabolism	Iron, myelin, sodium (^{23}Na-MRI and MRS);
	Tissue specific metabolites: e.g. *N*-acetylaspartate, in brain, α-hydroxyglutaric acid in glioma, triglycerides in liver, citrate in prostate;
	Energy metabolism: e.g. phosphocreatine, adenosine triphosphate, pH measurement;
	Metabolic pathways: e.g. glycolysis, tricarboxylic acid cycle

magnetic, physiologic and functional parameters *in vivo* (table 1). In conventional MRI, hydrogen nuclei of water molecules are the main source of signal. The signal is modulated by physical mechanisms that are either exploited to optimise soft tissue contrast or to explore or even quantify properties of tissues or organs. MRI is a truly 3D technique without depth limitation, but less time-consuming multi-slice 2D acquisitions can also be performed in any plane. The spatial resolution in preclinical studies ranges between 50 to 500 μm *in vivo* and depends on the available signal amplitude and acquisition duration in order to ensure an appropriate image quality. Due to this compromise, the temporal resolution is in the order of minutes and requires immobilisation of the animal.

2 Principles and setups

2.1 Physical principles

NMR is based on the excitation of nucleons of atoms with net magnetic moments placed in an external magnetic field B_0 by EM waves in the radiofrequency (RF) domain and the subsequent reemission of the absorbed energy. The resonance frequencies are determined by quantum mechanical properties (box 1) and the emitted EM wave therefore carries information about the molecular composition of the sample (exploited for MRS) or can be used to encode the 3D space (enabling MRI).

Box 1. Basic quantum physics of nuclear magnetic resonance.

Atomic nucleons (protons and neutrons), as well as electrons have a half-integer intrinsic angular momentum (i.e. spin quantum number of ½) and obey the Pauli exclusion principle stating that only one particle can occupy a given quantum state at one time. This principle determines how these particles are distributed among discrete energy states (Fermi–Dirac statistics). In a nucleus containing an odd number of protons (such as the hydrogen nucleus, ^{1}H) and/or an odd number of neutrons (such as carbon-13, ^{13}C), the total nuclear spin $\vec{I}$ in the ground state is an integer multiple of ½. A magnetic moment $\vec{\mu} = \gamma\vec{I}$ is associated with a non-zero nuclear spin, where the proportionality constant $\gamma = \frac{q}{2m}$ is the gyromagnetic ratio characteristic of each nucleus, with q being its electric charge and m its mass. The magnetic moment is only observable in an applied external magnetic field $\vec{B}_0$ in which it takes one of $2I+1$ orientations, with different potential energy levels (Zeeman splitting). The quantised energies E are given by the projection of the magnetic moment vector $\vec{\mu}$ onto the external magnetic field: $E = \vec{\mu} \cdot \vec{B}_0$. The interacting EM wave has the energy $E = h\nu$ with the resonance frequency ν allowing transition between these energy levels. The sum of the nuclear magnetic moments $\sum\vec{\mu}$ in a volume element (voxel) of the sample gives rise to a net local magnetisation vector $\vec{M}$. Transition between energy levels therefore corresponds to flipping $\vec{M}$ out of its equilibrium orientation aligned with $\vec{B}_0$.

During the acquisition sequences, the net magnetisation $\vec{M}$ of the sample is manipulated by precise control of spatial variations of the magnetic field (i.e. magnetic field gradients ΔB) and of the durations, amplitudes and frequency composition of the emitted pulses of EM radiation. The excited atomic nuclei relax by reemitting EM waves detected in an antenna by electromagnetic induction. This NMR signal is analysed for its frequency distribution which is determined by the local magnetic field experienced by the nuclei (B_0 plus applied ΔB plus internal molecular magnetic field distribution). The received signal amplitude depends on the quantity of the emitting nucleus but also on a collection of parameters given by its chemical environment that determine its relaxation. The most abundant nucleus (in the order of 100 mol L^{-1}) in biological tissues is the one of the hydrogens in the water molecule and most NMR applications distinguish two magnetic relaxation rates depending on the projections (longitudinal and transverse) of the magnetisation vector with respect to the orientation of the applied $\vec{B}_0$-field. Each biological tissue at a given magnetic field strength B_0 has characteristic longitudinal and transverse relaxation rates described by the longitudinal (T_1) and transverse (T_2) relaxation time constants, which in conjunction with sequence timing parameters result in different signal intensities and consequently MRI tissue contrast. Another accelerated transverse relaxation (with relaxation time constant T_2*) results from superimposed external magnetic field inhomogeneities that are not compensated for. Other mechanisms such as the motion of the studied atomic nuclei and the magnetic susceptibility of nearby tissue components add to the tissue contrast. All these contrast mechanisms occur simultaneously and constitute the strength of MRI since they can be differently weighted until the desired tissue contrast is obtained. However, for quantitative measures, they need to be disentangled by careful signal processing. Each of these contrast mechanisms can then be exploited in many ways to probe physical, magnetic, physiological, functional and molecular properties of the tissue but this requires appropriate physical models in order to yield relevant and quantitative measures. MRI contrast agents can be administered to either enhance tissue contrast based on their biodistribution or to perform pharmacokinetic studies to derive for instance perfusion or metabolic parameters (e.g. oxygen consumption). MRI contrast agents do not emit any MRI signal but all act by increasing the tissue susceptibility and the longitudinal and transverse relaxation rate of nearby water molecules.

2.2 Typical setups and state-of-the-art

NMR and MRI scanners are cumbersome equipment with specific requirements regarding site planning (supporting foundation, appropriate RF shielding, cooling system, absence of electric devices and ferromagnetic objects in the vicinity), power supply and security measures, and with a relatively high cost at purchase and installation and for maintenance and operation. In addition, advanced research applications may require sequence optimisation by a specialist with knowledge of NMR physics and signal analysis using mathematical models. The main components of the MRI scanners are (i) a usually superconducting electromagnet generating the high static B_0-field and weighing several tons, (ii) a magnetic field gradient coil system to correct for B_0-field inhomogeneities and to generate linear magnetic field variations for spatial encoding of the signal, (iii) a radiofrequency transmitter and receiver coil system acting as an antenna, (iv) a radiofrequency synthesiser, transmitter, amplifier and analyser, and (v) a sequence control, data acquisition, data storage, signal and image processing and display system. Modern clinical scanners operate at 0.5 to 3T. Such $\vec{B}_0$-field strengths are appropriate for imaging large animals. For small rodents, preclinical MRI systems with $B_0 \geqslant 4.7$T and dedicated RF coils are recommended, although compact easy-to-use 1T MRI scanners for the bench-top are also commercially available and sufficient for basic MRI contrast. The image quality and spatial resolution that can be achieved within a limited acquisition time is highly dependent on the available signal-to-noise ratio, which increases with magnetic field strength and with the RF coil performance. A substantial gain in signal-to-noise ratio can be obtained with RF coil systems based on phased arrays of individual surface coil elements as well as with cryogenic RF coils and preamplifiers.

3 Biomedical relevance

3.1 Application range and relevance

Table 1 lists examples of measurable parameters. Both morphologic and functional parameters are valuable for exploration of organs and organ systems in animal models (figure 1(a)). Preclinical applications of MRI and MRS range from phenotyping to non-invasive follow-up studies for monitoring disease progression and treatment effects or for developmental studies (figure 1(b)). Hybrid systems

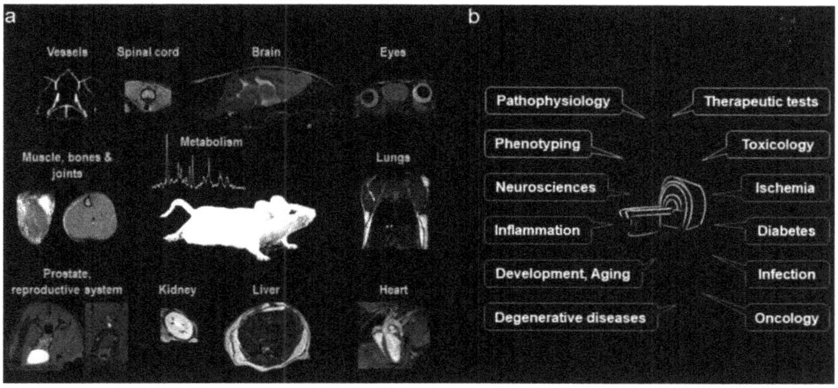

Figure 1. Examples of preclinical MRI applications. (a) Organs and systems that can be explored. MR images are not to scale. (b) Research topics and disorders that can be studied with MRI. Credit: TS Vergara Gomez, B Giannesini, F Kober, AT Perles-Barbacaru and A Viola from Center for Magnetic Resonance in Biology and Medicine (CRMBM), Marseille.

(e.g. PET/MRI) enable correlation of molecular imaging techniques with functional MRI measures for instance in pharmacological studies [1]. Various MRI contrast agents with different biodistributions and half-lives, as well as smart contrast agents targeting specific tissues, molecules or metabolic processes have been developed, making MRI more specific and increasing its application range [2].

3.1.1 Morphological / anatomical MRI

The high MRI contrast between tissues or lesions and the achievable spatial resolution make this modality useful for the volumetric and morphologic study of various body structures and lesions and their sub-compartments. Example applications are comparative studies between species or genetically modified animals, developmental and aging studies, monitoring of tumour progression and metastatic extensions as well as treatment effects. Within this category, 3D microscopic MRI requires ultra-high B_0-field preclinical scanners and involves longer acquisition durations in order to achieve high isotropic spatial resolutions (10 to 50 μm)^3. It is usually performed *ex vivo*, e.g. on excised and fixed tissues or entire embryos. The addition of often nonspecific contrast agents can improve the contrast and shorten acquisition times [3].

3.1.2 Quantitative MRI

Contrary to anatomical MRI, quantitative MRI aims at measuring one particular tissue parameter within each voxel by separating its contribution to the signal from all other contributions with as little bias as possible. In general, this involves multiple acquisitions for which one sequence parameter is varied followed by appropriate modelling of the signal behavior. The modelled parameters are then expected to be independent of sequence parameters and therefore comparable across acquisition methods and MRI equipment [4]. The relaxation time constants, the water diffusivity, the blood volume fraction and the blood flow are examples of physical and physiological parameters that can be mapped quantitatively over the entire tissue or organ. The measured physical parameters serve as quantitative biomarkers as they can indirectly inform about tissue composition, microstructure or pathological processes such as ischemia or inflammation [5]. For instance, myelin quantification, inferable from magnetisation transfer MRI, myelin water fraction, transverse relaxation time and tissue susceptibility mapping, is used to monitor demyelinating lesions in models of multiple sclerosis and the efficacy of drugs promoting remyelination. Quantitative parameters obtained by diffusion MRI are also used to assess myelin degradation and axonal damage.

3.1.3 Magnetic resonance angiography

Magnetic resonance angiography (MRA) selectively images functional (i.e. perfused) macroscopic vessels in order to study their branching and diameter to diagnose anatomic variants, occlusion or stenosis. It can be achieved with or without the administration of contrast agents. Without contrast agents, the sensitivity of the MRI signal to motion is exploited. Sequence parameters can be optimised to maximise the signal intensity difference between flowing blood water and quasi-static parenchymal water to study vascular morphology. Although for this time-of-flight (TOF) angiography technique, the blood signal intensity depends on the blood velocity within a certain range, another signal component, the phase, allows improved quantification of the blood velocity. The phase contrast (PC) angiography technique is based on the subtraction of acquisitions with and without flow encoding in each spatial direction. When contrast agents are administered intravenously and distribute within the blood compartment, their selective effect on the blood relaxation rate is exploited to generate vascular contrast. Figure 2 shows such acquisitions in a venous thrombosis model.

3.1.4 Magnetic resonance elastography (MRE)

The motion sensitivity of MRI is also exploited to measure the propagation speed of mechanically generated shear waves through tissue and infer the tissue stiffness from it. MRE-derived stiffness correlates with fibrosis stage and tumour cell density. Compared to ultrasound based elastography techniques, MRE is less operator-

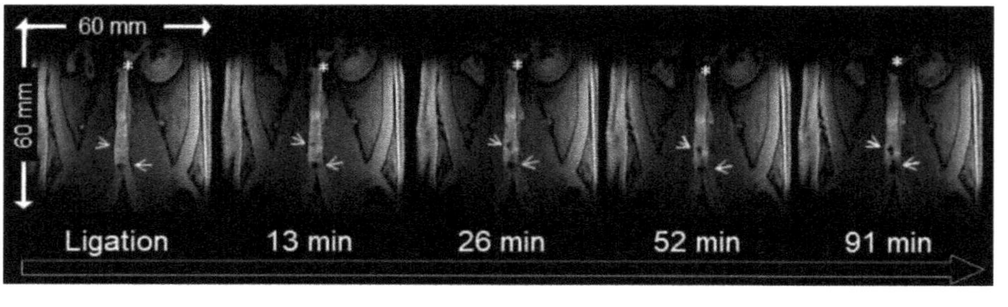

Figure 2. Repeated T_1-weighted MRI at 9.4T after contrast agent application (Gadospin F, Viscover, Germany) in a rat shows formation and growth of thrombi after ligation of the inferior vena cava (ligature at asterisk). (FLASH sequence, 100 ms repetition time, 1.9 ms echo time, 3 averages, 30 degree pulse flip angle, 14 coronal slices, 256 × 256 matrix, 0.234 × 0.256 × 1 mm^3 spatial resolution, 13 min scan time).

dependent and can generate 3D stiffness maps without being limited by depth or hampered by air and bone. However, confounding comorbidities, such as iron overload, can limit the reliability of MRE.

3.1.5 Cardiac MRI

Three options exist in order to image periodically moving organs with MRI acquisitions lasting several minutes: first, the organ can be immobilised for short time intervals while a part (segment) of the MRI data is rapidly acquired (i.e. breath hold in cooperative patients, or short intermittent interruptions of mechanical ventilation in anesthetised animals). Second, segments of data are acquired continuously while monitoring the motion of the organ simultaneously (e.g. by means of a navigator or electrocardiogram (ECG)), and data segments acquired within the same position over the motion cycle are retrospectively combined to yield an image without motion artefact. Third, the acquisitions can be prospectively synchronised by starting the acquisition of data segments needed to complete an image at identical positions in the motion cycle. In cardiac MRI, synchronisation with the ECG gives best results as long as the cardiac cycle is fairly regular. The ECG synchronised acquisitions at different time points within the cardiac cycle result in multiple images that can be visualised as a movie in false speed, a technique termed Cine MRI. In addition to the detection of lesions and morphological alterations, image processing involving segmentation of ventricle lumen and myocardium in multiple phases of the cardiac cycle enables the evaluation of functional parameters such as ejection fraction and contractility.

3.1.6 Perfusion MRI and pharmacological MRI

Fast serial MRI monitoring temporal signal changes in tissue (often the brain) during intravenous injection, distribution and elimination of a contrast agent bolus, allows deriving parameters related to microvascular perfusion [4]. The so-called first pass technique often relies on repeated T_2- or T_2*-weighted acquisitions, termed dynamic susceptibility contrast (DSC) MRI. After conversion of the signal intensity to tissue contrast agent concentration over time and deconvolution by the arterial input function (AIF), which is the contrast agent concentration curve over time in a vessel feeding the tissue, DSC-MRI yields the blood volume, the blood flow and the mean transit time of the contrast agent bolus. The dynamic contrast enhanced (DCE) technique uses serial T_1-weighted acquisitions to monitor the signal beyond the first pass. After conversion to contrast agent concentration over time and determination of the AIF, the data is fitted to an appropriate two or multi-compartment exchange model of the tissue in order to derive plasma volume v_p, the efflux rate (K^{trans}) of contrast agent into the interstitium and the extravascular distribution volume fraction v_e of the contrast agent.

The blood volume can also be assessed from the susceptibility-weighted signal difference before and after administration of a contrast agent restricted to the blood pool. By using blood pool contrast agents with a long blood half-life, such as ultrasmall particles of iron oxide, the acquired signal remains blood volume weighted over minutes to hours. Blood volume changes, such as occurring when brain tissue is activated, can then be monitored similarly to functional MRI.

Unlike such contrast agent-based techniques, arterial spin labelling (ASL) relies on magnetically labelled water as endogenous freely diffusible tracer perfusing the studied tissue. The labelling is achieved by preparing the magnetisation $\vec{M}$ of blood water in feeding vessels using pulses of EM radiation before it enters the tissue of interest. In order to derive the blood flow, a T_1-map, and the signal difference between a labelled acquisition and a control acquisition is required.

Perfusion MRI is used to asses ischemia, microangiopathy, tumour angiogenesis and treatment outcome. If blood flow is not the limiting factor, K^{trans} is proportional to the endothelial permeability, an important parameter in many pathological processes, in particular in brain tissue.

Pharmacological MRI (phMRI) consists in measuring one of these perfusion parameters before and during challenge with a vasoactive drug. For instance, the vascular reserve can be assessed by repeating the perfusion measurement after administration of acetazolamide [6]. When monitoring the cerebral blood volume or flow under pharmacological challenge with a psychoactive drug, pharmacological MRI is a means to assess activated or depressed brain structures.

3.1.7 Diffusion MRI and diffusion tensor imaging

The MRI signal is also sensitive to Brownian motion of water molecules resulting from thermal agitation. This incoherent motion allows probing the microstructure of biological tissue because the mean displacement of water within the time interval between excitation and signal reception (tens of milliseconds) is in the order of cellular structures (2–20 μm) [7]. Although diffusion affects the signal in almost all acquisitions, MRI equipment capable of generating high magnetic field gradients is needed to fully exploit this effect. Diffusion-weighted images and quantification of the apparent diffusion coefficient (ADC) from a couple of differently diffusion-weighted acquisitions [8] are widely used in biomedical imaging to evaluate the type of edema occurring in inflamed or ischemic tissue, cell density in space occupying lesions as well as therapeutic effects. A reduction of the ADC has also been observed in activated brain tissue and is likely due to cell swelling.

When diffusion is quantified in at least six independent directions (a technique termed diffusion tensor imaging, DTI), anisotropic tissue (e.g. muscle fibres and white matter) can be explored and characterised by quantitative parameters such as directional diffusivities and fractional anisotropy (FA). By determining the direction of least restricted diffusion in every image voxel, DTI post-processing algorithms can reveal fibre orientation and build a map of structural brain connectivity. Along with the functional connectivity maps, tractography is a valuable tool in neuroscience research. More complex signal analysis (such as diffusion kurtosis) and biophysical models of diffusion processes in white matter (such as the stick model of axons) derive additional quantitative parameters to characterise the tissue microstructure.

3.1.8 Magnetisation transfer (MT) MRI

MT occurs when off-resonance RF pulses are used to apply energy to the water molecules bound to macromolecules and therefore characterised by a very broad frequency spectrum. Part of this energy is transferred to the magnetisation of free water resulting in a signal decrease (expressed as the MT ratio, MTR) in tissues with a high macromolecule content. The MTR is sensitive to parameters of the off-resonance pulse. Quantitative MT imaging (QMT imaging) relies on biophysical models of the tissue and aims at quantifying the equilibrium magnetisation, fractions and relaxation rates of the free and bound water pool as well as the exchange rate between the pools. The MT effect increases the contrast between lesions and healthy parenchyma and is mainly used in neuroimaging. The MT induced signal decrease is also exploited for background suppression.

3.1.9 Molecular MRI

Smart contrast agents (often nanoparticles of iron oxide) can be functionalised to target specific interstitial, cytoplasmatic or cell membrane bound molecules and can be detected with contrast agent-sensitive MRI sequences such as T_2*-weighted or susceptibility imaging [2]. Application examples are *in vivo* tumour angiogenesis detection by targeting the endothelial growth factor receptor, evaluation of Alzheimer disease models by targeting amyloid plaques, detection of apoptosis, inflammation and enzymatic activity and tissue pH measurement. It is also feasible to monitor tissue migration of tumour cells or stem cells over time (cell tracking) by 'cellular' MRI after *in vitro* internalisation of nanoparticles of iron oxide and subsequent administration of such magnetically labelled cells into a living animal. This approach enables the study of pathophysiological mechanisms and the evaluation of treatment strategies.

3.1.10 Magnetic resonance spectroscopy and imaging

The characteristics of MR spectra such as the chemical shift and splitting of resonance frequencies, signal intensity and relaxation time constants are analysed by chemists and biologists to non-destructively identify molecules, determine their structure and investigate their dynamics [9]. With the advent of MRI scanners, MRS methods using the same equipment as for imaging were developed for the non-invasive investigation of metabolism in animals and humans. However, high B_0 permits increasing the quality of the spectrum and enhancing spectral resolution. MRS can detect small (<500 Da) mobile metabolites with tissue concentration in the millimolar range and is used in physiology, pathology and pharmacology research.

Localised MRS acquires the spectrum from a volume of interest (single voxel in the mm^3 range in preclinical studies), whereas the chemical shift imaging (CSI) technique also called MRS imaging (MRSI) enables acquisition of spectra from multiple contiguous voxels to generate metabolic maps.

In vivo MRS acquires the signal from biologically relevant atomic nuclei with net magnetic moments, mostly ^{1}H, ^{31}P and ^{13}C. While (^{1}H)-MRI detects water and lipid signals, ^{1}H-MRS necessitates water signal suppression to study tissue specific metabolites (e.g. neurotransmitters, N-acetylaspartate, choline-containing compounds in brain, α-hydroxyglutaric acid in glioma, triglycerides in liver, citrate in prostate).

^{31}P-MRS necessitates longer acquisition times than ^{1}H-MRS due to lower sensitivity. It is used to study bioenergetics (e.g. phosphocreatine, adenosine triphosphate, adenosine diphosphate), phosphomonoesters and phosphodiesters (generally increased in cancer) and to measure pH.

^{13}C-MRS is the least sensitive method and requires isotopic enrichment due to the low natural abundance of the ^{13}C nucleus. The enrichment can be obtained by intravenous injection of a stable ^{13}C labelled compound (e.g. glucose, acetate, pyruvate) or by dynamic nuclear polarisation. ^{13}C-MRS can be used to study almost all metabolic pathways (glycolysis, tricarboxylic acid cycle, pentose phosphate pathway, neuron/astrocyte cycle) and its main applications are in cancer research [10].

3.1.11 Functional MRI

Functional MRI (fMRI) aims at mapping task-related or cognitive state related activation of brain tissue by measuring a physiological process linked to neuronal activation. Diffusion MRI is sensitive to fractional changes of brain tissue compartments during activation. Perfusion MRI is sensitive to changes in either blood flow or blood volume accompanying activation. Such local hemodynamic changes occur due to neurovascular coupling required to supply the activated brain tissue with glucose and oxygen. Most fMRI studies, however, measure a signal change resulting from the combined effects of blood volume and blood flow increase and of the reduction of paramagnetic properties of oxygenated hemoglobin. This blood oxygenation level dependent (BOLD) signal is monitored with fast T_2*-weighted acquisitions. Brain tissue is believed to be activated when its temporal BOLD signal changes are highly correlated with the task design. With resting state fMRI (rs-fMRI), functional connectivity maps are obtained by analysing the correlation of task-unrelated BOLD signal fluctuations between brain regions. Functional MRI is described in more detail in a dedicated chapter of this compendium.

Another type of functional imaging, only feasible in preclinical studies due to potential long-term toxicity, is manganese enhanced MRI (MEMRI). After systemic or local administration, this paramagnetic agent is taken up by activated neurons as an analog to calcium. Due to anterograde and retrograde transport along the axon as well as across the synaptic cleft, manganese acts like a tracer of activated fibre tracts. The tracing remains visible on T_1-weighted acquisitions hours after the stimulation has ceased.

3.1.12 Susceptibility MRI

The magnetic susceptibility of a substance distorts the magnetic field around it and accelerates transverse relaxation and dephasing of nearby water. Susceptibility weighted imaging (SWI) exploits this effect to facilitate detection of paramagnetic and diamagnetic compounds in tissue, by analysing the phase of the signal in addition to its attenuation. SWI can detect microhemorrhages, small veins and other paramagnetic and superparamagnetic substances (such as contrast agents) and distinguish them from diamagnetic substances such as calcifications. Quantitative susceptibility mapping (QSM) aims at assessing local susceptibility of tissue hidden in the phase images by removing the effect caused by magnetic field distortion from other sources. QSM being independent of acquisition parameters enables quantification of tissue iron concentration, contrast agent accumulation and venous deoxyhemoglobin concentration. Susceptibility tensor imaging aims at quantifying susceptibility anisotropy and might enable assessment of myelin integrity.

3.1.13 Multiparametric MRI

The ability to obtain multiple relevant quantitative parameters from a single or a combination of MRI methods without harmful radiation greatly facilitates tissue characterisation and remains unparalleled by other biological imaging modalities.

3.2 Sample preparation

In preclinical *in vivo* MRI and MRS, the most convenient option to reduce motion and stress is to anesthetise the animal by prior injection or continuous inhalation. However, the impact of anesthesia on the physiology of the animal, on disease progression and on pharmacokinetics of studied substances, as well as on brain metabolism and functional networks has to be taken into account in the experimental design. The animal is placed into a suitable imaging cradle (figure 3), equipped with a nose cone for oxygen and anesthesia delivery, MRI-compatible sensors to monitor the animal's well-being (respiration, temperature, ECG, blood oxygenation and pressure), a heating system and the RF coil for imaging. Ear-bars, tooth-bars and adhesive tape are useful to avoid respiratory motion. If required, a contrast agent can be administered during acquisition via an intravenous or intraperitoneal catheter extended with microtubing for access from outside the magnet. In order to minimise total injection volume (max. 5 ml kg^{-1} i.v., 10 ml kg^{-1} i.p.) and to avoid air emboli, the extension line should be prefilled, and its dead volume minimised. An automatic injector is recommended for higher reproducibility. The contrast agent dose depends on the experimental protocol and the administration

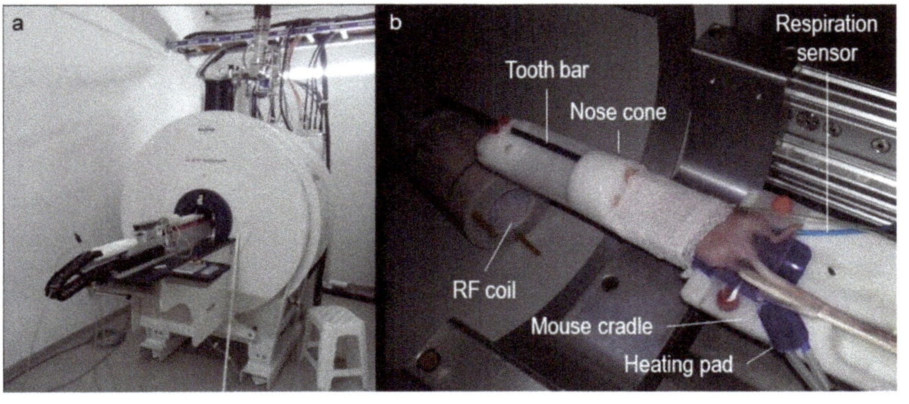

Figure 3. Preclinical MRI equipment. (a) B_0 = 7T Bruker PharmaScan. (b) Mouse setup for imaging.

route and is usually higher than in humans when using clinically approved contrast agents due to faster metabolism and elimination in rodents.

In vitro imaging of harvested organs or embryos requires avoidance of air tissue interfaces by immersion in a fixation liquid (e.g. formaldehyde) or perfluoropolyether fluorocarbon (hydrogen-free). Care should be taken to immobilise the sample (e.g. agarose embedding) and to avoid air bubbles producing susceptibility artefacts.

4 Parameters of image quality

The digitised MRI signal is acquired in the spatial frequency domain (k-space, figure 4). To reduce acquisition duration, the k-space is sampled in a limited number of discrete points (the k-space matrix) chosen prior to the acquisition, and often k-space points are omitted. The quality of the reconstructed image after inverse Fourier transformation depends on the data points acquired in k-space and the noise they contain. The k-space matrix has the same size as the image matrix. Low spatial frequencies determine the image contrast while high spatial frequencies determine the spatial resolution. To increase signal and average out the noise, k-space data points can be acquired multiple times. If one wishes to increase one image quality parameter (signal-to-noise ratio, contrast or spatial resolution) it comes with the cost of lengthened acquisition time. For instance, when spatial resolution is increased, the signal originates from smaller voxels and has a lower amplitude unless this is compensated for by increasing acquisition time or magnetic field strength. Another image quality parameter is the absence of artefacts. Careful preparation (sample immobilisation, correction of magnetic field homogeneity, RF-shielding) and sequence optimisation help reduce artefacts.

5 Data processing

The acquisition software also reconstructs images from the k-space data, involving Fourier transformation and additional data processing steps (e.g. filtering, zero-filling, distortion correction) and displays them. Qualitative contrast such as in morphological MRI is usually displayed in greyscale (figure 4). Parametric maps are

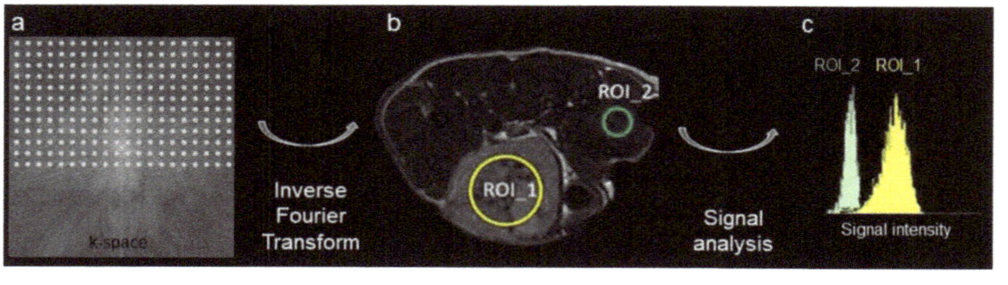

Figure 4. Processing of MRI acquisitions. (a) Data is acquired in k-space in discrete points (uniform grid overlay shown). (b) Morphological T_2-weighted MR image of a tumour bearing mouse. (c) Example of basic region of interest (ROI) analysis of signal intensity distributions in tumour (ROI_1) and muscle (ROI_2).

visualised using a grey scale or colour look-up-table accompanied by a bar specifying the values and the units of the parameter. TOF and PC angiography are best visualised by maximum intensity projection (MIP) where one can observe the vessels in the acquisition volume from different angles (projections). More complex post-processing algorithms based on pharmacokinetic or biophysical models often require dedicated commercial or free software (e.g. FSL, FreeSurfer) which may need further customisation. Serial imaging, multiparametric and multi-modal imaging may require image registration and image segmentation algorithms. Volumetric and morphologic comparisons between time points or experimental groups, and correlation between quantitative parameters can be analysed. MRS data processing also involves multiple steps and requires quantification methods in the time (e.g. AMARES, QUEST) or the frequency domain (e.g. LC-model).

6 Conclusions

6.1 Strength and limitations

Although MRI requires expensive equipment and expertise, the anatomical contrasts and detail and the multitude of quantitative parameters available, makes it a versatile tool for non-invasive *in vivo* or non-destructive 3D *in vitro* imaging. Many accessible physical and functional parameters correlate with pathophysiological mechanisms and can be acquired repeatedly without harm. MRI therefore contributes to the refinement of experiments and reduction of animals enrolled. Almost all MRI and MRS techniques are already performed in or are translatable to clinical research and enable comparisons of biomarkers between disease models and clinical pathologies.

Drawbacks of NMR are its relatively low sensitivity if NMR signals from atomic nuclei other than the water hydrogens are acquired, the need for complex physical models to obtain quantitative parameters and the compromise between acquisition duration, image quality and spatial resolution.

6.2 Future developments

Technical improvements in image quality and acquisition speed as well as in data post-processing are still being made, some of them relying on artificial intelligence. New contrast mechanisms and advanced quantitative MRI techniques are developed and tested for their biomedical relevance. More specific contrast agents are being developed to target particular tissues, cells or pathophysiological mechanisms. Theragnostic approaches are also explored.

Hybrid systems or inserts for simultaneous molecular imaging and MRI [1] were recently commericalised, in order to combine the functional specificity of nuclear medicine imaging (PET and SPECT) or bioluminescent imaging with the high tissue contrast and spatial resolution of MRI. Development of dual imaging agents for nuclear medicine and MRI is ongoing.

References and further reading

[1] Ross B and Gambhir S 2020 *Molecular Imaging—Principles and Practice* 2nd edn (Amsterdam: Elsevier)
[2] Schroeter A *et al* 2012 MRI *Small Animal Imaging—Basics and Practical Guide* ed F Kiessling, B J Pichler and P Hauff (Cham: Springer)
[3] Zhang X *et al* 2010 *Magn. Reson. Med.* **63** 1391–7
[4] Tofts P 2003 *Quantitative MRI of the Brain: Measuring Changes Caused by Disease* (Chichester: Wiley)
[5] Yankeelov T E, Pickens D R and Price R R 2011 *Quantitative MRI in cancer* 1st edn ed W R Hendee (Boca Raton, FL: CRC Press)
[6] Perles-Barbacaru T A *et al* 2012 *NMR Biomed.* **25** 498–505
[7] Le Bihan D and Iima M 2015 *PLoS Biol.* **13** e1002203
[8] Le Bihan D 2013 *Radiology* **268** 318–22
[9] de Graaf R A 2018 *In vivo NMR Spectroscopy – Principles and Techniques* 3rd edn (Chichester: Wiley)
[10] Lai M, Vassallo I, Lanz B, Poitry-Yamate C, Hamou M F, Cudalbu C, Gruetter R and Hegi M E 2018 *Int. J. Cancer* **143** 127–38

Chapter II.7.b

Functional magnetic resonance imaging

Xin Yu and Patricia Pais-Roldán

Functional magnetic resonance imaging (fMRI) allows measuring hemodynamic changes that are relevant to neural function non-invasively, with whole-brain coverage. The setup used to acquire the 4D-functional images is the same as for anatomical MRI, but several stimulation-related devices are added to study evoked functional responses in animals. Usually, a whole-brain volume is imaged every few seconds for a duration of several minutes, which results in a 4D matrix, i.e. one time course at each voxel within the brain. Besides evaluating the activation of voxels relevant to particular circuits (e.g. somatosensory cortex in response to peripheral electrical stimulation), it is possible to acquire resting-state scans (>10 min, without stimulation) and evaluate connectivity networks. In the past years, a substantial effort has been made to merge the fMRI platform with other methodologies, like cell-specific optogenetics or calcium imaging through optical fibres, which allows complementary analysis to better understand the brain function. This chapter provides an overview of fMRI contrasts, sequences, common setups and applications, and it will guide the reader through the main steps that are needed to conduct an fMRI experiment in small animals.

1 Introduction

Neuronal activity results in increased local blood flow, which leads to higher ratio of oxy-deoxy hemoglobin in the adjacent capillaries, and in local vasodilation. These hemodynamic changes are the source of most fMRI contrasts (e.g. T2-weighted sequences track changes in the blood oxygenation) [1]. Although the typical fMRI scheme measures changes that attain to the vascular compartment, it has been demonstrated that, although indirectly, this fMRI signal reliably tracks the activity of neuronal populations.

doi:10.1088/978-0-7503-3747-2ch10

In order to detect functional changes, fMRI volumes are repeatedly acquired within short time intervals, which results in fMRI time courses with relatively high temporal resolution to detect hemodynamic responses (in the order of 0.5–2 s for common studies covering the whole animal brain). In animals, the platform combining fMRI with cell-specific optogenetic stimulation constitutes one of the main tools to dissect brain circuits [2]. Besides, the inclusion of animal models of disease in neuroimaging labs, together with the recent interest for understanding the brain state, has boosted the use of resting state fMRI (rs-fMRI) to evaluate network connectivity patterns linked to specific neurological states.

2 Principles and setups

2.1 Physical principles

The source of the fMRI signal depends on the sequence used during its acquisition. fMRI sequences have been developed to optimise the detection of particular hemodynamic events linked to neuronal activity (i.e. to exploit a specific contrast that allows differentiation between baseline and activation conditions). Often, several sequences exist that provide a decent fMRI contrast, and the choice for one of them is made based on the application (e.g. faster sequences for studies focused on the temporal dynamics of the signal). In this chapter, common fMRI contrasts are described, such as blood oxygen level dependent (BOLD) [3, 4], cerebral blood volume (CBV) [5–7] or cerebral blood flow (CBF) [6].

BOLD fMRI takes advantage of the variable magnetic properties of the iron-based protein hemoglobin that is present in the blood cells. When bound to oxygen, oxy-Hb is mostly diamagnetic, and in the absence of oxygen, the deoxy-Hb remains paramagnetic. This difference in the magnetic properties of the hemoglobin confers the blood an oxygen-dependent magnetisation constant that is easily detected in T2-weighted MRI. When neurons fire, the increased blood flow towards the activated region leads to a higher oxygenation of the blood and, hence, to a higher ratio of oxygenated hemoglobin in the local capillary bed. Therefore, BOLD fMRI is characterised by higher MR signals in active areas (i.e. longer T2/T2* constants in T2-weighted images).

Because oxygen lengthens both, the T2* and T2 constants of blood, the BOLD effect can be assessed using Gradient Echo (GE) or Spin Echo (SE) sequences, with the first one being more sensitive to the hemodynamic changes but less accurate in terms of the potential location of the neuronal source. At high magnetic fields (>7T MRI scanners), SE sequences could be preferred for applications that are especially sensitive to the location of the signal (e.g. laminar analysis), since the T2* from large pial veins obscures the changes that are attributable to T2, which are more relevant to the local neuronal activations. Given the fast-sampling rate, GE-BOLD fMRI is the preferred methodology to study whole brain function. Some common fMRI sequences are listed in table 1.

Another indicator of neuronal activation is the dilation of arterioles supplying blood to the surrounding capillary bed, which is less spatially-biased and shows faster temporal dynamics than the BOLD signal. Vessel dilation can be tracked in

Table 1. fMRI sequences. Abbreviations: arterial spin labelling (ASL), echo planar imaging (EPI), fast low angle shot (FLASH), gradient echo (GE), monocrystalline iron oxide nanoparticle (MION), specific absorption rate (SAR), spin echo (SE), vascular occupancy (VASO).

MRI sequence	fMRI contrast	Advantages	Disadvantages
GE-EPI	T2* BOLD	Fast High SNR Non-invasive	Unspecific
GE-EPI + MION/ ferrous ion	CBV	High SNR Specific to arterioles	Infusion of exogenous iron compounds
GE-EPI + inversion recovery pulse to null blood water (VASO)	CBV	Specific to arterioles Non-invasive	Low sensitivity
SE-EPI	T2 BOLD	Specific to capillary bed Non-invasive	Low SNR High SAR
FLASH	T2* BOLD	High Spatial Resolution High SNR No distortions Non-invasive	Slower
EPI with ASL	CBF	Better spatial localisation Non-invasive	Low temporal resolution

fMRI with sequences that are sensitive to the vascular compartment, with or without employment of a contrast agent. A non-invasive strategy consists of adding an inversion recovery pulse to an EPI sequence to null the blood signal. Because water in the blood exhibits specific relaxation constants that are different from those in other tissues, it is possible to apply the RF excitation pulse at the exact moment when the blood signal is completely zero, which results in a blood-free image. This technique, referred to as vascular space occupancy (VASO), provides a signal that is proportional to '1—blood volume', therefore, a reduced signal is expected in a voxel holding neuronal activation. In animals, VASO is typically substituted by an invasive but more efficient technique which involves intravenous infusion of a ferrous agent, typically monocrystalline iron oxide nanocalloid (MION). This kind of CBV measure relies on the fact that ferrous materials shorten T2* significantly, which leads to a signal dropout in T2*-weighted images that is proportional to the concentration of the iron-based molecule. Hence, increases of blood volume upon neuronal firing are observed as a decrease in the T2*-weighted signal acquired with GE-EPI sequences. BOLD and iron-based CBV can complement each other in fMRI experiments to distinguish venules (positive BOLD) from arterioles (negative iron-based CBV), which results in arteriole-venule brain maps [7].

A third method to track neuronal activity is based on measuring CBF, which increases locally upon neuronal firing. A common CBF fMRI sequence performs arterial spin labelling (ASL) with saturation or inversion pulses, which allows distinguishing the moving protons in the blood from the static protons in the tissue water without using exogenous tracers. In addition, CBF can be estimated by applying bipolar gradients and measuring the resulting phase shift, which is proportional to the velocity of the blood flow.

Given that the fMRI signal is an indirect measure of neuronal activity, the temporal dynamics of this signal in a given voxel are different from those observed in electrophysiological recordings (i.e. local field potentials or single/multiunit activity). An extracellular electrode located in the cerebral cortex can detect neuronal responses that occur within a few milliseconds; in contrast, the BOLD signal recorded from a cortical voxel in response to a stimulus event peaks at 5–6 s. Despite the different nature of the signals, the relationship between fMRI (e.g. BOLD) and local field potential recordings has been demonstrated, and a standard hemodynamic response function (HRF) has been proposed that is used in fMRI analysis. However, it has been shown that the specific relationship between the firing of a neuron and the vascular changes that follow, i.e. the shape of this HRF, can vary depending on the age or the neurological state of the subject. The mechanisms underlying the coupling between the vasculature and the neuronal function (the 'neuro-vascular coupling') constitute a topic of active research [8]. Potential players mediating this process are astrocytes (with end-feet in neurons and in capillaries), pericytes, or specific VASO-active factors and neurotransmitters released from neurons or astrocytes.

2.2 Typical setups

In addition to the MR-related hardware (see chapter II.7.a *Nuclear magnetic resonance spectroscopy and imaging*, section 2.2), fMRI experiments typically include a series of devices to monitor the physiological state of the subject. Examples of such devices are:

- A **blood pressure transducer**.
  In terminal animal fMRI experiments, a catheter is placed surgically in the femoral artery, which allows reliable measurement of the blood pressure during the whole imaging time. The catheter is connected, in the other end, to a blood pressure transducer. This procedure also allows one to take blood samples during the experiment that can be analysed on site with a blood analyser. Alternatively, a non-invasive blood pressure recording system can be placed on one of the paws of the animal; however external devices can be easily detached during an fMRI experiment and may fail to report blood pressure changes. The heart rate can be automatically calculated from the blood pressure waves.
- A **respiratory circuit connected to an artificial ventilator**.
  Due to the high sensitivity of fMRI to motion, many fMRI experiments are conducted in paralyzed animals that need ventilatory support. The animal artificial ventilator, which provides a mixture of air and oxygen (and inhalatory anesthetics, when applicable), guarantees continuous gas supply to the animal and allows the investigator to control the ventilatory parameters (pressure, volume, rate), which can be used to adjust the blood

oxygenation and, hence, influence the BOLD signal. Ventilated animals require intubation at the beginning of the experiment. When animals are not intubated (i.e. when no paralyzer agent is used and passive ventilation is not required), a respiratory circuit is still built that supplies a specific gas mixture to the animal snout (a cone mask allows free respiration) to enhance oxygenation and provide inhalatory anesthesia, if necessary.

- An **end-tidal carbon dioxide (et-CO_2) transducer**.
  In both, intubated and non-intubated animals, the respiratory circuit usually involves a tube that conducts the exhaled air from the animal towards a capnograph. This allows tracking the production and clearance of CO_2 in real time, which is an indicator of the anesthesia depth and helps keeping the animal in the right physiological state.
- A **chest motion transducer**.
  In small animals, this device consists of a small sensor pad that is located under the belly of the animal (placed in prone position), allowing for non-invasive monitoring of the chest motion relevant to respiration. Respiratory rate can be calculated easily from this recording. In humans, the equivalent sensor comes in the form of a belt that is attached to the chest of the subject.
- A **temperature probe**.
  The temperature is an important variable that affects the physiological state of anesthetised animals and it must be carefully controlled over the imaging period. To guarantee constant corporal temperature (37.5 degrees for rats), a feedback loop is generated using a rectal temperature probe and a heating pad placed under the animal body.
- An **infusion pump**.
  Many animal experiments require continuous infusion of an intravenous anesthetic. In these cases, catheters are surgically placed through the femoral vein and connected to an infusion pump placed in the room next to the MR scanner. The speed of the infusion (usually a combination of anesthetic and paralyzer agent) can be modified during the course of an experiment to reach a stable physiological state. Attention should be paid to the maximum volume per kilogram and per hour that may be intravenously injected into the animal (e.g. for rats, maximum 4 mL/kg/h).
- A **recording system**.
  Most of the transducers described above (e.g. blood pressure transducer, chest motion detector, etc) are connected to analog-to-digital converters and read by software that allows the investigator to easily monitor all the physiological variables in the MR control room. In human scanners, the physiological recordings can normally be displayed directly on the MRI acquisition system.
- A **stimulation device** (evoked fMRI).
  Common stimulation devices for animals include needles for somatosensory stimulation (inserted between the phalanges for paw stimulation or placed over

Table 2. Typical stimulation parameters.

Type of stimulation	Electrical somatosensory	Optogenetic
Pulse width	300 μs	5 ms
Frequency	1–15 Hz	1–100 Hz
Train length	1–60 s	1–60 s

the whisker pad for whisker pad stimulation), light panels for visual stimulation or, in optogenetic experiments, optical fibres that are directly implanted into a particular brain area. In all cases, the other end of the stimulation element is connected to a pulse generator, which is a device which sends electrical pulses to the stimulation elements, according to the programmed stimulation parameters (see table 2 for an example of typical settings). The pulse generator receives its primary input from the MRI scanner, which triggers the stimulation onset according to the parameters set in the MRI console. In this way, the stimulation is synchronised with the fMRI acquisition. In humans, the most common form of stimulation is visual, and for this purpose, a screen is usually located at the back of the scanner room, which the subject can observe through a mirror that is mounted on the head coil.

3 Biomedical relevance

3.1 Application range

Functional MRI is typically conducted to assess either the brain function in response to stimulation (evoked fMRI) or the brain connectivity during rest (rs-fMRI). In animals, evoked-fMRI allows one to verify the integrity of well-known pathways (e.g. somatosensory pathways) or to dissect more complex or unknown brain circuitry using brain stimulation (usually, optogenetics). In humans, evoked paradigms usually involve motor tasks (e.g. finger tapping), or relatively complex visual stimulation aimed at detecting memory, language or psychological aspects of the brain. The rs-fMRI paradigm consists of acquiring fMRI volumes during relatively long periods of time (> 10 min) without the influence of any stimulation, and it aims at identifying resting-state networks that support certain neurological states or particular activation patterns that may reflect changes in the brain state.

Both, evoked or rs-fMRI can be performed in anesthetised or awake animals. Although imaging the brain of awake healthy animals promises a more direct translation of methods and findings to the human, awake paradigms involve coping with the stress of animals and prevention of motion (not only the voluntary movement of limbs but also the spontaneous breathing). Ongoing efforts are being made to improve the awake rodent fMRI platform to fully resolve the problematic stress and motion artefacts.

The combination of rodent fMRI with other imaging techniques (e.g. calcium recordings) and powerful stimulation methods like optogenetics, has amplified the capabilities of fMRI in preclinical research to study the animal brain across multiple scales [2, 9]. Some applications include the investigation on the mechanisms underlying the neurovascular coupling, the resting state oscillations or the identification of specific circuits involved in particular behavioral responses.

Besides their use in neuroimaging, many functional MRI sequences can be used to characterise vasculature-related pathologies that may emerge outside the brain. Examples include the delineation of tumours from their surrounding tissue based on a different BOLD signal resulting from increased blood flow and metabolic rate, or the differentiation of breast tumours based on the different perfusion of the tissue, using ASL [10].

3.2 Sample preparation

The preparation procedures for animal fMRI largely depend on whether or not anesthesia will be used during the fMRI acquisition. Awake animals are not subjected to any surgery (with the exception of chronic brain implants in multimodal studies), but they require long periods of training to habituate them to the MRI environment. In the more common case where animals undergo a terminal fMRI experiment under anesthesia, some surgical procedures are usually performed to allow adequate monitoring and adjustment of the physiological state, which may include the following:

- **Endotracheal intubation**. Rats are usually intubated using 14–16G flexible needles which are inserted through the animal mouth and into the trachea. Proper ventilation through the tube should be verified before continuing with other surgical procedures. A common mistake consists of inserting the intubation tube through the pharynx, which is connected to the stomach and not to the lungs and leads to respiratory insufficiency.
- **Femoral artery and femoral vein catheterisation**. PE-50 tubing filled with heparin diluted in PBS is usually employed to catheterise the vessels.
- **Ear filling**. In order to avoid field inhomogeneities at air–tissue interfaces, a fluoride paste can be used to fill the inner ear, which contains air and leads to distortions in the images. To access the inner ear, PE-50 tubing can be cut sharply at one end and connected to a syringe filled with the paste at the other end. Then the tubing is conducted towards the tympanic membrane, penetrates it and is left in place while pushing the syringe plunge to inject the paste into the cavity. This allows imaging the whole brain without distortion in the temporal lobe.
- **Intracerebral implantation**. This procedure involves craniotomy and stereotactic implantation of an optical fibre or MR-compatible electrode that may be used for stimulation or recording.

Concerning anesthesia, it is important to note that, since fMRI relies on a good neurovascular coupling, the use of anesthetics that alter the hemodynamics can largely interfere with the fMRI responses in terms of area activated and magnitude and duration of the responses. Although most anesthetics interfere, to a certain extent, with the cardiovascular system, some of them are known to have a greater influence in fMRI studies. For instance, isoflurane is a vasodilator and decreases the

Table 3. Anesthetics commonly used in fMRI. Abbreviations: α-amino-3-hydroxy-5-methyl-4-isoxazolepropionic acid (AMPA), gamma aminobutyric acid (GABA), glycine (Gly), Nicotinic acetylcholine receptor (N-Ach), N-methyl-D-aspartate (NMDA).

Anesthetic	Main receptors	Advantages	Disadvantages
Alpha-chloralose	GABA-A agonist	Preserved fMRI responses Stable light anesthesia. Low cardio-respiratory depression	Toxic (not suitable for survival studies)
Urethan (ethyl carbamate)	N-ACh agonist/ GABA-A agonist, Gly agonist NMDA antagonist, AMPA antagonist	Preserved fMRI responses	Toxic (not suitable for survival studies) and carcinogenic.
Medetomidine	Alpha-2 agonist	Preserved fMRI responses Suitable for survival studies	Needs fine control during the experiment for stability.
Isoflurane	NMDA antagonist Gly agonist GABA-A agonist	Suitable for survival studies	Altered hemodynamics (poorer fMRI signal)
Propofol	GABA-A agonist	Suitable for survival studies Used also in humans	Altered hemodynamics (poorer fMRI signal)

BOLD response significantly, therefore, although it is suitable for survival studies, it is usually replaced by other anesthetics, like alpha-chloralose, in terminal studies, which provide fMRI results closer to those expected during wakefulness. Table 3 provides a summary of the most common anesthetics used for fMRI.

Despite their known influence on the physiological state, several general anesthetics allow one to keep a stable state during long scanning periods, which remains difficult to achieve in awake animals. Besides, at appropriate regimes, most neuronal responses are preserved under anesthesia, which can be detected with fMRI, and when combined with paralyzer agents, anesthesia significantly dampens the influence of motion, which remains one of the biggest concerns in fMRI. Nevertheless, since different anesthetics may yield different fMRI responses, it is a good practice to reproduce any novel result with at least two different anesthetic agents, when possible, to rule out a potential dependence of the observed effect on particular anesthetic receptors.

4 Parameters of image quality

4.1 Signal-to-noise ratio

Given that voxels of an fMRI experiment consist of time courses instead of single values, the common signal-to-noise-ratio (SNR), typically calculated as the ratio between the mean intensity in a brain region (signal) divided by the standard deviation of intensities in a background area (noise), can be replaced by temporal-SNR (tSNR), defined, for a given voxel, as the mean value of its time course divided by its standard deviation. However, in evoked fMRI, the signal standard deviation is correlated with the extent of activation, and hence, another metric is usually considered in evoked data: contrast-to-noise ratio, defined as the amplitude of the signal in an activated voxel (difference between peak response and baseline) divided by the standard deviation in a voxel of no interest.

4.2 Motion and field distortion

One of the key variables influencing the reliability of fMRI results is motion of the animal head/neck during acquisition. In paralyzed animals, motion should be negligible. In contrast, non-paralyzed anesthetised animals can exhibit motion due to active breathing. Especially in awake animals, large mean head displacements could lead to untrustworthy data sets. Besides head motion, active breathing in non-paralyzed animals may produce dynamic field inhomogeneities, which lead to diverse image distortions, to which EPI is highly susceptible. Static field distortions can be partially corrected before acquisition with appropriate shimming; however, the dynamic field distortions that appear in relation with the chest movement cannot be corrected. One way of quantifying the resulting contamination post-processing is to identify its dominant frequency in an fMRI time course and calculate its amplitude throughout the brain, then, a band-stop filter could help reducing the noise component. To reduce motion artefacts, the temporal displacement of voxels can be regressed out of the functional time courses. Additionally, motion censoring can be included in the analysis to ignore volumes with displacements bigger than a specified threshold. However, because post-processing methods are not perfect, when possible, paralyzer agents are used in animal fMRI to reduce the motion content *in situ* during the experiment.

4.3 Spatial/temporal resolution

Choosing a particular voxel size for fMRI involves a trade-off between local specificity and SNR (lower resolution or bigger voxels contain more vessels, which give stronger signals at the expense of poorer localisations). A high temporal resolution (e.g. >0.5 s) allows one to delineate the dynamics of the fMRI response at a fine scale, and it may help reducing the aliasing of respiratory signals. However, high temporal resolution is achieved with shorter TRs, which, by definition, reduces the magnitude of the MR signal, decreasing the SNR. Higher spatial resolutions allow finer identification of functional units (e.g. cortical layers). However, these tend to involve longer TRs and hence, poorer temporal resolutions, to maintain the same field of view and SNR.

4.4 fMRI statistical parameters

Parameters that can be reported in an evoked fMRI experiment include the t-statistic per voxel (the significance of a certain temporal activation pattern), the beta value per voxel (coefficient of stimulus regression during general linear modelling (GLM)), or the number of voxels surviving a certain threshold of t-statistic or beta within the brain. Under comparable physiological conditions, these measures typically improve as the number of stimulation blocks is increased. For this reason, a higher number of trials per subject and a higher number of subjects is recommendable to increase the statistical significance. In rs-fMRI, when several subjects have been imaged and are co-registered to the same template, a t-statistic can be calculated from the resulting connectivity maps. Importantly, when computing voxel-wise statistics, the high number of tests performed may yield false positive results (this is known as a multiple comparisons problem, i.e. the more voxels subjected to the test, the higher chance of including a randomly activated voxel as a statistically significant cluster). To correct for this, several approaches have been proposed, which consist of either increasing the statistical threshold (decreasing the p-value proportionally to the number of voxels, e.g. Bonferroni correction or, less strict, false discovery rate) or imposing conditions such as a minimum number of voxels per cluster (cluster threshold estimation based on simulations—more computationally expensive).

4.5 Physiological parameters

The chances of evoking neuronal activity upon stimulation, as well as the efficiency of the neurovascular coupling, depend largely on the physiological state of the animal. To guarantee a good physiological state, the temperature needs to be controlled (normally using a closed-loop heating system) and the blood gases are tested and adjusted if necessary during the experiment. Animals that are intubated, paralyzed and subjected to intravenous injection of an anesthetic have a higher chance of producing reliable data, as they present low motion (due to paralysation) and their physiological state can be corrected by altering the ventilation parameters and the infusion rate of the anesthesia. As a general rule, low blood oxygenation can be corrected by increasing the ventilatory rate, the partial oxygen concentration in the breathing air or, under conditions of low et-CO_2 (if monitored), decreasing the anesthetic infusion rate. The opposite condition (high oxygen in blood) could be reverted by decreasing the ventilation rate (this will also increase CO_2) or, if accompanied by high blood pressure, increasing the infusion rate of the anesthetic. Some experience in controlling the physiology of the animal and familiarisation with the physiological equipment are required to perform successful fMRI.

5 Data processing and visualisation

The data acquired during an fMRI experiment are usually processed in two main steps: pre-processing and experiment-related analysis. The pre-processing pipeline includes treatment of the data motivated to improve image quality (e.g. reduce motion, align data sets, filter undesired frequencies, etc), while the experiment-related analysis generates the final results (e.g. assess evoked-related responses to generate an activation map or identify resting-state networks). Figure 1 and table 4

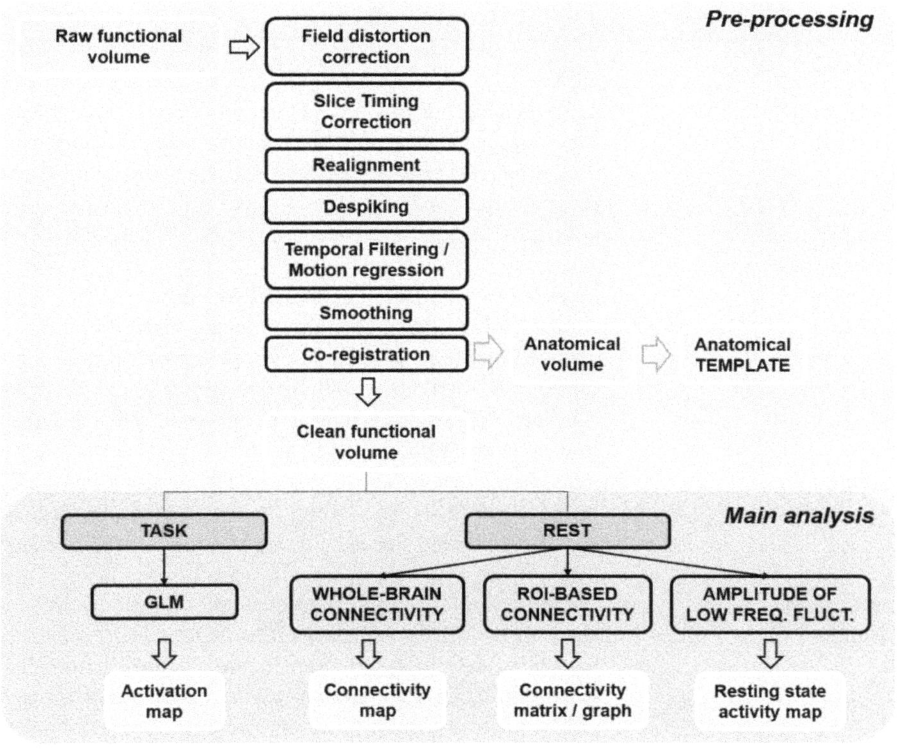

Figure 1. General pipeline to process fMRI images.

include the main steps of a pre-processing pipeline. Some additional steps include the following.

5.1 Masking

It is advised to create a mask to limit the analysis to voxels within the brain. Most of the functions during the processing allow the user to input this mask to reduce the computational load. Additionally, in some cases, the use of a mask is necessary, e.g. when producing a global connectivity map, when the connectivity of every voxel with every other voxel is calculated.

5.2 Global mean removal

Although the mean fMRI signal has been typically associated with noise, recent studies have shown that this global mean might have a relationship with the brain state. In addition, removing this signal can result in false negative correlations between voxels and therefore, global signal removal is not recommended as a default step in rs-fMRI analysis.

Table 4. Steps for the pre-processing of fMRI. (*): Strictly necessary.

Pre-processing step	Purpose
Field distortion correction	Correct the image distortions that are produced by permanent field inhomogeneities (to apply this correction step, a fieldmap or one additional EPI volume acquired with opposite phase-encoding direction are required).
Slice timing correction (in 2D acquisitions)	Compensate for the difference in acquisition times of consecutive slices.
Realignment*	Correct for motion from one volume to the next.
Despiking	Eliminate sudden signal changes (potential artefacts).
Temporal filtering/ motion regression	Eliminate noise (e.g. hardware noise, physiological noise or motion). Rs-fMRI is usually filtered with 0.01–0.1 Hz bandpass filtering. Regression of motion parameters further penalises motion in the data.
Smoothing	Increase homogeneity (averaging signal from neighbor voxels) and improve group statistics. Note that smoothing also decreases spatial specificity, therefore it might not be suitable for some applications.
Co-registration (with anatomical scan)	Space normalisation brings the functional data set to the same space as the anatomical data set for anatomical localisation of the functional results. In order to avoid blurring, in high-resolution studies, the anatomical scan should be the one that suffers the transformation and is moved to the functional space.

Several software packages exist to perform most of the analysis and visualisation of fMRI data. Some of the most common ones are: AFNI, SPM (Matlab based), FSL or Freesurfer.

5.2.1 Experiment-related analysis and visualisation

Evoked fMRI results are usually reported as an activation map. To generate this map, the experimenter assumes, for a particular voxel, that

$$Signal = regressors \times beta + error, \tag{1}$$

where ‘signal’ is the recorded fMRI signal, regressors are the expected responses to stimulation (created by convolving the stimulation paradigm by a hemodynamic response function), beta is the weight or coefficient for each regressor (to be calculated) and error is the amount of signal that cannot be explained by any of the included regressors. By solving this equation (referred to as the GLM), one beta value can be obtained for each voxel within the brain. During visualisation, an anatomical image is usually loaded as underlay and a functional map (e.g. a beta map) is overlapped on it to identify areas of activation. To have a better understanding of the temporal dynamics, functional time courses can be averaged within a certain activation area and plotted as a vector. From the fMRI time course, several parameters can be calculated, such as the time to peak, the onset time or the peak response.

In common resting-state fMRI, model-free analysis is performed, usually aimed at investigating the connectivity (assessed usually as temporal correlation of time courses) between brain regions. Some resting state analysis methods include the following:

- **Seed-based connectivity**. One voxel is selected as seed, and its time course is correlated with the rest of the brain. The result is a seed-correlation map, where the value of each voxel refers to its connectivity with the selected seed.
- **ROI-based connectivity**. Several (n) ROIs are chosen *a priori*. The time courses of all voxels in each ROI are averaged together. An $n \times n$ correlation matrix is created by computing the correlation between the average time course of each ROI with every other ROI. A correlation threshold can be set to filter only reliable connections, which may be plotted as a connectivity graph.
- **Centrality measures (e.g. degree centrality, eigenvector centrality)**. These approaches assess the connectivity of each voxel with every other voxel in the brain. For each voxel, its correlation values with every other voxel are averaged (the weight of each correlation might be different, depending on the method). The result is a map where the value at each voxel represents how well that voxel is connected. This method provides an unbiased measure of connectivity (i.e. no seed or ROI needs to be chosen *a priori*), but as a drawback, it does not provide information about specific connections within the brain.
- **Independent component analysis (ICA)**. This method allocates voxels into different categories according to their time course. Each category is characterised by a unique temporal feature, which identifies specific networks (one network = 1 independent component). Importantly, many of the independent components that are automatically identified may depict noise; therefore, some experience is required to correctly identify neural networks based on ICA.
- **Fractional amplitude of low frequency fluctuations (fALFF)**. This method measures the amount of activity per voxel, within a particular frequency band (usually, 0.01–0.1 Hz). For each voxel, the relative power of the resting state fluctuations (e.g. 0.01–0.1 Hz) with respect to the power of the whole spectrum is calculated, which results in a map indicative of the resting-state-relevant activation.

Besides static connectivity, dynamic connectivity can be studied to track different brain states during the course of a scan, and other measures such as causality effects, e.g. Granger-based or, with a priory information, dynamic causal modelling, can also been investigated to test relationships between regions within particular circuits of the brain.

6 Conclusions

6.1 Strength and limitations

fMRI constitutes a powerful imaging tool to investigate whole-brain function indirectly. Among the strengths of fMRI are the following:

- It allows whole brain coverage (better source identification compared to other functional techniques, e.g. local field potential recordings).
- It provides better temporal and spatial resolution than other whole brain imaging techniques (e.g. compared to positron emission tomography).
- The imaging procedure itself is not invasive.
- It provides a key neurovascular coupling perspective of brain function with translational potential.
- Due to the complexity of MRI, it is possible to develop customised sequences.

However, it also presents several important limitations:

- It provides an indirect measure (relies on perfect and constant neurovascular coupling, and the HRF is unknown).
- It needs specific expertise on both hardwire design (e.g. RF coils) and software implementation (e.g. the echo planar imaging (EPI) sequence modification).
- It is not readily used in routine neuroscience labs and needs access to an animal MRI facility.

6.2 Future developments

The animal fMRI platform has demonstrated its potential in a wide range of applications combining optical fibre recordings (e.g. calcium or glutamate sensors), deep brain stimulation (e.g. optogenetics) or pupillometry, among others. Besides the hemodynamic fMRI for brain functional mapping, new molecular sensors are being developed to measure brain function-related neuromodulators (e.g. serotonin, dopamine) or even the sodium/potassium-related neuronal activity directly with MRI. Last but not least, given the large pool of new transgenic animal models developed, awake animal fMRI studies may soon become the primary practice to decipher the brain function.

References and further reading

[1] Glover G H 2011 *Neurosurg. Clin. N. Am.* **22** 133–9 vii

[2] Albers F *et al* 2018 *Mol. Imaging Biol.* **20** 171–82

[3] Logothetis N K and Wandell B A 2004 *Annu. Rev. Physiol.* **66** 735–69

[4] Ogawa S *et al* 1990 *PNAS* **87** 9868–72

[5] Huber L *et al* 2017 *Neuron* **96** 1253–63 e7

[6] Huber L, Uludag K and Moller H E 2019 *Neuroimage* **197** 742–60

[7] He Y *et al* 2018 *Neuron* **97** 925–39 e5

[8] Iadecola C 2017 *Neuron* **96** 17–42

[9] Wang M *et al* 2018 *PNAS* **115** E1647–56

[10] Fiordelisi M F *et al* 2019 *J. Clin. Med.* **8** 1883

Chapter II.7.c

Hyperpolarised ^{13}C magnetic resonance spectroscopic imaging

Kevin Brindle, Arnaud Comment, Felix Kreis and Adam Gaunt

When ^{13}C nuclear spins are brought into a hyperpolarised (HP) nuclear spin state, molecules isotopically labelled with this carbon isotope can be used as highly sensitive MRI detectable biomarkers. The distribution of HP ^{13}C-labelled substrates and their metabolic products can be mapped *in vivo* with magnetic resonance (MR) spectroscopic imaging techniques. The method can be used to probe biology and biochemistry in a noninvasive manner and has recently been translated into the clinic. Hardware requirements for HP ^{13}C MR include an MRI scanner equipped with ^{13}C transmit/receive capability and a hyperpolariser to generate the HP state in the ^{13}C-labelled substrate. The technique can be used to investigate various metabolic activities including glycolysis, TCA cycle activity, fatty acid oxidation and gluconeogenesis as well as other parameters that characterise living organisms, including tissue pH, cell membrane transport, redox state, cell death (necrosis) and perfusion. Because of this variety of applications, it is an attractive addition to conventional proton MRI. It has the potential, in some cases, to replace or complement the assessment of glycolytic activity using the clinically established technique of 2-deoxy-2-[^{18}F]fluoroglucose–positron emission tomography (FDG-PET). Here we describe the underlying physical concepts behind the technique and give examples of applications with some of the HP ^{13}C labelled substrates that are available currently.

1 Introduction

Dynamic ^{13}C MR measurements of metabolic fluxes using ^{13}C-labelled substrates are aided by the broad chemical shift range of ^{13}C when compared to ^{1}H, which facilitates discrimination of individual metabolite signals (figure 1(a)). However, at thermal equilibrium of the nuclear spin state, ^{13}C MR measurements lack sensitivity

doi:10.1088/978-0-7503-3747-2ch11

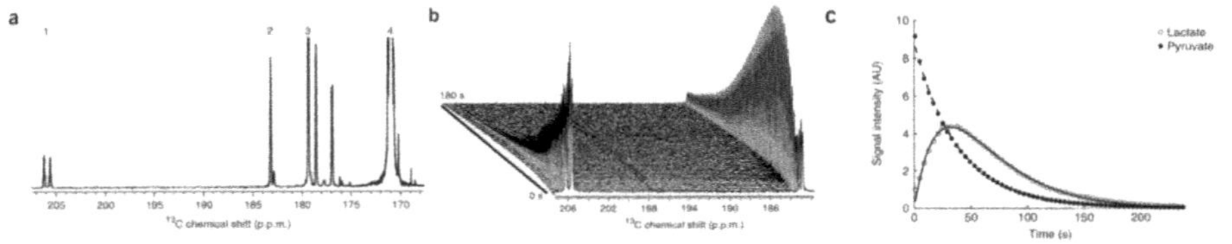

Figure 1. (a) Spectrum of an EL4 cell suspension after the addition of HP [1–^{13}C]pyruvate: 1, natural abundance [2–^{13}C]pyruvate; 2, [1–^{13}C]lactate; 3, [1–^{13}C]pyruvate hydrate; 4, [1–^{13}C]pyruvate. (b) Sequential spectra acquired during the first 3 min after addition of HP [1–^{13}C]pyruvate to the cell suspension. (c) Example of fits of the [1–^{13}C]pyruvate and [1–^{13}C]lactate peak intensities to a kinetic model for the exchange of ^{13}C label. The lactate peak intensity has been multiplied by a factor of 100. Reproduced with permission of Springer from [6], copyright (2007).

and can only probe metabolic transformations with a relatively poor temporal resolution (>10 min) and very low spatial resolution. Nuclear spin hyperpolarisation can increase the ^{13}C signal well beyond the inherently low thermal equilibrium level by transferring spin polarisation from unpaired electrons on a radical compound to the ^{13}C nuclear spins. Currently, the most effective and commonly used hyperpolarisation method in clinical and preclinical biomedical experiments is dissolution dynamic nuclear polarisation (DNP), where the HP ^{13}C-labelled substrate is cooled to ~1 K in a polariser magnet of several Tesla and following polarisation is rapidly thawed by dissolving it in superheated water. This 'one-shot' technique has been used successfully to measure the cellular metabolism of numerous substrates [1–4], in particular in tumours [5].

2 Principles and setups

2.1 Physical principles

The ^{13}C-enriched substrate to be polarised is mixed with a stable radical and rapidly frozen in order to form an amorphous solid. The unpaired electron spins in the radical become completely polarised at ~1 K in a strong static magnetic field, typically 3–7 T. Microwave irradiation at a frequency close to the electron spin resonance (ESR) frequency of the unpaired electron spins is then used to promote cross-polarisation between the electron spins and neighboring ^{13}C spins. The microwave frequency can be modulated to transfer polarisation from a wider range of ESR frequencies. The DNP process takes place inside an apparatus typically called a hyperpolariser (figure 2). At high magnetic field and low temperature, the solid-state longitudinal relaxation time (T_1) of the nuclear spins is on the order of hours, while the electrons relax in less than a second. Therefore, the electron spin polarisation can be pumped from the electron and retained on the nuclear spins.

The HP sample is then rapidly thawed and transferred from the polariser to the measurement system. Once in the liquid state, the polarisation returns irreversibly towards the thermal equilibrium value through spin-lattice (longitudinal) relaxation, with a time constant given by the T_1 relaxation time. ^{13}C T_1s are typically on the

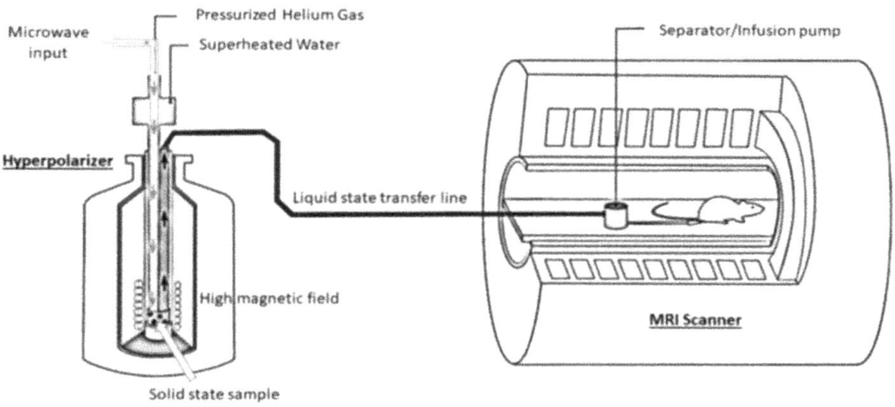

Figure 2. Representation of a dissolution DNP experimental setup for preclinical HP ^{13}C MR experiments. The sample is frozen and polarised at a high magnetic field in the hyperpolariser. Dissolution is then carried out in the high magnetic field and the liquid-state sample is shuttled rapidly to the measurement magnet and injected, in this example into a rodent.

order of 20–40 s *in vivo* (figures 1(b) and (c)). Therefore, to observe the HP ^{13}C MR signal, it is essential to minimise the time between dissolution and measurements. Moving the liquid sample through areas of zero field or very high field gradients will destroy some or all of the polarisation. The application of RF pulses to allow detection of the HP magnetisation has to be carefully managed as this also drives the spin system out of the HP state resulting in irreversible loss of the HP signal. Typically, frequency selective small flip angle pulses are used during the ^{13}C HP MR acquisitions to reduce the loss of the HP state.

2.2 Radicals

Trityl radicals are preferred as the source of the unpaired electron spins in clinical experiments due to their narrow ESR linewidths, which results in higher levels of nuclear spin polarisation. For clinical studies, a radical that aggregates due to a change in pH is used so that it can be removed by filtration before the HP solution is injected into a patient. In non-human experiments it is permissible to use other radicals such as TEMPO which are more cost effective. However, these often provide inferior levels of polarisation due to the broad ESR lineshape exhibited by these nitroxide radicals. It is also common in preclinical experiments to add trace amounts of gadolinium (Gd) to the sample, which can sometimes improve the efficiency of the polarisation. Although, the effect of the Gd is highly dependent on factors such as field strength, sample composition, concentration and temperature to name but a few. For some substrates, such as keto-acids, it is possible to generate a radical directly from the substrate itself using photo-irradiation at low temperatures [7, 8]. At high temperatures these photo-generated radicals quench and the molecule recombines, leaving the sample radical-free at the point of injection, avoiding any need for filtration.

2.3 Typical setups and state-of-the-art

In a typical HP ^{13}C MR experiment, the hyperpolariser is placed remotely with respect to the MR system and the HP solution has to be transferred rapidly between the two, post-dissolution. The original hyperpolariser design employed a pumped bath cryostat operating at 1.2 K and 3.35 T. More recently, a multi-sample polariser (SPINLab™) based on a 5 T magnet and closed-cycle helium system operating at approximately 0.8 K was developed for clinical applications [3]. This system can achieve ^{13}C polarisation levels of 55 % with [1–^{13}C] pyruvic acid using a trityl radical. In recent years the rising cost of helium has meant that a pumped-bath cryostat polariser setup has become prohibitively expensive to run. Therefore, new systems have been developed using various closed-cycle helium refrigerators and liquid helium-free magnets.

3 Biomedical relevance

3.1 Application range and relevance

HP ^{13}C MR can provide datasets containing spectral, temporal and 3D spatial information. The technique has been used to interrogate many different aspects of cell biochemistry *in vivo*. The ability to detect noninvasively and in real-time downstream metabolites produced *in vivo* from injected substrates has made this technology particularly attractive for applications in the field of cancer research [5, 6]. HP ^{13}C MR has also been used to image a variety of other pathologies and tissues including ischemia, cardiac metabolism, acute renal tubular necrosis, diabetes and kidney disease [1–4].

3.2 Hyperpolarised ^{13}C-labelled cell substrates

Three important substrates and their applications are described in the following to illustrate the versatility of HP ^{13}C MR.

3.2.1 Pyruvate to probe glycolysis

HP ^{13}C MR has been used to great effect to probe the glycolytic pathway with [1–^{13}C]pyruvate [1–6]. The degree of lactate labelling following administration of [1–^{13}C]pyruvate is related to the lactate pool size and therefore provides an indicator of the aerobic glycolysis (the ‘Warburg effect’) that is often displayed by tumours (figure 3). The labelled lactate-to-pyruvate ratio within a tumour can provide information about the stage of the tumour and the effectiveness of a therapy. Monitoring treatment response with HP ^{13}C MR has several advantages over other cancer imaging techniques. Unlike conventional (proton) MRI, which reports on the changes in tumour volume that may only become manifest weeks after successful treatment, HP ^{13}C-pyruvate can detect changes in metabolism that can occur within days of treatment. In a clinical scenario, this could minimise the duration of ineffective treatments. The clinically well-established technique of using PET to image uptake of the glucose analog, FDG-PET, can fail to give diagnostically useful information about tumours in regions where FDG uptake is inherently high, such as

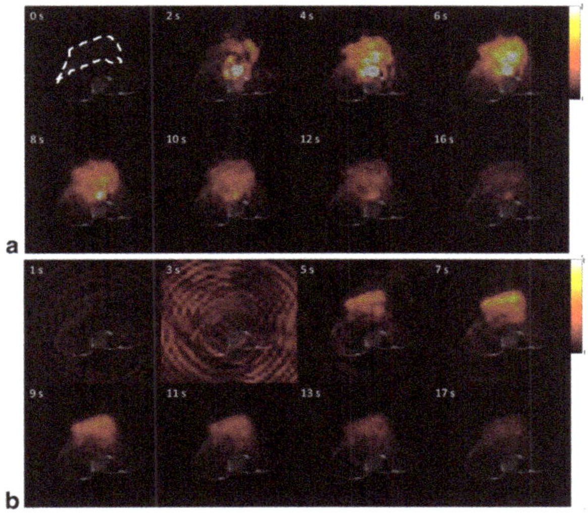

Figure 3. MR spectroscopic images of HP [1–^{13}C]pyruvate in an EL4 tumour-bearing mouse. ^{1}H T_2-weighted images (in greyscale) overlaid with (a) HP [1–^{13}C]pyruvate and (b) [1–^{13}C]lactate images. Republished with permission of John Wiley & Sons, from [9], reprinted under a Creative Commons License (CC BY 4.0).

the brain, or if the tumour is adjacent to a region with high uptake, for example a prostate tumour adjacent to the bladder. In contrast, HP ^{13}C-pyruvate can detect aberrant metabolism in these tissues. To date, pyruvate is the only HP ^{13}C-molecule that has been translated to human studies [1–5].

3.2.2 Fumarate to probe cellular necrosis

Following the onset of cell death and subsequent breakdown of the plasma membrane, the intracellular enzyme fumarase can leak out into the extracellular space. This can be detected by the production of [1,4–$^{13}C_2$]malate in the fumarase-catalyzed hydration of injected HP [1,4–$^{13}C_2$]fumarate. The labelled malate-to-fumarate ratio is a marker of cell death and could be used as an indicator of the potency of cancer treatment [1–5].

3.2.3 Urea to assess tissue perfusion

Non-metabolised HP ^{13}C-molecules can be used to image perfusion. A ^{13}C-labelled perfusion marker needs to have a long T_1 *in vivo*, be non-toxic and exhibit little or no cell uptake within the lifetime of the HP state. All these requirements are fulfilled by [^{13}C]urea, which has been successfully used for this purpose [2–5]. When the two amide nitrogen atoms are isotopically labelled with ^{15}N, the ^{13}C T_2 in whole blood is dramatically increased from 250 ms to about 5 s at 3T and 37 °C providing an opportunity to use longer trains of refocusing pulses in the imaging pulse sequence, which can improve image signal-to-noise ratio (SNR). ^{13}C T_1 is also affected, with *ex vivo* values typically ~ 44 s and 46 s at 3T for molecules with and without ^{15}N labelling respectively.

3.2.4 Further applications

There are many more ^{13}C-labelled compounds that can be hyperpolarised. These can be used, for example, to probe TCA cycle activity (HP [2–^{13}C]pyruvate results in the production of [5–^{13}C]glutamate), pH (HP ^{13}C-labelled zymonic acid changes its chemical shift as a function of pH), ketone body metabolism (HP [1–^{13}C]butyrate labels acetoacetate), or glutamine metabolism (HP [1–^{13}C]glutamine labels 2-hydroxyglutarate 2-HG). For further reading on this topic, the authors suggest several review articles [1–5].

4 Parameters of image quality

Due to the vast array of image acquisition techniques available in MRI, there are numerous possibilities for acquiring HP ^{13}C images. However, a typical strategy is to collect a train of rapidly acquired images over a time course that is on the timescale of the polarisation lifetime to collect spectral, spatial and temporal information. The achievable image resolution is limited by the available SNR, which is determined by such factors as the number of molecules per voxel, the B_0 field, the sensitivity of the receiver coil and the level of polarisation. This is especially true for the typical single-shot acquisitions used in HP ^{13}C MR where small flip angles are employed to prolong the lifetime of the HP signal. To achieve acceptable SNR, it is often necessary to sacrifice temporal information and sum over all acquired time points in the acquisition.

Although general acquisition parameters for MRI are beyond the scope of this chapter, some specific circumstances such as acquisition modes that accelerate imaging are noteworthy. Due to the short lifetime of the HP signal, it is often necessary to employ rapid imaging techniques that either traverse k-space in a non-Cartesian fashion or allow sections of k-space to be omitted entirely. Images are acquired using a range of pulse sequences, from standard EPI with Cartesian sampling to more exotic methods such as non-linear sampling of k-space with added under-sampling techniques.

Figure 4 shows how spiral traversal of k-space is implemented to acquire a time course of rapidly acquired images. The 3D trajectory is determined by the non-linear gradient waveforms that navigate the x,y plane of the k-space and then gradient 'blips' determine the z position [9]. This sequence allows several images to be acquired on a timescale that is well below that of the lifetime of the ^{13}C polarisation (figure 3). Spatially and frequency selective pulses can be used to excite a specific resonance in the target voxel while preserving the off-resonance polarisation at other spatial and chemical shift locations.

5 Data processing

Visualisation of the HP ^{13}C images is aided by a further acquisition of high-resolution ^{1}H images of tissue anatomy, on which the lower resolution ^{13}C images can be overlaid. The low sensitivity of ^{13}C detection (even in the HP state) means that there are usually significantly fewer voxels collected during the ^{13}C acquisition than in a typical ^{1}H acquisition and therefore co-registration requires digital

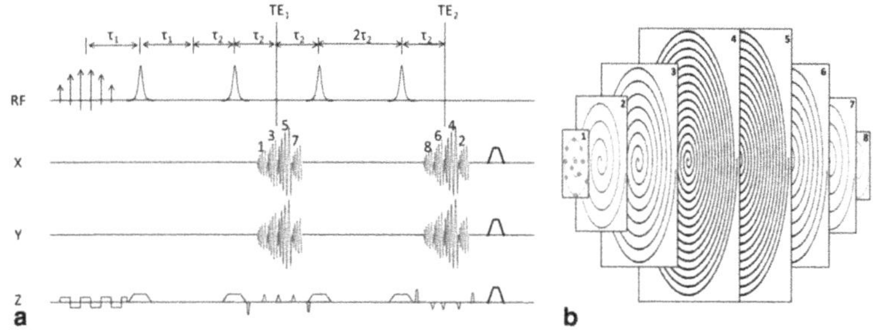

Figure 4. (a) A single-shot 3D HP ^{13}C MRI pulse sequence. (b) 3D k-space trajectory where the blue dots in the first slice are an example of the possible k-space points that could be sampled. The pulse sequence includes a spectral-spatial excitation pulse and two pairs of adiabatic inversion pulses. The whole of k-space is acquired as a stack of interleaved spirals. Republished with permission of John Wiley & Sons, from [9], reprinted under a Creative Commons License (CC BY 4.0).

enhancement of the ^{13}C image resolution. An example of HP ^{13}C images acquired with the pulse sequence shown in figure 4 and then co-registered with higher resolution ^{1}H images, is shown in figure 3.

6 Conclusions

6.1 Strength and limitations

HP ^{13}C MR is unique in its ability to image the metabolism of labelled substrates at relatively high temporal and spatial resolution *in vivo*. However, it is difficult to relate signal intensity to concentration since this is a function of the initial polarisation and subsequent T_1 relaxation, which may vary in different areas of the body. The need for additional equipment in the form of a hyperpolariser, which can be expensive to purchase and operate, and ^{13}C RF coils for the MRI scanner add to the complexity of the HP ^{13}C experiment. However, the vast signal enhancement provides a wealth of data that can only be achieved with this technique.

6.2 Future developments

To facilitate the use of HP ^{13}C MR, a technique is being developed that could make 'on-site' polarisers obsolete. The idea is to have a central production facility that can prepare transportable HP ^{13}C-molecules stored at cryogenic temperatures, where the polarisation lifetime can be several hours, enabling delivery to surrounding MR centers. This would bring the cost per dose down, improve reproducibility and remove the financial burden of operating a polariser next to local MR scanners. A technique that could allow these developments is the use of non-persistent radicals that are photo-induced in keto-acids and can be eliminated at cryogenic temperatures in order to extend the T_1 of the polarised ^{13}C-molecules [7, 8]. The increasing cost of liquid helium may mean that it is necessary to adopt 'dry' magnet systems such as that proposed by Cheng *et al* [10]. Systems and methods such as these are heavily automated and relatively inexpensive to run, therefore further increasing the accessibility of hyperpolarisation for biomedical applications.

References and further reading

[1] Comment A and Merritt M E 2014 *Biochem.* **53** 7333–57

[2] Wang Z J, Ohliger M A, Larson P E Z, Gordon J W, Bok R A, Slater J, Villanueva-Meyer J E, Hess C P, Kurhanewicz J and Vigneron D B 2019 *Radiology* **291** 273–84

[3] Kurhanewicz J *et al* 2019 *Neoplasia* **21** 1–16

[4] Keshari K R and Wilson D M 2014 *Chem. Soc. Rev.* **43** 1627–59

[5] Hesketh R L and Brindle K M 2018 *Curr. Opin. Chem. Biol.* **45** 187–94

[6] Day S E, Kettunen M I, Gallagher F a, Hu D-E, Lerche M, Wolber J, Golman K, Ardenkjaer-Larsen J H and Brindle K M 2007 *Nat. Med.* **13** 1382–7

[7] Marco-Rius I, Cheng T, Gaunt A P, Patel S, Kreis F, Capozzi A, Wright A J, Brindle K M, Ouari O and Comment A 2018 *J. Am. Chem. Soc.* **140** 14455–63

[8] Capozzi A, Cheng T, Boero G, Roussel C and Comment A 2017 *Nat. Commun.* **8** 1–7

[9] Wang J, Wright A J, Hu D E, Hesketh R and Brindle K M 2017 *Magn. Reson. Med.* **77** 740–52

[10] Cheng T, Gaunt A P, Marco-Rius I, Gehrung M, Chen A P and van der Klink J 2020 *NMR Biomed.* **33** e4264

Section 8

Magnetic particle imaging

Imaging Modalities for Biological and Preclinical Research: A Compendium, Volume 2
Parts II–IV: *In vivo* preclinical imaging, multimodality imaging and outlook
Andreas Walter, Julia G Mannheim and Carmel J Caruana

Chapter II.8

Magnetic particle imaging

André Behrends, Anna Bakenecker, Mandy Ahlborg, Ksenija Gräfe, Jonas Schumacher, Jan Stelzner, Franz Wegner, Alexander Neumann, Yvonne Blancke Soares, Kerstin Lüdtke-Buzug, Xin Chen, Huimin Wei, Ankit Malhotra, Thomas Friedrich and Thorsten M Buzug

This chapter aims to give an overview on the topic of magnetic particle imaging and its possibilities in diagnostics and therapy. An elementary discussion on particle properties and physical principles is proceeded by the concepts of signal generation, particle detection and spatial encoding. A general description of the hardware setups and the reconstruction methods will give an impression of common challenges and their solutions. Furthermore, the medical relevance, especially the possibility to combine diagnostics and therapy is discussed. The chapter will be completed by a debate on strengths and limitations as well as future directions of the magnetic particle imaging modality.

1 Introduction

Magnetic particle imaging (MPI) was first introduced in 2005 by Jürgen Weizenecker and Bernhard Gleich [1] and has since continued to attract attention. MPI relies on the nonlinear magnetic properties of magnetic nanoparticles (MNPs) and is therefore a tracer-based imaging modality. It is a promising candidate for real-time, high-resolution imaging. Today's preclinical scanners can image 50 volumes of $20 \times 20 \times 10$ mm^3 per second with a spatial resolution of 1 mm to 8 mm. Due to its working principle MPI is suitable for different cross-modality approaches like MPI-MRI or MPI-CT. As it is a highly sensitive technique, imaging applications cover a wide range of applications such as vascular imaging or cell tracking. Currently, MPI is primarily a preclinical imaging modality but first attempts to elevate MPI to clinical applications have been made. Recently, besides imaging, theranostic cross-modality applications such as targeted drug delivery,

doi:10.1088/978-0-7503-3747-2ch12

manipulation of magnetic objects or magnetic hyperthermia have become the focus of research.

2 Principles and setups

In the following subsections, the properties of the particles, the physical principles of the imaging technique, and the instrumentation of an MPI scanner will be discussed.

2.1 Particle properties

The magnetic behavior of MNPs is the key property that is commonly used for medical applications. This is especially true for imaging applications like MPI, where the particles serve as tracer material. The particles have to be stable in water at pH 7 in a physiological matrix. This depends on the charge, surface chemistry, and dimension of the particles. The stability of the particles is mainly influenced by the coating material of the MNP's core. MNPs used as tracer for MPI have a magnetically relevant core size between 10 nm and 50 nm. For *in vivo* applications, the MNPs must be coated with a biocompatible polymer to prevent agglomeration or biodegradation when the MNPs are exposed to a biological system. The coating of the MNP's core with biocompatible polymers will also provide a binding to active ingredients, drugs or receptors through adsorption or covalent attachment to the MNPs. An important parameter determining the biocompatibility of these tracers is the material of the MNP core. The most commonly used tracers for MPI are superparamagnetic iron oxide nanoparticles (SPIONs) made of magnetite (Fe_3O_4) which is a mixture of iron (II) oxide and iron (III) oxide. Particles can be fabricated by either chemical synthesis [2], lithographic processes or in other ways. After the completion of the synthesis, various separation steps are carried out in order to obtain a defined particle size distribution. Some SPIONs were already clinically approved for their use in MRI, which shows the feasibility of translating MPI to clinical applications regarding the tracer. However, tracer-usage in MPI must be approved before clinical application.

2.2 Physical principles

MPI leverages the nonlinear magnetic response of magnetic materials to an external alternating magnetic field. As such, MNPs are used as a tracer material in a spatially varying, alternating magnetic field and a spatially encoded magnetic response from the particles is obtained. Due to this encoding, the distribution of the particle concentration within the encoding area, referred to as field of view (FOV) can be reconstructed. In general, the encoding of the external magnetic field is achieved by superimposing an alternating magnetic field $\boldsymbol{H}_{\mathrm{DF}}(\boldsymbol{r}, t)$ to excite the particles, with a static gradient field $\boldsymbol{H}_{\mathrm{S}}(\boldsymbol{r})$. The total magnetic field at point $\boldsymbol{r}$ at time t is given as

$$\boldsymbol{H}(\boldsymbol{r}, t) = \boldsymbol{H}_{\mathrm{DF}}(\boldsymbol{r}, t) + \boldsymbol{H}_{\mathrm{S}}(\boldsymbol{r}) = \boldsymbol{H}_{\mathrm{DF}}(t) + \boldsymbol{G}\boldsymbol{r} \tag{3}$$

with $\boldsymbol{G}$ being the gradient matrix assuming a linear gradient and $\boldsymbol{H}_{\mathrm{DF}}(\boldsymbol{r}, t)$ simplifying to $\boldsymbol{H}_{\mathrm{DF}}(t)$ if the field is homogeneous. $\boldsymbol{H}_{\mathrm{S}}(\boldsymbol{r})$ is called the selection field (SF), which has a field free region (FFR). $\boldsymbol{H}_{\mathrm{DF}}(t)$ is typically called the drive field (DF) or excitation field, which excites the MNPs and moves the FFR. Furthermore, $\boldsymbol{H}_{\mathrm{DF}}(t)$ might also include a varying or constant focus field to shift the FOV and thus increase the imaging area. Depending on the gradient field, the FFR is either a point or line classifying MPI scanners either as a field free point (FFP) scanner or a field free line (FFL) scanner [3]. This allows the reformulation of equation (3) in the FFP case as:

$$\boldsymbol{H}(\boldsymbol{r},\, t) = \boldsymbol{G}(\boldsymbol{r}_{\mathrm{FFP}}(t) - \boldsymbol{r}), \tag{4}$$

where $\boldsymbol{r}_{\mathrm{FFP}}(t)$ is the trajectory of the FFP in space. The trajectory varies for different scanners and further classifies them. Typical trajectories are the Cartesian and the Lissajous trajectory (figure 1).

The signal detected by an MPI scanner is an induced voltage due to a change of magnetic flux in one of the scanner's detection coils. Neglecting any spatial dependency of the sensitivity and the dimensions of the detection coils, the induced voltage might be written as:

$$\boldsymbol{U}_{\mathrm{ind}} = -\frac{\partial}{\partial t}\int \boldsymbol{B}(\boldsymbol{r},\, t)dA = -\mu_0 \int\left(\frac{\partial}{\partial t}\boldsymbol{H}(\boldsymbol{r},\, t) + \frac{\partial}{\partial t}\boldsymbol{M}(\boldsymbol{r},\, t)\right)dA, \tag{5}$$

with μ_0 being the permeability of free space. Since $\frac{\partial}{\partial t}\boldsymbol{H}(\boldsymbol{r},\, t)$ contains the drive field it is orders of magnitude bigger than the signal generated by the particles $\frac{\partial}{\partial t}\boldsymbol{M}(\boldsymbol{r},\, t)$ and must be attenuated (see subsection 2.3.1). $\boldsymbol{U}_{\mathrm{ind}}$ contains higher harmonics of the excitation frequency, since the magnetisation $\boldsymbol{M}(\boldsymbol{B}(\boldsymbol{r},\, t))$ of the MNPs is nonlinear. In figure 2 the basic concept of MPI is illustrated.

The function describing the superparamagnetic behavior is called Langevin function and in a graph plotting magnetisation with respect to the magnetic field, the curve resembles an S-shape with a nearly linear slope at lower fields and a constant behavior at higher fields. Note, that real particles used in MPI never behave superparamagnetically and show hysteresis due to frequencies above 10 kHz typically used by an MPI scanner. When the magnetic moments are aligned with

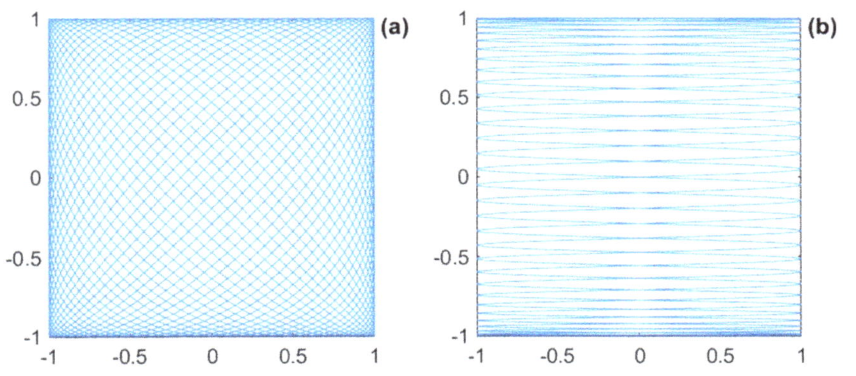

Figure 1. Common trajectories of the FFP, the Lissajous (a) and Cartesian (b) trajectory. Animation available at https://iopscience.iop.org/book/978-0-7503-3747-2.

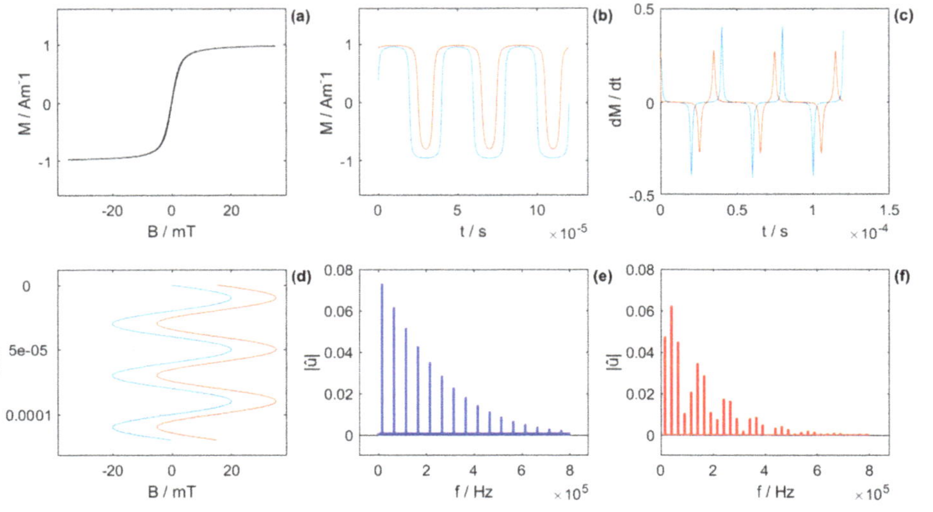

Figure 2. Driven by the magnetic field (d) and due to the nonlinear magnetisation (a) the particle signal is distorted (b) and the induced signal (c) contains higher harmonics, visible in the signal spectra (e, f). The particle behavior is shown for particles in the FFR (blue) and in the vicinity of the FFR (red). For every location in the FOV a unique spectrum is generated. Animation available at https://iopscience.iop.org/book/978-0-7503-3747-2.

the selection field the particles are saturated. Saturated particles will not contribute to the signal and are not visible within MPI. In general, the particles are saturated if the particles are outside the FOV. In the vicinity of the FFR, the particles are giving a magnetic response and thus an induced signal in the receive coils. The signal depends on the position of the particles within the trajectory and can be interpreted as a fingerprint of the particles at that position (figure 2). This fingerprint is then ultimately used in the reconstruction to obtain an image of the particle distribution within the FOV (see section 3).

2.3 Instrumentation

The following subsections present common setups for imaging (MPI) and spectrometry (MPS), as well as their components and their purpose.

2.3.1 Imaging setup

As described in section 2.2, the MNPs are excited by magnetic fields and their change in magnetic moment is detected by means of different coil types. Various scanner topologies have been proposed for both FFP and FFL imaging, but the underlying excitation signal chain for both imaging and spectrometry is quite general (figure 3).

The excitation signal is typically comprised of a purely sinusoidal varying current where the amplitude and frequency must be chosen to avoid peripheral nerve stimulation (PNS) in the lower frequency range and tissue heating due to specific absorption rate (SAR) in the higher frequency range. The usual frequency range in MPI is between 10 kHz and 100 kHz and is safe to use with respect to PNS and

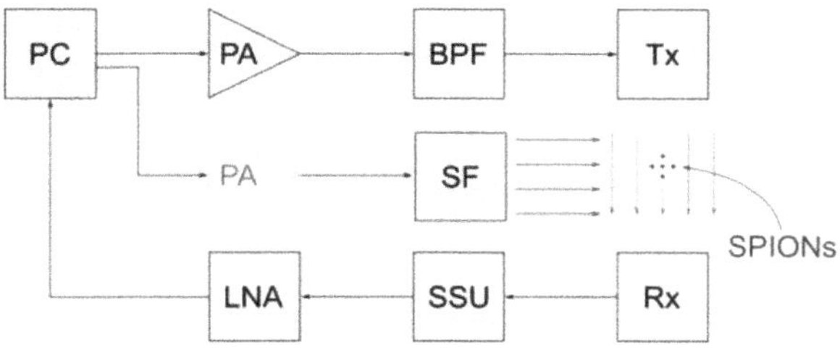

Figure 3. Basic signal chain of an MPI/MPS system.

SAR. The excitation signal is amplified by a power amplifier (PA) and filtered by a band-pass filter (BPF). The magnetic field is then created by transmit coils (Tx), which are often impedance matched in order to optimise the power transfer.

Typically, the drive field is produced by Helmholtz coils, whereas the selection field is produced by Anti-Helmholtz coils or permanent magnets. To maintain an open bore and low power consumption usually solenoid coils are used for magnetic fields along the bore and saddle coils for fields perpendicular to the bore. The power consumption heavily depends on the hardware setup and can reach several tens of kilowatts. A lot of research is being done in order to find power loss optimised coil topologies, especially for a large FOV fitting a human subject. For acquiring the MNPs response to the excitation field, commonly receive coils (Rx) are used which rely on the Faraday law of induction. Due to the feedthrough of the excitation frequency (equation (5)), which is several orders of magnitude higher than the particle response, a signal separation unit (SSU) is necessary. A common solution is given by band-stop filters, which attenuate the excitation frequency and therefore MPI relies on the broadband nature of the particle signal. A further option is a cancellation approach, realised, for instance, by gradiometric receive or send coils or a complete replication of the field generator. Cancellation is based on the principle of adding two equal voltages with opposing polarities, yielding a strongly reduced feedthrough of both the excitation frequency and interferences from e.g. the PA. After signal separation, the remaining signal is fed to a low noise amplifier (LNA) for adapting the signal amplitude to the input amplitude of the converter used for digitisation.

2.3.2 Spectrometry

Magnetic particle spectrometry (MPS) devices can be seen as zero-dimensional magnetic particle scanners, where spatial encoding is not employed. Therefore, no selection field is applied inside an MPS system. Apart from that, the concept of hardware design between an MPS system and a scanner is quite similar (see figure 3). In order to synthesise MPI optimised MNPs, MPS is used to analyse the characteristics of the MNPs by their spectral response. An MPS device is more sensitive, more compact and consumes less power compared to an MPI scanner. In multidimensional MPS several offset fields are used to emulate the magnetic field, including the selection field, inside an MPI scanner. Thus, it is capable of recording system matrices of MPI scanners, which are used for the reconstruction of the images.

3 Data processing

In the following subsections the MPI reconstruction problem will be formulated and the common ways for image reconstruction, system matrix reconstruction and x-space reconstruction will be presented.

3.1 MPI problem formulation

In MPI, the receive signal includes information about the presence and distribution of the MNPs in the FOV. Considering just the particle contribution of the induced signal and the receive coil sensitivity $\sigma^{Rx}(\boldsymbol{r})$, as well as the local concentration of the particles $c(\boldsymbol{r}, t) \approx c(\boldsymbol{r})$ equation (5) can be rewritten as [4]

$$\boldsymbol{U}(t) = -\mu_0 \int_{\Omega} c(\boldsymbol{r}) \frac{\partial}{\partial t} \boldsymbol{M}(\boldsymbol{r}, t) \sigma^{\mathrm{Rx}}(\boldsymbol{r}) d^3\boldsymbol{r}. \tag{6}$$

After applying a Fourier transform, the signal can be described as follows:

$$\hat{\mathbf{U}}_k = -\frac{\mu_0}{T} \int_{\Omega} \int_0^T c(\boldsymbol{r}) \frac{\partial}{\partial t} \boldsymbol{M}(\boldsymbol{r}, t) \sigma^{\mathrm{Rx}}(\boldsymbol{r}) e^{\frac{-2\pi i k t}{T}} dt d^3\boldsymbol{r}. \tag{7}$$

Due to the nonlinear behavior of the MNPs, higher harmonics of the excitation frequency are present in the signal spectrum, which contain information about small structures. In order to reconstruct the particle signal, two different approaches are common in MPI. While the system matrix approach is suitable for complex trajectories and multiple excitation fields, the x-space approach offers a simple and fast reconstruction scheme without the need of a time consuming calibration.

3.2 System matrix reconstruction

One possibility to reconstruct MPI images is by using a system matrix that calibrates the imaging setup. To determine a system matrix, for each individual position in the FOV the receive signal of a particle sample is recorded. This measurement strongly depends on the used particles and different scanning parameters. Thus, a system matrix can only be used for a limited range of parameter settings and a new calibration is necessary if different particles or imaging sequences are used. This process is very time consuming. That is why different concepts to reduce the calibration time are investigated. With the system matrix $\boldsymbol{S}$, the unknown particle concentration $\boldsymbol{c}$ can be calculated by solving the discretised version of equation (7)

$$\boldsymbol{S}\boldsymbol{c} = \boldsymbol{U}, \tag{8}$$

where $\boldsymbol{U}$ contains the induced signal of an unknown particle distribution. Various mathematical concepts can be used to solve this ill-posed inverse problem. In MPI, the iterative Kaczmarz algorithm in combination with a Tikhonov regularisation is typically used. This reconstruction concept allows the reconstruction of any MPI data. However, due to large system matrices the reconstruction speed is slow and further concepts are necessary to achieve a fast reconstruction.

3.3 X-space reconstruction

If an ideal MPI setup is assumed, it is possible to formulate a direct reconstruction scheme known as x-space reconstruction [5], in which the time signal is mapped to the spatial domain. Assuming one homogeneous drive field, the 1D signal equation can be written as:

$$U(t) = -\mu_0 \sigma^{Rx} \dot{H}_{\mathrm{DF}}(t) c(x_{\mathrm{FFP}}(t)). \tag{9}$$

Then, the particle concentration can be calculated via

$$c(x) = \frac{U(t)}{-\mu_0 \sigma^{Rx} \dot{H}_{\mathrm{DF}}(t)}. \tag{10}$$

The x-space reconstruction provides a fast method for reconstructing the data without the need of tedious calibration scans, but it is limited to simple trajectories.

4 Biomedical relevance

The medical relevance of MPI can be divided into diagnostic scenarios with a focus on precise image acquisition and therapeutic scenarios for medical treatment. The great potential of MPI is the direct combination of both aspects in a single device, which enables a variety of different theranostic applications. This subsection presents a brief selection of the potentials of MPI and its therapeutic extensions.

4.1 Diagnostic scenarios

The unique combination of high temporal and spatial resolution as well as the absence of nephrotoxic contrast agents and ionising radiation proves MPI to be a very promising 3D imaging modality for cardiovascular imaging and the guidance of vascular interventions. The real-time capability of MPI was first shown in 2009, when imaging of a beating mouse heart was realised (figure 4(a)). Besides its real-time imaging capability, MPI is able to resolve anatomical structures like the vascular system (figure 4(b)).

MPI-visible and MPI-compatible interventional instrument designs enable MPI to locate and track devices in real-time. Furthermore, the concepts of MPI guided angioplasty and stent implantations were proven in *in vitro* studies. Due to the quantitative characteristic of MPI, the quantification of stenosis is possible. First results suggest that MPI may overcome the disadvantages of MRI and CT regarding stent lumen visualisation, which is of highest clinical relevance. The detection of axillary sentinel lymph nodes in breast cancer patients is another potential medical application scenario from the field of tumour staging. The sentinel lymph node is located at the first position after the tumour in the lymphatic system. After localisation of the sentinel lymph node a biopsy and histological examination is performed for cancer staging. Another diagnostic setting is a method called colour MPI, which allows the colour coding of different tracer material, tracer viscosity or tracer temperature.

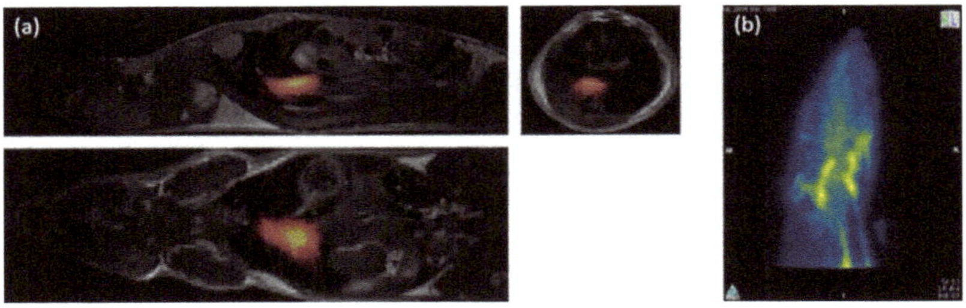

Figure 4. In (a) the fusion of MRI and real-time MPI data for different slices is shown. Various observations such as the arrival of the bolus in the vena cava, the filling of the right atrium, the left and right ventricles, the pulmonary arteries and the vessel tree can be made in the sequences (I–III). Reused from: [6], copyright IOP Publishing. Reproduced with permission. All rights reserved. In (b) the head of a rat is rendered. Anatomical structures can be recognised by resolving the spatial distribution of MNPs. Reused from: [7], copyright (2017) with permission of Springer. Animation available at https://iopscience.iop.org/book/978-0-7503-3747-2.

4.2 Therapeutic scenarios

One promising therapeutic extension to MPI is magnetic fluid hyperthermia (MFH), which utilises MNPs as a conduit for targeted energy deposition into the body to heat and destroy diseased tissue. MPI and MFH can be integrated together to seamlessly switch between imaging and therapeutic mode. Combining MPI and MFH can provide spatially targeted MFH with high resolution. In the presence of the selection field, only the particles in the vicinity of the FFR, where the MNPs are unsaturated, can respond to an alternating excitation field and generate heat, whereas the magnetisation of the particles in the saturated regions barely changes and no heat will be deposited. The spatial resolution of heating scales similarly to the MPI spatial resolution since the localisation mechanism is the same. As discussed in subsection 4.1, colour MPI is capable of mapping the temperature of the tracer, allowing for evaluation of MFH treatment. Moreover the magnetic fields of an MPI scanner can be used for applying a magnetic force or torque to the MNPs. Magnetic objects or particles, featuring a magnetic moment, experience a force pointing away from the FFR. Inside the homogeneous magnetic field, generated by the focus field, the magnetic moment of objects tends to align with the magnetic field. When applying alternating currents to the focus field coils, a homogeneous magnetic field with rotating field vector can be generated. A magnetic object follows this rotation and a forward movement is induced, when it features a helical shape inducing a forward motion. Thus, it can be navigated in an untethered manner towards regions of the body which are difficult to access (figure 5). Hence, the dual use of MPI for magnetomotion and visualisation can be realised in a single device. Thus, targeted drug delivery can be realised when MNPs are bound to a therapeutic or localised MFH is performed.

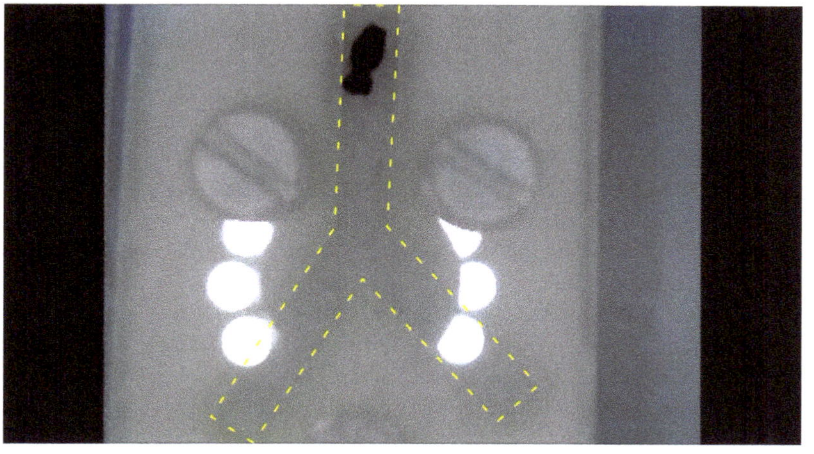

Figure 5. A helical shaped magnetic object is steered in an untethered way through a branching vessel phantom using focus fields. Reprinted from [8], Copyright (2019), with permission from Elsevier. Animation available at https://iopscience.iop.org/book/978-0-7503-3747-2.

5 Conclusions

5.1 Strength and limitations

MPI is a young imaging modality typically characterised by properties such as real-time imaging, a high sensitivity and (sub-)millimeter spatial resolution. To this point, it is possible to detect smallest quantities of MNPs in the picogram range, real-time imaging has been successfully shown and (sub-)millimeter spatial resolution can be achieved. However, a major difficulty in MPI is scalability. Upscaling to human-sized scanners increases power consumption dramatically and can, at this time, only be achieved by accepting compromises in image quality.

5.2 Future developments

Current developments include the enhancement of imaging quality by improving reconstruction algorithms, MPI-specific particle synthesis and hardware optimisation, as well as overcoming the problem of upscaling. Additionally, the quantifiability of particle concentrations is possible but still challenging in complex application scenarios. Further fields of research lie in cross-modality applications such as combined use with magnetic manipulation, magnetic fluid hyperthermia or targeted drug delivery. Another focus in future work is given to the acquisition methodology for secondary information such as the temperature or viscosity of the particle matrix. The change of temperature in a particle matrix can be used to verify the treatment of an MFH application, whereas changes in the viscosity of the particle matrix, may indicate changes in tissue structure, resulting, for example, from degeneration.

References and further reading

[1] Gleich B and Weizenecker J 2005 *Nature* **435** 1214–7

[2] Ferguson R M *et al* 2014 *IEEE Trans. Med. Imaging* **34** 1077–84

[3] Weizenecker J, Gleich B and Borgert J 2008 *J. Phys. D: Appl. Phys.* **41** 105009

[4] Rahmer J, Weizenecker J, Gleich B and Borgert J 2009 *BMC Med. Imaging* **9** 4

[5] Goodwill P W and Conolly S M 2010 *IEEE Trans. Med. Imaging* **29** 1851–9

[6] Weizenecker J *et al* 2009 *Phys. Med. Biol.* **54** L1

[7] Zheng B *et al* 2017 *Magnetic Particle Imaging Design and Applications of Nanoparticles in Biomedical Imaging* ed J Bulte and M Modo (Cham: Springer)

[8] Bakenecker A C *et al* 2019 *J. Magn. Magn. Mater.* **473** 495–500

[9] Kratz H, Taupitz M, de Schellenberger A A, Kosch O, Eberbeck D, Wagner S, Trahms L, Hamm B and Schnorr J 2018 *PLoS One* **13** e0190214

[10] Knopp T, Gdaniec N and Möddel M 2017 *Phys. Med. Biol.* **62** R124–78

[11] Grüttner M, Knopp T, Franke J, Heidenreich M, Rahmer J, Halkola A, Kaethner C, Borgert J and Buzug T M 2013 *Biomed. Tech.* **58** 583–91

[12] Bakenecker A C, Ahlborg M, Debbeler C, Kaethner C, Buzug T M and Lüdtke-Buzug K 2018 *Innov. Surg. Sci.* **3** 179–92

[13] Wu L C *et al* 2019 *AJNR Am. J. Neuroradiol.* **40** 206–12

Section 9

Computed tomography

IOP Publishing

Imaging Modalities for Biological and Preclinical Research:
A Compendium, Volume 2

Parts II–IV: *In vivo* preclinical imaging, multimodality imaging and outlook

Andreas Walter, Julia G Mannheim and Carmel J Caruana

Chapter II.9

Computed tomography

Maria Gomez-Lazaro, David Haberthür and Ruslan Hlushchuk

Computed tomography (CT) is a non-destructive imaging modality which allows three-dimensional visualisation of humans, big animals, biological samples and small animals (referred to as microCT). The possibility to obtain three-dimensional high-resolution datasets without destroying the integrity of the specimen appoints microCT as a technique alternative to serial histology in both *in vivo* and *ex vivo* applications. CT in general is particularly well suited for straightforward imaging of bone, fat and lung tissue. Other soft tissues and vasculature can also be imaged, but with the usage of radiopaque contrast agents.

Traditionally, CT and microCT applications are focused on the evaluation of bone morphometry and densitometry. In recent years, the range of possible applications has remarkably expanded and, nowadays, includes the characterisation of heart morphometry and function, visualisation and analysis of the vasculature (from the level of the organism to organ and down to the finest capillaries, the latter only possible with microCT *ex vivo* imaging), quantification of fat tissue, liver and lung structure, assessment of tumour load in organs as well as the characterisation of kidney morphometry and function.

The purpose of this chapter is to introduce the operating principles for CT and microCT in biomedical research. In this chapter, we describe biomedical applications together with the corresponding sample/animal preparation procedures, guiding principles for optimal scanning parameters and their influence on the quality of the obtained datasets, as well as recommendations for data processing workflows. At the end of this chapter, the advantages and limitations of CT as well as its prospective application in human and preclinical imaging are discussed.

doi:10.1088/978-0-7503-3747-2ch13

1 Introduction

CT imaging relies on the differential attenuation of X-ray photons by matter. In practice, the number and energy of X-ray photons leaving the X-ray source decreases on their way to the detector as they interact with the sample. The number of X-rays detected is proportional to the attenuation characteristics of the object in the X-ray beam. By applying back-projection algorithms to the X-ray projection images obtained at different angles, a three-dimensional dataset of the sample is reconstructed (figure 1). Since this technology is able to retrieve high-resolution internal structural information of humans and animals (*in vivo*) and samples (*ex vivo*) without the need to destroy the three-dimensional structure of the sample, its use has been increasing over the years. It is possible to conduct true longitudinal studies in preclinical research for the characterisation of animal phenotype and pathologies or monitor disease progression, contributing to the reduction of the total number of animals used per study. CT image acquisition is very efficient for bone, lung, and body fat without the need for prestaining. Unfortunately, vasculature and other soft tissues display similar attenuation in CT imaging and cannot be easily differentiated without contrast agents.

2 Principles and setups

2.1 Physical principles

CT scanners for human patients and microCT systems are based on the same physical principles. However, microCT systems are able to obtain images with higher resolution although for smaller objects. CT uses a polychromatic X-ray source, typically producing X-ray photons with different wavelengths; although the use of monochromatic X-rays is possible with synchrotron radiation microCT. The energy (E) for each photon is described by the expression (equation (1)):

$$E = h\nu hc\lambda^{-1} \tag{1}$$

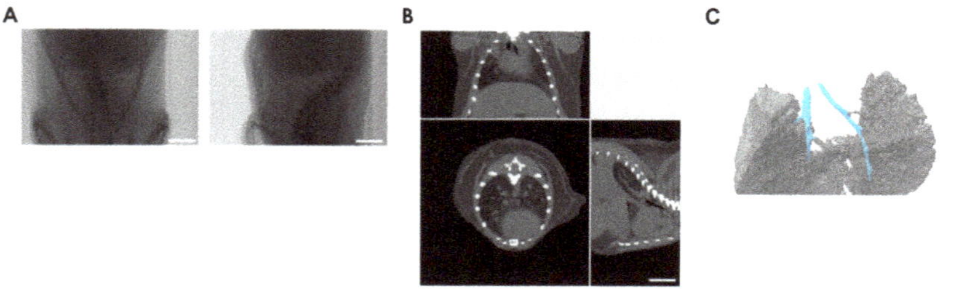

Figure 1. (a) 2D X-ray projection images of murine lung acquired at two different angles using the SkyScan1276 system (Bruker microCT, Kontich, Belgium). (b) Example of cross-sectional views of the lung obtained after reconstruction using NRecon software (Bruker, version 1.7.4.2) and DataViewer (Bruker, version 1.5.6.2), for reconstruction and data view, respectively. (c) Three-dimensional visualisation of the lung generated by the CTAnalyser software from Bruker (version 1.17.7.2+) and CTVol (Bruker, version 2.3.2.0), for model creation and viewing respectively. Scale bar 5 mm.

where h is Planck's constant (6.33×10^{-34} J·s); c is the speed of light (3×10^8 ms^{-1}); and λ the wavelength of the X-ray photon. X-ray photons with longer wavelengths have lower energy than those with shorter wavelengths.

In the X-ray tube, electrons are produced by heating a filament (usually tungsten, the cathode) and then accelerated towards the anode by the tube voltage. The X-ray radiation is then produced by the deceleration of the electrons in the anode, which is generally made of solid metal. In practice, the energy of the X-ray spectrum is determined by the acceleration voltage, while its intensity is regulated by the anode current. The number of X-ray photons arriving at the detector will depend on the differential absorption at the sample. X-ray photon interaction with matter might be explained by: photoelectric interaction; Compton scattering; coherent interaction; pair production; and photodisintegration [1]. The first three occur in the energy range of diagnostic CT and preclinical CT, while pair production and photodisintegration occur in the therapeutic energy range (between 10 kV and 300 kV).

The X-ray photons are attenuated by the interaction within the sample and are then converted to visible light by a scintillating material in the detector. The charge-coupled device (CCD) camera or complementary metal oxide semiconductors (CMOS) sensor in the detector digitises this visible light by producing an electric signal proportional to the number of X-ray photons (and their respective energy) hitting on a two-dimensional array. The specific value of each pixel will be then related to the electric signal produced and the dynamic range of the detector. The intensity of the X-ray reaching the detector (I) will depend on the initial intensity (I_0); the attenuation coefficient of the material (μ); and its thickness (d) (equation (2)):

$$I = I_0 exp(-\mu d) \tag{2}$$

Final attenuation is determined not only by the composition of the sample, but also its mass density [1].

2.2 Typical setups and state-of-the-art

There are two main microCT system designs (figure 2): (i) with a rotating sample holder in-between both a stationary detector and an X-ray source (rotating specimen) for high resolution applications; and (ii) with a fixed sample holder in-between a rotating detector and an X-ray source (rotating gantry), more appropriate for *in vivo* applications in order to minimise animal movement.

Main parts of a standard microCT scanner include: an X-ray tube; an X-ray detector; a collimator used to reduce the irradiation of the areas that are not covered by the detector; a set of energy filters; sample holders; and a stepping motor for either gantry or specimen holder rotation. The *in vivo* scanners for preclinical imaging are usually equipped with further elements like: a heating system to avoid loss of animal body temperature, an external system for volatile anesthesia, and a system to monitor physiological signals, which might be used for image gating.

A generic setup for CT scanners for humans includes the CT table for allocating the patient, an X-ray tube with a high-voltage generator, a detector, collimators to

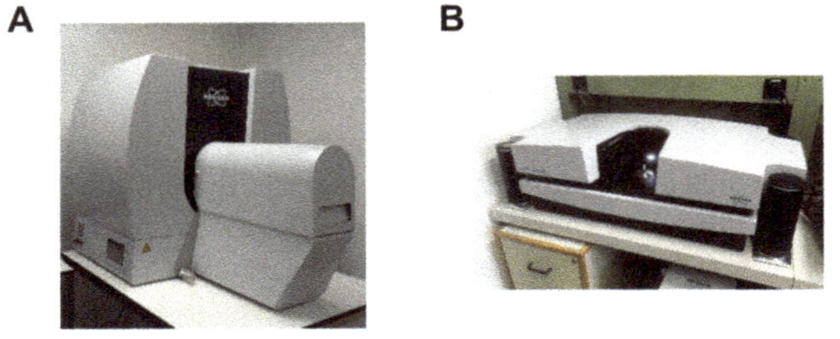

Figure 2. Images from commercially available equipment from Bruker microCT. The *in vivo* scanner SkyScan1276 (a) and the *ex vivo* scanner SkyScan1272 (b).

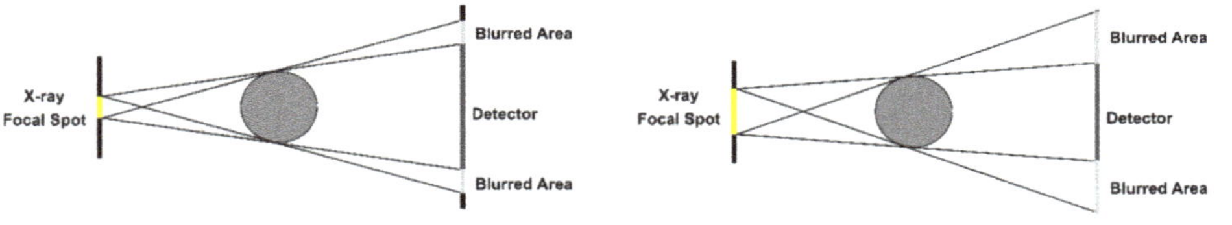

Figure 3. The effect of the X-ray focal spot size onto the image quality.

reduce the X-ray dose received by the patient and to define the slice thickness of the image, and additional filters. During the scan the CT table moves at a constant speed while the gantry rotates. The specific protocol for scanning the patient will define the collimator and detector apertures, X-ray tube voltage and current, scan mode together with CT table speed movement and gantry rotation speed [6].

X-ray tubes for commercially available microCT scanners are mini- and micro-focused X-ray sources with a focal spot size ranging from 20 μm to less than 5 μm in diameter, while working in ranges from 10–160 kV. In the case of nanoCT the focus spot size can be smaller than 0.5 μm. The size of the focal spot influences the so-called penumbra, a blurred (partially shadowed) area around the objects projections, which leads to a less sharp image. A smaller focal spot reduces the blurring effect, leading to an improvement in maximum achievable detail resolution (figure 3) [1]. In the case of a clinical CT, the X-ray tube might operate at 120 kVp and 200–800 mA, with possibility to choose between two focal spot sizes (smaller for high resolution and larger for high-flux applications) [6].

Further, the detection of the X-ray photons can be either direct or indirect (through a scintillator). X-ray detectors may be based on: amorphous silicon (aSi)-based flat panels, CCD or CMOS detectors. While CCD detectors display the best image quality (signal-to-noise ratio, SNR), quantum efficiency, and dynamic range, the readout of the data is slower than with CMOS detectors, which can be read out faster due to their crystalline form, enabling faster scans with lower radiation dose for *in vivo* scans of animals. Improvements of CMOS sensors developed specifically for scientific applications (sCMOS) have increased the image fidelity to the CCD's level. Flat panel detectors might present higher noise. However, they present much larger dimensions and sensitivity to X-rays than CMOS detectors [2]. For the CT scanners, X-ray detectors can be either solid state scintillators coupled with photodiodes or inert gas detectors [6].

To reduce the radiation dose inflicted on the human subject or the animal, it is thus important to filter lower energy photons when performing *in vivo* imaging, since they are absorbed by the body and do not contribute to the resultant image. Such X-ray energy filters are commonly made of aluminum with different thickness (i.e. 0.5 and 1 mm) and mixtures of aluminum and copper. In the case of CT scans an additional type of filter can be used, the so-called bowtie filter, which has a concave shape to alter the X-ray beam intensity according to the oval-shape of the human body.

3 Biomedical relevance

The ability to obtain internal 3D structural information is the greatest advantage of this technology for both *in vivo* and *ex vivo* applications. After imaging, the sample can be further processed for use in other non-destructive imaging techniques (optical imaging, magnetic resonance imaging (MRI)) and/or destructive methods such as classical histology or other biochemical analysis (i.e. RNA and protein content analysis). This technology offers the opportunity to perform *in vivo* longitudinal studies obtaining high-resolution information.

3.1 Application range and relevance

The classic research area for microCT is bone research due to the high X-ray attenuation properties of bone, allowing for the highest spatial resolution with shorter acquisition times. Bone tissue is particularly well suited for microCT imaging due to its characteristic composition (hydroxyapatite (~45%), organic part (~40%) and water (~15%)). Typically, the analysis of bone includes morphometric analysis of trabecular and cortical bone [3] and additionally its density without the need of any contrast agent or sample pretreatment. MicroCT has been used for the phenotypic characterisation and monitorisation of therapeutic interventions in animal models of osteogenesis, ovariectomy, osteoporosis, bone cancer, and metastases [7], as well as studies of bone resorption and remodelling in regeneration and bioengineered therapies [8]. A further insight of mechanical properties of bones can be obtained by the application of finite element analysis to the microCT images.

Due to its low density, lungs are easily imaged and subsequently segmented. Applications of lung imaging include longitudinal monitoring of lung carcinoma and lung fibrosis. Allowing for phenotypic characterisation of lung cancer by quantification of the volume of metastases and also lung function can be monitored by quantification of total lung volume during expiration and inspiration phases. With the use of contrast agents, evaluation of lung vasculature is also possible, although smaller vessels such as capillaries (4 to 10 μm in diameter) can only be evaluated in *ex vivo* protocols.

Whole adipose tissue evaluation is another application for *in vivo* microCT without resorting to contrast agents, with the possibility to quantify total adipose tissue in animal models of obesity or diabetes for phenotypic characterisation and monitorisation of directed therapies. With the images acquired, it is possible to differentiate between subcutaneous and visceral adipose tissue. The latter being more important in metabolic disorders associated with type 2 diabetes.

Most of the applications on cardiovascular imaging are only possible in microCT imaging by resorting to the use of contrast. *In vivo* cardiac imaging allows the morphological quantification of wall thickness and additionally, functional information comparing the heart ejection volume. Huge limitations for *in vivo* imaging of vasculature are the contrast agent circulating time, the amount of contrast in the image related to agent concentration, the maximum X-ray dose for living animals and the practically inevitable movement, which together limits image contrast and the maximum resolution, respectively. With these limitations in *in vivo* applications it is only possible to image larger caliber vasculature. These limitations also apply to CT imaging for cardiovascular imaging on human patients. If smaller vessels are to be assessed, *ex vivo* imaging protocols for sample preparation and image acquisition need to be employed [1]. Despite the almost microscopic resolution provided by the current microCT equipment, the fine vasculature cannot be differentiated from the surrounding soft tissue without a radiopaque contrast agent. Various agents have been used to visualise the vessels of the liver, heart, lungs as well as other organs. Until rather recently, the combination of vascular corrosion casting (involves tissue maceration) and subsequent microCT visualisation was the state-of-the-art microvascular imaging. Only recently has the correlative *ex vivo* microangioCT approach with μAngiofil as polymer-based contrast agent been introduced (see figure 4). The mentioned approach enables investigation of the vasculature from the whole organ to the capillary level without tissue maceration. Another advantage of microangioCT with μAngiofil is its blue colour, which enables direct visual control of the perfusion success (figure 4(b)). The intravascular solidified μAngiofil does not interfere with the eventual further processing for histology (see figure 4(e)). Furthermore, its inherent fluorescence greatly facilitates the search for the corresponding site of interest within the microangioCT dataset [4].

Due to the optimised perfusion technique, even the finest capillaries can be filled continuously. Moreover, its solidification brings extra stability to the sample during rather long microCT-scanning sessions. Even such gentle organs as the eye can be easily scanned and visualised (see video 1 available at https://iopscience.iop.org/book/978-0-7503-3747-2). Furthermore, the X-ray absorption level of the μAngiofil is rather high and distinct from that of bone tissue, enabling reliable visualisation of the vasculature within the bone tissue without decalcification (see video 2 available at https://iopscience.iop.org/book/978-0-7503-3747-2) and even in close proximity to the metal implants (see figure 5).

Imaging of abdominal organs is also possible with the use of contrast agents. Liver metastasis can be studied by using liver-specific contrast agents which accumulate in hepatocytes. Spleen also accumulates contrast agents circulating in the vasculature. In the case of kidney, it is possible to quantify their volume, length and thickness by using contrast agents. However, if structural detail is needed, *ex vivo* imaging allows the evaluation of glomerular volumes, tubular sections of the nephron and vasculature. Renal function can also be evaluated by imaging at different times by employing a contrast agent which is cleared renally. In the case of

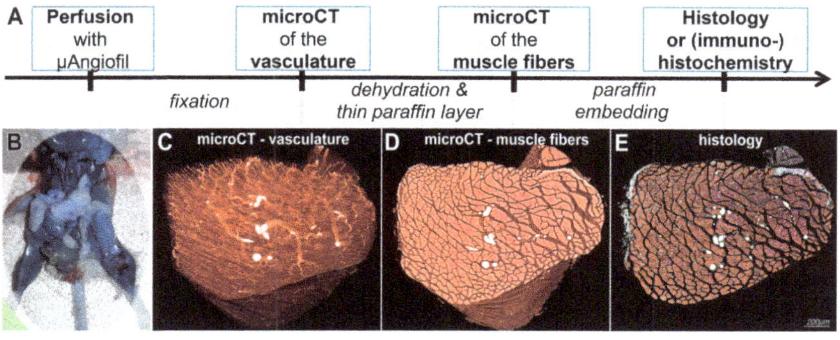

Figure 4. Correlative imaging approach to visualise the vasculature and muscle fibre arrangement by microCT and histology. (a) Workflow. (b) Photograph of a successfully perfused mouse with blue-coloured tissues. (c) Contrast-enhanced microCT-based visualisation of the vasculature of plantaris muscle in 3D. (d) Muscle fibres of the same plantaris muscle in 3D after dehydration step and covering the sample with a thin paraffin layer. (e) Corresponding (to c and d) histological cross-section stained with Masson trichrome. Larger blood vessels were manually highlighted in white. Adapted from [10], copyright (2017) with permission of Springer, and [11] with permission of the American Physiological Society.

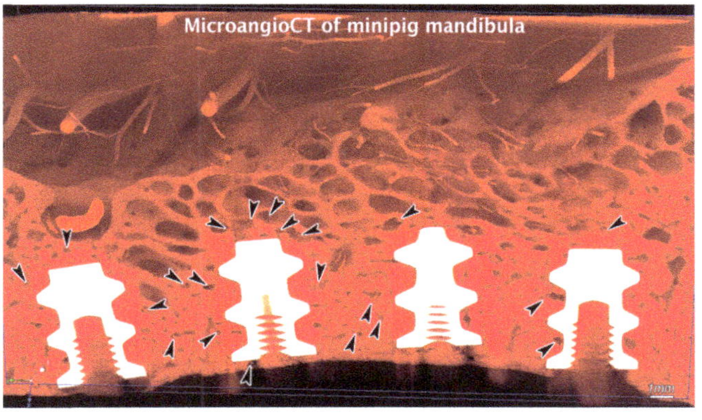

Figure 5. MicroangioCT of the vasculature of the minipig mandibula. Scanner: multiscale laboratory nano-CT system 2214 (Bruker microCT, Kontich, Belgium). The contrast agent μAngiofil has a different, from bone and metal, X-ray absorption index, which enables simultaneous visualisation and differentiation of those structures (arrowheads indicate the microvessels in the vicinity of implanted metal parts).

the gastrointestinal tract, the evaluation of colourectal cancer in animal models of the disease is possible when using simultaneously a contrast agent injected intra-peritoneally together with anal delivery (imaging by negative contrasting).

In the case of CT for human subjects, the applications might be divided into: thoracic imaging (with examples of pulmonary embolism, acute pulmonary edema, acute aortic syndrome amongst others); gastrointestinal and genitourinary imaging (i.e. acute gastrointestinal bleeding, bowel perforation, gallstone ileus, acute pancreatitis, renal and liver transplant dysfunction, amongst others); neurology and non-traumatic spinal imaging (i.e. stroke, carotid artery dissection, cerebral venous sinus thrombosis, herpes simplex encephalitis, amongst others); pediatric imaging (i.e. intussusception, bowel malformation, acute otitis media, amongst others); trauma imaging (i.e. cardiac injury, pneumothorax, lung contusion and lung laceration, amongst others); and interventional and vascular imaging and iatrogenic complications (i.e. acute arterial ischemia, endotracheal tube misplacement, endovascular stent endo leak, amongst others) [9].

3.2 Sample preparation

Depending on the sample and the desired quantification, sample preparation will be needed. However, as discussed before, for every sample, proper immobilisation in the sample holder is crucial to avoid imaging artefacts due to sample movement. In the case of *ex vivo* bone imaging, a fixation step should be avoided if measurement of bone density is of interest, while for morphometric analysis previous fixation will not affect the results. Additional for *ex vivo* hydrated samples, if longer scans are needed, dehydration needs to be prevented in order to avoid sample displacement or structure alteration due to shrinkage.

In the case that the target tissue for imaging does not present inherent differential attenuation it is possible to stain the specific tissue resorting to the use of contrast agents. For *ex vivo* imaging, normally a first fixation step will precede the incubation with the specific contrast agent. A final washing step might be needed to remove excess of contrast agent or not bound contrast agent. For *in vivo* imaging, a major difference on available contrast agents is their corresponding body clearing time (conventional iodinated contrast agents are rapidly eliminated by the kidneys while blood-pool contrast agents remain in the circulation for longer periods). Depending on the target tissue a contrast agent will be injected in animals intraperitoneally or intravenously. For example, to image tumours of the gut a combination of intraperitoneal injection and rectal administration has been described.

In the case of CT for human patients, the use of contrast agents while not necessary in all applications might be beneficial for data interpretation. The route for the administration of contrast agents in humans can be the intravenous injection (i.e. small iodine-based contrast agents) for example for imaging the vasculature and evaluating blood flow, or the oral administration (i.e. barium-based contrast agents) for imaging the gastrointestinal tract (an enema is also an option). However, some patients might develop adverse effects to the intravenous injection of contrast agents, further discussed in [9].

4 Parameters of image quality

Image quality in CT is represented by contrast, noise and sharpness. In CT, measurements and data visualisation are performed on the 3D reconstructed images. However, the quality of the 3D images depends on the quality of the signal obtained in the X-ray projections. A good SNR is to select the percentage of minimal transmission of 10%–20% (difference between the maximal and minimal values of the X-rays detected by the detector in the shadow image). However, the quality of the CT scan is normally evaluated in the reconstructed dataset, and additional artefacts can arise from the reconstruction process.

As noted previously, the spatial resolution of every system is influenced by the size of the focal spot of the X-ray tube, the detector pixel size, the scattering of the optical photons within the scintillator and the SNR. Resolution can be defined as the distance between two objects at which they can still be identified as independent. It is usually given as distance (in the case of microCT as μm) or as spatial frequency (line pairs per mm). As a rule of thumb, the voxel size should be at least three times smaller than the structures to be resolved through imaging. To evaluate the performance of a microCT the point spread function (PSF) can be used. In the case of microCT the PSF can be derived from two other measurements: the line spread function (LSF) and the edge spread function (ESF) (equations (3) and (4)). In the LSF, the signal from the detector in one line is collected by using a high density object with a slit (with the slit opening size equal to the size of the point size diameter). The ESF is obtained by imaging the edge of a high density object. Not only can the PSF be obtained through the ESF but also the LSF by the following formulae (assuming that the PSF of the system is symmetric):

$$LSF(x) = d(ESF(x))/dx \tag{3}$$

$$LSF(x) = \int_{-\infty}^{\infty} PSF(x, y)dy \tag{4}$$

In CT the spatial resolution is more commonly characterised over a range of dimensions by using the so-called modulation transfer function (MTF). The MTF is a measurement on how much contrast of the sample is preserved in the image obtained by the detector. It can be obtained by imaging a dense object consisting of spaced lead foils. By normalising the intensity profile perpendicular to the edges of the foils we can obtain the ESF. Then the MTF can be calculated by obtaining the Fourier transform of the ESF's first derivative or the Fourier transform of the LSF (equation (5)).

$$MTF(k_x) = F\{d/dxESF(x)\} = F\{LSF\} \tag{5}$$

with k_x being the spatial frequency in the x direction. The MTF is normalised to one and given in line-pairs per mm (LP/mm). In preclinical and clinical CT, it is possible to detect structures of spatial resolution bigger than the 10% MTF. Noise in microCT images can be regarded as contrast-to-noise ratio (CNR) and SNR. CNR is obtained by equation (6).

$$CNR = |\mu_0 - \mu_b|/\sigma_b \tag{6}$$

where, μ_0 is the mean grey value of the object, μ_b is the corresponding mean grey value of the background, and σ_b the standard deviation of the grey values.

In the case of the SNR it can be calculated as specified in equation (7):

$$SNR = \mu_0/\sigma_0 \tag{7}$$

being μ_0 and σ_0 the mean grey pixel value and their corresponding standard deviation of the specific region.

4.1 Radiation dose

Due to the use of X-rays, specific governmental safety regulations apply to the use of this technology. Non-medical microCT are designated as industrial, thus following the safety regulations for industrial radiology. Most of the commercially available microCT are highly shielded and radiation arising from them is minimal, protecting the operator. However, when used for *in vivo* applications, radiation is a concern for human patients and animal welfare. In the case of humans, to minimise the risk for developing radiation-induced cancer, the radiation dose should be kept as reasonably achievable. The dose received by the animal and its consequences on animal welfare, will depend on: the parameters of the scan (X-ray energy and intensity, and scan time); on the scanning frequency; on the total dose received; on the animal strain; and its age. Although living animals present some intrinsic capacity to recover/repair from certain levels of damage induced by ionising radiation, in longitudinal studies scan frequency should be longer than one scan per week to allow for damaged repair [6]. In practice, to lower the radiation dose received by the animal we can decrease the scanning time (less projection images (by increasing the rotation angle) and/or less number of images acquired in the same position (reducing the average acquisition)), and the use of filters to remove X-ray photons with shorter wavelengths.

Most strategies to minimise the radiation dose received by the patients are related to increasing the efficiency of the CT scanners by increasing the quantum efficiency of the detectors; reducing the dose by using pre-patient collimators, and the reduction of the scattered radiation received with the post patient collimators; use of beam-shaping filters; limitation of the scanned body area; and the optimisation of the radiation according to the patient's size.

4.2 Control of artefacts and image quality

Artefacts can appear due to the process of image acquisition (physics); imperfections in the scanner; movement of the sample during the scanning process; presence of very high-density material such as metal implants; and artefacts in the reconstruction process. Some of these errors might be completely or partially corrected by the commercial available software from each scanner.

Image contrast in CT imaging is governed by the difference attenuation of different materials/tissues (radiographic contrast) but also it is influenced by the detector's ability to translate the different X-ray inflow into different electronic signals. X-ray images present inherent noise which affects the quality and resolution of the final image due to X-ray beam intensity being not uniform; non-homogeneous intensity arriving at the detector; the stochastic nature of X-ray production and attenuation (Poisson distribution); the electronic noise of the detector; and non-uniform sensitivity of the pixels in the detector which might produce different pixel values from equally intense photons. The correction for the different inhomogeneities is called flat field image, which is acquired with the detector with X-rays on and without any sample. Ideally, a detector might give a black image when there are no X-rays, however, in reality, there is always some noise produced by the electronics (electronic noise or dark current). Dark current images are collected to reduce this electronic noise. Since this noise is independent of any optical photon (static noise), X-ray images are corrected for this noise. Additionally, microCT images present an inherent noise due to the limited power of the X-ray source.

The emitted X-rays are polychromatic, and as they interact with the sample, the lower energy photons are absorbed, causing the so-called beam hardening (increasing its mean intensity). In practice, the edges of the object or the interfaces between objects with very different attenuations appear brighter, changing the true attenuation value in that area. The so called ring artefacts appear in the reconstructed images as circles centered on the rotation axis. They are mostly related to wrongly calibrated pixels in the detector. Another source of artefacts is photon starvation, which can appear in areas displaying high attenuation when the beam is closer to the horizontal and suffers greatest attenuation giving rise to noisy images.

For *in vivo* microCT scanners, the animal's movement during scanning is diminished by using holders to fix the animal position during the scanning process. Although living animals are anesthetised during the scanning procedure, they are moving due to breathing, heart beating and gut peristaltic movement. In the case of human patients, they need to control their breathing and hold their breath for a few seconds during image acquisition to limit the image blurred. There are two approaches to correct these motion-derived errors in microCT: prospective or retrospective gating. In the prospective gating strategy, image acquisition is triggered by a physiological signal (i.e. respiratory phase or ECG (electrocardiogram) cycles) leading to longer image acquisition periods. While in the retrospective scan, several images are acquired at each angular projection, which later will be sorted based on image data or physiological signals.

5 Data processing

A typical workflow for microCT data processing and analysis includes: (1) reconstruction of the 3D dataset from the 2D projections. This 3D information is obtained by back-projection of the images into the sample space (other reconstruction algorithms might be applied [5]). Classically, this is done by reformatting the same line from all projections into a so-called sinogram, where all the horizontal lines in one sinogram correspond to the same horizontal line in all projections. One single sinogram is then back-projected into one reconstruction, which is one horizontal slice of the 3D dataset. Exhaustive details of the process can be found in [5]. In the reconstructed dataset the grey value intensity of each voxel is proportional to the mean linear attenuation coefficient of the sample at the specific location. In clinical CT it is common to express this attenuation in the so-called Hounsfield units (HU), which are the normalised attenuation related to water (0 HU) and air (−1000 HU). Corresponding attenuation for soft tissue falls within that range. However, bone as a denser material presents a value above 1000 HU. (2) Visualisation of the

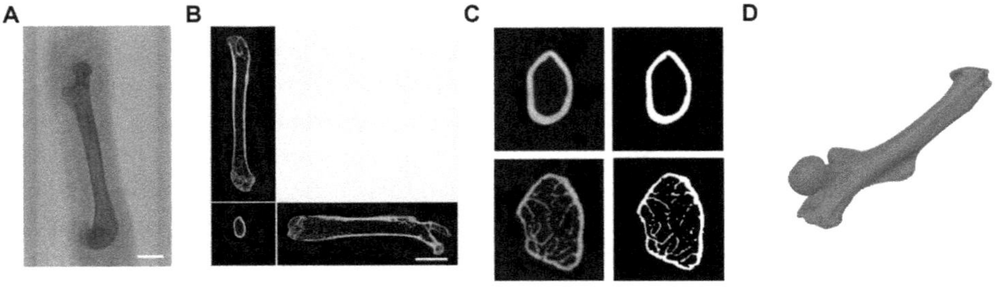

Figure 6. A typical workflow for image processing in microCT. (a) A shadow projection image obtained from mouse femur. Scale bar 2 mm. (b) Reconstructed dataset projections for orientation adjustment of the 3D dataset. Scale bar 3 mm. (c) Segmentation of trabecular and cortical bone for quantification of morphometric and desitometric parameters. (d) 3D model created from the 3D reconstructed images of the mouse femur.

3D dataset by: slice-by-slice, slice-by-slice orthogonal projections or rendered 3D volume. (3) Data analysis (morphometry and densitometry) (figure 6).

Standard workflow for data analysis includes: (i) selection of the region of interest (ROI) in the 3D reconstructed dataset; (ii) segmentation of features of interest, either through grey value thresholding or manual segmentation, calculation of features like volume, thickness and tortuosity after segmentation; and (iii) visualisation.

6 Conclusions

6.1 Strength and limitations

The major advantage of this technology is that CT is a non-invasive imaging modality which allows for the acquisition of information on the internal structures in humans, animals and samples, with high spatial and temporal resolution (the latter for specific applications such as rat cardiac imaging). Depending on the attenuation characteristics of the sample no preprocessing of the sample is needed. Interestingly, internal anatomical structures of humans and small animals can be imaged for disease phenotyping and monitorisation of therapeutic interventions. However, the major limitation for *in vivo* imaging, is the radiation dose received by human patients and animals. This imposes a limitation in maximally achievable resolution and the SNR in the resulting images. The latter will depend on the imaging scheme (i.e. longitudinal studies or single imaging before animal euthanasia, limitation of the body area scanned in patients). Additionally, poor soft tissue contrast makes it necessary to use contrast agents for both *in vivo* as well as *ex vivo* applications.

Although nano-computed tomography has been developed and allows for a resolution limit in the range of 50–400 nm, this technology is not (yet) applicable *in vivo* due to the long acquisition times. The mentioned higher resolutions are achieved with the use of X-ray sources with smaller focal spots. A drawback of imaging at these high resolutions is the small fields of view that can be achieved and the difficulty of performing imaging with either specialised lab setups or endstations at synchrotron sources.

6.2 Future developments

Developments of new detectors with higher sensitivity, spatial and temporal resolution might contribute to the reduction of the radiation dose. We anticipate that in the near future the biggest advances of this technology will be the development of new contrast agents capable of bigger contrast allowing for the reduction of the image noise. Developing of contrast agents with minimal adverse effects might also expand the applicability of this technology in the human clinics. Traditionally, microCT has been used for morphometrical and anatomical analysis. However, with the development of new contrast agents functional studies are now emerging as complementary tools to add functional information to morphological data. Examples are the use of contrast agents for quantification of tumour blood volume and blood vessel leakage. Besides, amongst these tools are the progress in developing nanoparticle contrast agents with specific ligands that target the contrast agent to specific cellular receptors aiding in the molecular characterisation of tumours or phenotypes.

Additionally, the development of new contrast agents allowing for multimodal imaging capacities constitutes a great advantage. With the integration of different imaging modalities acquisition of complementary information (anatomical, functional, and molecular) will be possible. Moreover, these combinations will contribute to a significant reduction in the number of experimentation animals used in preclinical studies. Nowadays, some dual modality contrast agents for CT imaging exist in the market such as microCT/Optoacoustic agent, but there is still much to improve and develop.

Acknowledgements

MGL is the Head Coordinator of the Bioimaging i3S Scientific Platforms member of the national infrastructure PPBI—Portuguese Platform of Bioimaging (PPBI-POCI-01-0145-FEDER-022122), and acknowledge the support of the Portuguese funds—FCT-UID/BIM/04293/2020.

References and further reading

[1] Buzug T M 2008 *Computed Tomography: From Photon Statistics to Modern Cone-Beam CT* (Berlin: Springer)

[2] Salplachta J, Zikmund T, Horvath M, Takeda Y, Omote K, Pina L and Kaiser J 2019 *9th Conf. on Industrial Computed Tomography (Padova, Italy) (iCT 2019)*

[3] Bouxsein M L, Boyd S K, Christiansen B A, Guldberg R E, Jepsen K J and Müller R 2010 *J. Bone Miner. Res.* **25** 1468–86

[4] Hlushchuk R, Haberthür D and Djonov V 2019 *Vascul. Pharmacol.* **112** 2–7

[5] Hsieh J 2003 Computed Tomography: Principles, Design, Artifacts, and Recent Advances (SPIE Monographs vol 188) (Hoboken, NJ: Wiley)

[6] Cavanaugh D, Johnson E, Price R E, Kurie J, Travis E L and Cody D D 2004 *Mol. Imaging* **3** 55–62

[7] Clark D P and Badea C T 2014 *Phys. Med.* **30** 619–34

[8] Liu H, Jiao Y, Zhou W, Bai S, Feng Z, Dong Y, Liu Q, Feng X and Zhao Y 2018 *J. Transl. Med.* **16** 137

[9] Lewis G, Patel H, Modi S and Hussain S 2016 *On Call Radiology* (Boca Raton, FL: CRC Press)

[10] Schaad L, Hlushchuk R, Barré S, Gianni-Barrera R, Haberthür, Banfi A and Djonov V 2017 *Sci. Rep.* **7** 41842

[11] Hlushchuk R *et al* 2018 *Am. J. Physiol.-Renal Physiol.* **314** F493–9

Section 10

Nuclear imaging

IOP Publishing

Imaging Modalities for Biological and Preclinical Research: A Compendium, Volume 2

Parts II–IV: *In vivo* preclinical imaging, multimodality imaging and outlook

Andreas Walter, Julia G Mannheim and Carmel J Caruana

Chapter II.10

Positron emission tomography

Claudia Kuntner-Hannes, Thomas Wanek and Wolfgang Wadsak

Positron emission tomography (PET) is a molecular imaging method based on the administration of radiolabelled substances and external measurement of emitted photons. It provides functional three-dimensional images of the living body visualising metabolic processes and is used in clinical practice and clinical/preclinical research. PET allows a quantitative measurement of the radiolabelled substance within the body over time and as such enables the study of biological function in both health and disease. Clinical PET is mostly used in oncology, neurology, and cardiology, whereas preclinical PET has found applications in drug development, as well.

1 Introduction

Positron emission tomography (PET) is a nuclear medical imaging technique whereby a biologically active compound labelled with a positron emitting radioisotope (tracer) is introduced into the body typically by injection. This compound accumulates in the patient and the pattern of its following radioactive emission is used to assess the temporal and spatial distribution of the radioisotope and hence the radiolabelled compound. The reconstructed image represents the distribution of the radiolabelled compound (e.g. drug, pharmaceutical, endogenous substance) *in vivo*. Hence, PET enables the determination of certain metabolic processes and the corresponding rate at which these processes take place. Therefore, PET can determine if the biological function of an organ is altered, while many other forms of medical imaging—such as computer tomography (CT)—are mainly used to determine the anatomical structure of the organ. Thus, PET is most frequently used to receive information about the biological function of organs and diseases [1]. In oncology (cancer) the metabolic rates give information on whether the tissue is cancerous and how it responds to treatments. A major advantage of PET is that it

doi:10.1088/978-0-7503-3747-2ch14

does not only allow visual inspection and interpretation of the acquired data but also quantification of the images.

2 Principles and setups

2.1 Radiopharmaceuticals/radiotracers

Radiopharmaceuticals, also known as radiotracers, are radiolabelled drugs for *in vivo* application in clinical or preclinical nuclear medicine. In general, radiotracers consist of two parts: (i) a molecular backbone responsible for transport, uptake and excretion of the radiotracer within the organism (pharmacokinetics and pharmacodynamics) and (ii) a radionuclide that has properties that either allow the accurate detection from outside of the organism for subsequent visualisation (e.g. annihilation photons) or allow the local delivery of high amounts of energy through radioactive decay for internal radionuclide therapy (e.g. alpha- or beta-minus-emitters) [2].

Radiotracers are administered at very low amounts so as to not perturb normal physiology but to still allow *in vivo* studies under physiological conditions. For the organism itself, radiotracers are indistinguishable from their non-radioactive counterparts. Hence, they are also treated equally on a molecular level allowing one to follow-up the fate of the radiotracer through measurement of the emitted radiation coming from the attached radionuclide. However, it has to be considered that the identical non-radioactive molecules will 'dilute' the radiotracer and therefore potentially block the signal internally. Hence, the so-called molar activity (or apparent molar activity in the case of radiometal labelling) needs to be sufficient (e.g. >100 GBq μmol^{-1}; dependent on the availability of target molecules and half-life of the radionuclide). This is especially important in preclinical PET imaging, since the volume of administration is very limited in mice and rats and a significant saturation of target sites *in vivo* cannot be ruled out when molar activities of the used radiotracers are too low. Commonly used positron emitting radionuclides that are used in human and small-animal PET are summarised in table 1.

The by far most commonly used PET radiopharmaceutical is 2-deoxy-2-[^{18}F]fluoro-D-glucose ([^{18}F]FDG) [3]. Starting from the late 1970s, it has been in use to visualise glucose-dependent molecular mechanisms and pathways, predominantly in

Table 1. Commonly used PET radionuclides and their half lives.

Nuclide	Production	Half life
C-11	$^{14}N(p,\alpha)^{11}C$, cyclotron	20 min
N-13	$^{16}O(p,\alpha)^{13}N$, cyclotron	10 min
F-18	$^{18}O(p,n)^{18}F$, cyclotron	110 min
Ga-68	$^{68}Ge/^{68}Ga$-Generator	68 min
Rb-82	$^{82}Sr/^{82}Rb$-Generator	1.3 min
Zr-89	$^{89}Y(p,n)^{89}Zr$, cyclotron	78.4 h
I-124	$^{24}Te(p,n)^{124}I$, cyclotron	4.2 d

oncology (staging, therapy monitoring), cardiology (vitality of the myocardium) and neurology (dementias, movement disorders). In the last years, ^{68}Ga-labelled radiotracers became more interesting as they can be prepared independently from the availability of a nearby cyclotron using a radionuclide generator directly on site. Especially, [^{68}Ga]Ga-PSMA-11, a compound targeting the prostate specific membrane antigen (PSMA), is now used broadly for tumour staging in prostate cancer. Apart from that, a plethora of ^{18}F- and ^{11}C-labelled radiotracers are described and used in preclinical or clinical research. ^{11}C has the advantage of being able to insert a radiolabel without alteration of the original molecule as carbon is present in all organic compounds. Also, the low positron energy allows for high spatial resolution in imaging. On the other hand, the short half-life limits its usability, both in radiosynthesis and application [4]. Larger compounds such as antibodies usually exhibit slower pharmacokinetics and are thus radiolabelled with longer-lived radionuclides such as ^{89}Zr or ^{124}I to enable imaging at later time points (e.g. 24 h after administration).

2.2 Physical principles

PET radionuclides typically decay via positron emission also called beta+ (β^+) decay. This means that in the unstable parent nucleus a proton (p) is transformed into a neutron (n), a positron (e^+) and a neutrino (ν_e). The positron, which is the antiparticle of an ordinary electron, is emitted from the nucleus and travels some distance (depending on the kinetic energy). This distance is typically within a few millimetres of the site of its origin in the body tissue. When the positron is nearly at rest it attracts an electron (e^-) and shortly forms a so called 'positronium', where the positron is acting as the 'nucleus' and has a lifetime of around 10^{-10} s. Thereafter, the positron combines with the electron in an annihilation reaction, in which the mass of the two particles is converted into energy (annihilation photons or also called gamma-rays). Both particles possess a mass equivalent to 511 keV and thus 2 annihilation photons with 511 keV each are emitted. Due to the conservation of momentum, these annihilation photons are emitted in opposite directions 180° ± 0.5° (see figure 1).

In a PET scanner multiple detector rings consisting of pixelated detectors are arranged around the object of interest for detecting the emitted annihilation photons [5]. The detection principle is based on coincidence detection, which means that if two annihilation photons are detected in opposing detectors within a short time (called the coincidence time window, around 10 ns) and energy window, the event it is counted as a valid (or also called 'true' coincidence) event and a line-of-response (LOR) is defined between these two detectors. Coincidence logic is used to analyse the signals from the opposing detectors. Although the annihilation photons are emitted simultaneously, a small but finite coincidence window width is applied to allow for different distances of travel by the two annihilation photons from the annihilation event to the detector. This small difference in arrival time is used in time-of-flight (TOF) PET scanners for improving signal-to-noise ratios (SNRs).

Within the multiple detector rings and the pixelated detectors forming the ring, a high amount of coincidence events is possible. All coincidence events are stored

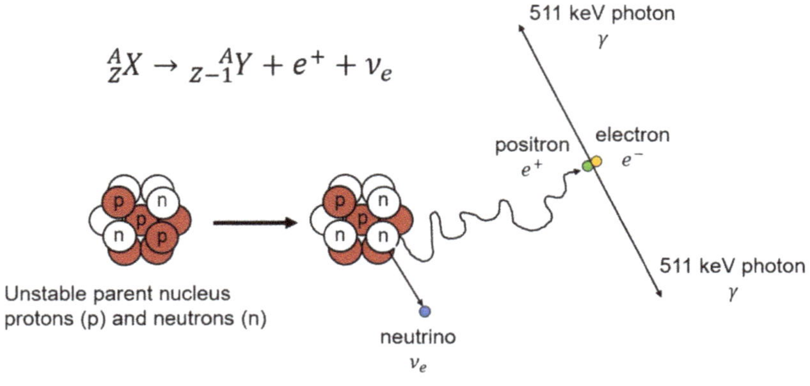

Figure 1. Schematic drawing of the positron decay process leading to the annihilation event with the emission of two 511 keV annihilation photons (gamma-rays).

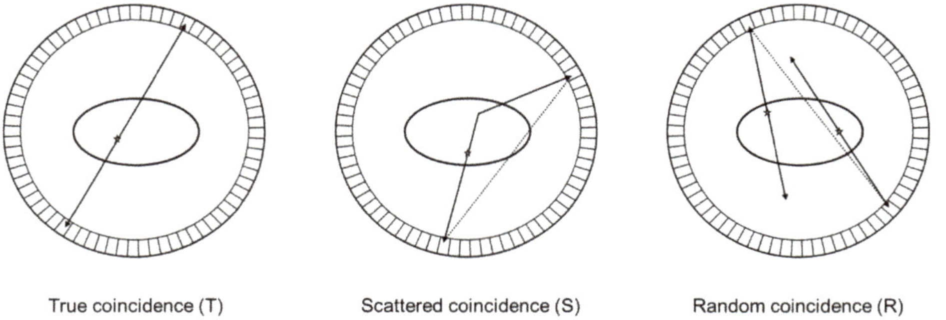

Figure 2. True coincidence event (left), scattered coincidence even (middle) and random coincidence event (right). Scattered and random coincidence events lead to incorrect positional information and contribute to a relatively uniform background consequently decreasing the image contrast.

including a digital time stamp in a file also known as list mode file. By using dedicated reconstruction algorithms three-dimensional images are then formed based on the list mode file.

In PET, different types of coincidence detection events can appear within the specified coincidence time window. All these events are called *prompt coincidences*. In figure 2 three types of prompt coincidence events are depicted. The *true coincidence (T)* describes the coincidence event where both annihilation photons produced from the same annihilation event are detected. In the *scatter coincidence (S)* event, one or both annihilation photons produced from the same annihilation event undergo scattering and are detected in another detector. Scattering can occur within the subject but also with components from the scanner itself. Scatter coincidences are more severe in clinical than in preclinical imaging and the incorrect information of the position leads to an increase in the background signal of the image. This results in a loss of contrast and a decrease in the SNR. Finally, *random coincidences (R)* occur when annihilation photons from two unrelated annihilation events are detected within the coincidence time window. The random coincidence rate increases as the square of the amount of administered radioactivity and decreases in proportion to the width of the coincidence time window.

For quantitative PET imaging, methods for scatter and random corrections are typically applied in the image acquisition and reconstruction process.

2.3 PET detectors

Most commonly used PET detectors are built up of scintillating crystals connected to photodetectors. These photodetectors are either photomultiplier tubes (PMTs), or semi-conductor-based detectors such as avalanche photodiodes (APDs) or silicon PMTs (SiPM). Typically, an array of pixelated scintillating crystals is connected to one or more photodetectors (referred to as block detectors). The motivation of this type of detector is that the scintillation crystals are allowing a measurement of the detected energy of the gamma-ray. Thereby, scattered annihilation photons which would only increase the background signal in the reconstructed images can be discarded based on the lower energy. A scintillating crystal absorbs the energy of the gamma-ray and re-emits the energy in the form of visible light. This visible light is then detected by the photodetector and converted into an electric signal. The mostly used scintillators in PET scanners are nowadays cerium-doped lutetium(-yttrium) oxyorthosilicate crystals also known as LSO:Ce or LYSO:Ce. In preclinical systems, other scintillators such as BGO or GSO:Ce are implemented, as well. Table 2 lists the mostly used inorganic scintillators in PET detectors.

2.3.1 Time-of-flight-PET

Scintillators with extremely short decay time constants offer the possibility of TOF-PET, in which opposing detectors measure the difference in the arrival time of the two annihilation photons. Thus, the position of the annihilation event can be localised along the line connecting the two detectors and this information can be used for reconstruction to enhance image quality. Depending on the timing resolution of the PET detector, TOF-PET reduces the LOR from around 70 cm to about 15 cm (assuming 500 ps timing resolution) or lower [7]. Thus TOF-PET significantly improves SNR as illustrated in figure 3. Due to recent improvements in new scintillators and fast digital SiPMs photodetectors, it is expected that TOF

Table 2. Characteristics of selected inorganic scintillation crystals used in clinical and preclinical PET scanners [6].

Scintillator	LSO	LYSO	GSO	BGO
Chemical formula	Lu_2SiO_5:Ce	$LuYSiO_5$:Ce	Gd_2SiO_5:Ce	$Bi_4Ge_3O_{12}$
Density (g cm^{-3})	7.40	7.10	6.71	7.13
Effective atomic number	66	64	60	75
Decay time (ns)	42	48	60	300
Light yield (photons keV^{-1})	25	32	8	9
Emission wavelength (nm)	420	420	440	480

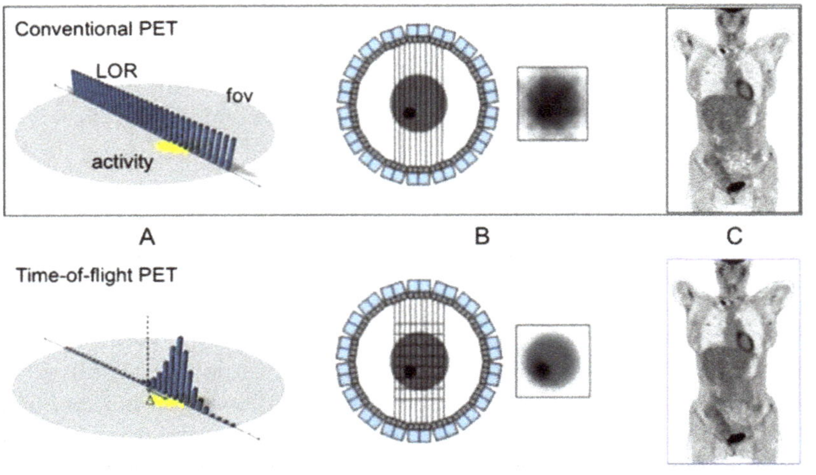

Figure 3. Schematic illustration of the differences between conventional PET (top) and TOF-PET (bottom). In conventional PET, there is equal probability (illustrated by the blue bars) for the recorded event along the LOR (a). In TOF-PET, the time difference between the arrival of the two annihilation photons, is used to create a probability distribution locating the recorded event at distance Δ from the axis of the tomography (dashed line). A timing resolution of 600 ps, for example, results in a FWHM of about 9 cm. The probability distribution and knowledge about the timing resolution of the detector system can be incorporated into the reconstruction process, illustrated here for a single projection (b). (c) Reconstructed PET emission images illustrate the improved SNR in TOF-PET for a patient with BMI 24. (Image is reproduced from [7] under the terms of the Creative Commons Attribution 2.0 International License (https://creativecommons.org/licenses/by/2.0)).

timing resolution will reach 100 ps or even less. With 100 ps timing resolution the uncertainty of determining the origin of the annihilation point along the LOR would be reduced to 3 cm. TOF-PET is only available for clinical scanners, as for application in preclinical scanners an even faster timing resolution would be needed (~ 30 ps). In addition, due to the small size of rodents the scatter fraction is lower and thus SNR is typically higher in preclinical images.

2.3.2 Depth-of-interaction

The limiting factor for uniform spatial resolution throughout the whole (axial and radial) field-of-view (FOV) of a PET scanner is the length (or thickness) of the scintillation crystals. A substantial length of the scintillator is required to efficiently stop the 511 keV annihilation photons of around 1–3 cm. However, the relatively thick detector elements lead to a degradation of the radial spatial resolution also known as the depth-of-interaction (DOI) effect. In PET scanner systems, two different designs are used to account for DOI effects. The first design is based on so called 'phoswich' detectors, where different layers of scintillator materials are used to define the DOI. The identification of the crystal, which absorbed the annihilation photon, is mostly based on differences in light yield or decay time. The second design to account for DOI effects is based on the use of monolithic crystals. In this case the light distribution of the scintillation photons after absorption of the annihilation photon is used to identify the position but also the depth-of-interaction. The DOI concept is implemented in selected preclinical and clinical PET scanners.

2.4 Typical setups and state-of-the-art

Commercial PET systems are available for clinical and preclinical application. However, new technologies are continuously developed. For an extensive review covering new detector technologies and also new algorithms for image reconstruction in clinical and preclinical imaging systems, please see [8] for details.

2.4.1 Preclinical PET

Preclinical detector design shows a higher variability as compared to clinical scanners. The majority of the preclinical PET scanners is based on the block detector design using pixelated scintillators either in one layer or as a phoswich arrangement. In addition, the continuous block detector design with DOI capability is also used in some scanners. The different block detector designs are illustrated in figure 4. In most of the scanners the detectors are arranged in a ring around the subject. Typical preclinical block detectors are based on a matrix of pixelated scintillators (~ $1.2 \times 1.2 \times 10$ mm^3) coupled to position-sensitive PMTs or SiPMs. In preclinical PET the transaxial FOV ranges from 7–13 cm and the axial FOV ranges from 10–15 cm. Spatial resolution is around 1 mm FWHM and sensitivity is in the range of 3%–12%. Preclinical PET scanners are either stand-alone systems or hybrid systems. As for clinical PET, preclinical PET/CT, PET/MR and SPECT/PET/CT scanners are available and described in detail in chapters III.2.a, III.2.b and III.2.c. In figure 5(a) a preclinical PET/CT scanner is illustrated. Other vendors from preclinical PET(/CT or MR) scanners include Bruker, Inviscan, Mediso, Mi Labs, Molecubes, MR solutions, Sedecal, Sofie biosciences, and TriFoil Imaging (this list does not claim completeness).

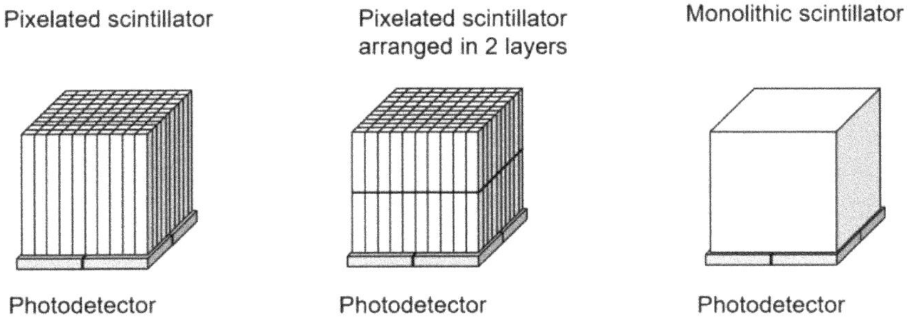

Figure 4. Illustration of the different PET block detector designs. (Left) pixelated scintillator (10 × 10 matrix) coupled to an array (2 × 2) of photodetectors. (Middle) pixelated scintillator matrix arranged in two layers (phoswich geometry) coupled to the photodetector array. (Right) monolithic scintillator coupled to the photodetector array.

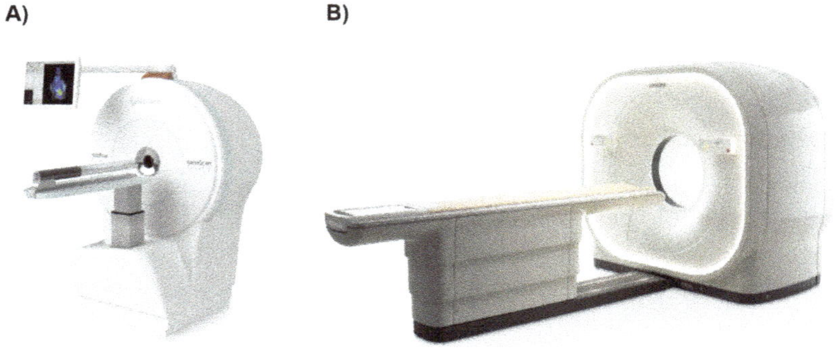

Figure 5. Example of a preclinical PET/CT (a) and clinical PET/CT (b) scanner. (Photo courtesy of Mediso and Philips Healthcare).

2.4.2 Clinical PET

Clinical PET scanners are based on the block detector design forming a ring around the patient. To enhance the detection efficiency multiple rings attached to one another are used. Typical block detectors consist of an array of 13×13 scintillation crystals coupled to four photodetectors (either PMTs or SiPMs) with scintillation crystals having a size of around $4 \times 4 \times 20$ mm^3. Bore sizes of the PET scanner are in the range of 60–90 cm and axial FOVs range from 16–25 cm. Spatial resolution at 1 cm radial offset is around 4 mm FWHM and sensitivity (referring to the percentage of detected annihilation photons) is in the range of 5–100 kcps MBq^{-1} (0.5%–10%). Most clinical PET scanners are TOF-PET systems with timing resolutions of around 250–400 ps.

It also has to be mentioned, that for clinical PET only hybrid scanners (in combination with computed tomography (CT) or magnetic resonance imaging (MRI)) are nowadays available. In figure 5(b) a clinical PET/CT scanner is illustrated. Other vendors from clinical PET/CT or PET/MR scanners include GE, Kindsway Biotech, Mediso, Philips, Siemens, Toshiba, and United Imaging (this list does not claim completeness) [9]. More details on PET/CT and PET/MR can be found in chapters III.2.a, III.2.b and III.2.c.

2.4.3 Clinical total-body PET

In clinical practice, total-body PET images are acquired by either using multiple bed positions or by using continuous bed motion to cover the whole body. The disadvantage of both methods is that dynamic total-body imaging at the same imaging time point is not possible and additional scan time is prolonged. In 2017 a total-body PET scanner (EXPLORER) was introduced by Zhang and Cherry [10], which has an axial length of 2 m. This scanner has a 40-fold higher sensitivity as compared to conventional scanners and thus allows a much shorter scan time, or the use of small radiotracer activities leading to lower radiation doses. In addition, it also allows dynamic total-body scanning opening new areas of clinical research in drug development.

2.5 Image reconstruction

The major purpose of PET is the quantitative determination of the underlying biological changes in the chemistry and physiology of injected compounds labelled with radionuclides. In this case the function to be reconstructed is the distribution of radioactivity in the body and the measurements are used to estimate the total activity along lines of known locations. Image reconstruction procedures can be divided into analytical and iterative approaches. While analytical reconstruction methods calculate the radiotracer distribution directly from the measured projections, iterative reconstructions start with an estimate of the image and refine this estimate iteratively by comparison to the measured data. The classical analytical reconstruction algorithm is filtered-backprojection (FBP). The advantage of analytic reconstruction methods is that they offer a direct mathematical solution for the formation of an image [11].

Iterative methods are based on a more accurate description of the imaging process resulting in a more complicated mathematical solution requiring multiple steps to obtain an image. Iterative methods offer improvements over the analytical approaches as they can account for the noise structure in the observations and can also incorporate a model of the scanner leading to better spatial resolution. Often used iterative reconstruction algorithms are based on the maximum-likelihood expectation-maximisation (MLEM) method.

In PET typically the measured raw data (in the form of list mode files) is first grouped into projection images, called sinograms. The sinograms are sorted by the angle of each view and can also be grouped into defined time frames for dynamic imaging studies. Then, in the reconstruction step different corrections are applied to the sinogram such as normalisation, deadtime correction, decay correction, attenuation correction and scatter correction before obtaining the final image.

3 Biomedical relevance

3.1 Application range and relevance

PET is a versatile imaging modality and its application range is defined by the applied radiotracer. In daily clinical practice, PET imaging is typically applied in the fields of oncology, neurology and cardiology.

Especially for preclinical research, the property of PET to non-invasively quantitate the distribution of radiolabelled compounds *in vivo* combined with the capability of radiolabelling of (nearly) any substance of interest, offers a broad spectrum of PET applications. Thus, preclinical PET is often applied in drug development (pharmacokinetic studies, pharmacodynamic studies, drug–drug interactions), therapy monitoring, and also to investigate the biological mechanisms behind a certain disease (e.g. mechanism of transporter proteins, systematic studies to elucidate the origin of Alzheimer's disease, labelling of cells to investigate immunotherapy). The major advantage is the availability and continuous development of animal models mimicking specific parts of human disease. These range from cardiac and pulmonary diseases, aging, infectious diseases, obesity, diabetes, inflammation, neurological diseases, joint diseases, visual disorders, cancer, hypertension, genetic diseases, to diseases of abuse.

3.2 Subject preparation

PET is an *in vivo* imaging technology and thus requires living subjects, which remain still during the scan process. As such, human and animal preparation for the scanning procedure vary slightly.

3.2.1 Patient preparation

Patient preparation for PET is strongly dependent on the radiopharmaceutical in use and the area of interest within the organism. In clinical practice patients are typically undergoing a PET/CT examination. [^{18}F]FDG is used in 90% of the regular PET examinations for tumour detection and localisation. For [^{18}F]FDG scans, patients are typically asked to follow a limited carbohydrate diet for the previous 24 h before the examination. Additionally, usually they have to fast for at least 6 h and do not perform strenuous exercise prior to the examination. In general, patients should be asked to drink a lot of water (not sparkling!) to ensure efficient renal clearance of excess radiopharmaceutical and therefore minimise the radiation burden to the kidneys. They should not be on steroid medication for 24 h and should not receive dextrose infusion for 12 h before the scan. Then the radiotracer is injected over the arm vein and patients are asked to wait in the uptake room, which should be quiet, dimly lit and warm. During the uptake period, patients should remain as calm as possible, without moving or talking, to minimise muscle uptake. Directly before the scan, patients should visit the toilet to empty the bladder and are then positioned on the exam table and given instructions to remain still (PET) or hold the breath (CT) during the acquisition. The 'classical' FDG-PET scan is performed around 45–60 min after intravenous injection. It lasts around 20–30 min with 7–8 bed positions (2–3 min acquisition/bed position). Furthermore, claustrophobia can be an issue, especially when PET/MRI is used. In PET/CT the gantry is normally wide enough and the whole scanner short enough to minimise the feeling of being trapped. Depending on the scanning region, the arms of the patients are either positioned in arms down or arms up orientation. More information on patient preparation can be found in a recent review [12].

3.2.2 Animal preparation

PET acquisitions in general usually last from 10 up to 90 min; therefore, to obtain high resolution images, the subject must not move during data acquisition (to avoid motion artefacts). In contrast to humans, animals are therefore typically anesthetised for a PET scan. Anaesthesia is introduced either by using inhalant anaesthetic agents (e.g. isoflurane, sevoflurane) mixed in oxygen or medical air, or by the injection of an anaesthetic agent (e.g. ketamine/xylazine, propofol). Depending on the respective PET application, animals are already anesthetised during radiotracer injection, remain under anaesthesia during the uptake process and are then placed on the PET scanner bed for data acquisition ('static scan'). Alternatively, radiotracer injection and PET data acquisition are initiated simultaneously after the animal has been positioned in the PET field-of-view ('dynamic scan'), so that the uptake or distribution phase of the administered compound is recorded. In most of the cases, inhalant anaesthesia is used as it is better tolerated by the animals, is relatively easy to adjust, has faster induction and recovery times, does not involve injection and also allows multiple scanning/day, even though the number of anaesthesia periods that can be safely applied on animals during a day is limited. Most importantly for rodents, external body warming has to be applied during the anaesthesia periods to protect the animals from hypothermia.

A typical workflow for animal preparation using inhalant anaesthesia is:

Animals are taken out of the cage and positioned in a warmed induction box where anaesthesia is induced. When the animals are unconscious, they are taken out of the box and positioned on the animal bed (imaging chamber), where warming and anaesthesia is continued. Then a small catheter is inserted into the tail vein of the animal and flushed with saline to ensure correct positioning in the vein. Subsequently, a syringe containing the radiotracer is connected to the catheter and the radiotracer is administered. Depending on the study setup, the animal is either placed back into a warmed area for the uptake period or the PET acquisition started concomitantly with the tracer administration. Injection durations are depending on the radiotracer characteristics and typically range from some seconds to some minutes up to continuous infusions.

4 Parameters of image quality

The versatility of PET applications raises specific challenges to the experimenter in terms of chemistry, technology, methodology and biology aspects [13] (figure 6). All these factors together significantly influence PET image quality, or to be more specific, image quantitation and study reproducibility. The important question for the examiner prior to initiation of any PET study is to identify the factors potentially influencing the study that are relevant for the respective research question and to standardise these factors as well as reasonably possible. For clinical investigations some of these influencing factors have been assessed and summarised in respective guidelines (e.g. EANM/EARL FDG-PET/CT accreditation [14]). In the preclinical field, however, the establishment of such specific guidelines has proven to be extremely challenging, best exemplified by a recent published multicentre preclinical study on [^{18}F]FDG biodistribution in healthy mice [15]. Consequently, accurate reporting is of utmost importance for preclinical PET studies and specific guidelines on how to report are available in the literature [16].

4.1 Chemical aspect influencing image quantification

PET studies require the use of radiolabelled compounds, so-called radiotracers, that are the primary source of image contrast. As the commonly used positron emitting radionuclides have very short half-lives (e.g.: ^{11}C: 20.4 min; ^{18}F: 109.8 min), these radiotracers have to be synthetised shortly prior to tracer administration. Consequently, radiotracer synthesis introduces concerns regarding the chemical

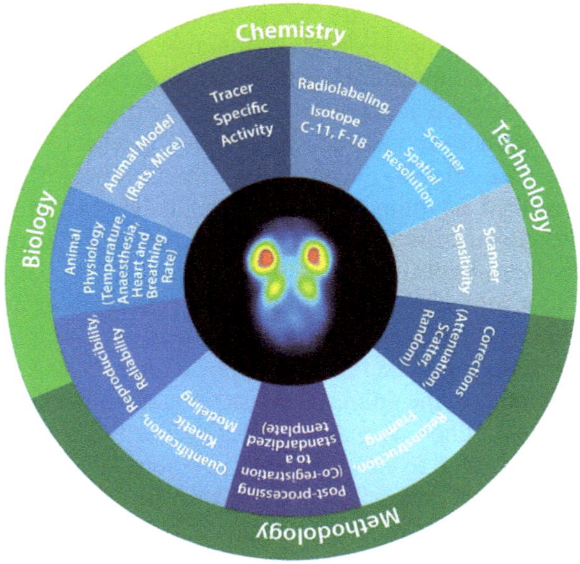

Figure 6. Technical, methodological, biological, and radiochemical aspects influencing the quantification accuracy of PET studies. (Image is reproduced from [13] with permission of Springer, under the terms of the Creative Commons Attribution 4.0 International License (http://creativecommons.org/licenses/by/4.0/)).

aspects of a study. Radiolabelling of a substance needs to be optimised to obtain high radiochemical yields with high purity of the radiotracer in a reasonable amount of time. In addition, due to the limited sensitivity of the scanner, the injected radioactivity should be sufficient to obtain reasonable count rates. However, a high injected radioactivity amount might contain an increased amount of non-labelled compound, which is dependent on the molar or specific activity of the radiotracer (ratio of radiotracer activity to compound mass) and can vary widely for each synthesis. Especially in neuroimaging, high injected radioactivity with low molar activity might lead to receptor occupancy and thus to competition of the radiotracer and the non-labelled substance at the binding site producing 'false-negative' images out of receptor saturation. However, there are multiple applications in which the molar activity of a radiotracer has a lower impact (e.g. transporter substrate studies) on the study result. Preclinical PET studies bear the challenge of the very low blood volume and also lower receptor densities present in mice and rats, compared to humans. Additionally, only very small amounts of volume can be safely injected into animals (e.g. 100 μL per mouse).

4.2 Technological aspect influencing image quantification

Scanner design itself has an influence on image quality and quantification. However, these usually tend to not vary in imaging studies within an institution as such studies are most likely performed on the same instrumentation. The technological aspects become relevant when comparing studies across different institutions or if the PET instrumentation itself changes, e.g. potential scanner drift. The most influencing factors are scanner sensitivity and spatial resolution. As described above, scanner sensitivity for conventional clinical PET and preclinical PET scanners is around 2%–10%. To increase the count rate (using a given scanner) it is possible to either increase the radioactivity of the injected radiotracer or to use longer scan times. However, these two options are not always feasible as increased radioactivity also leads to an increase absorbed dose in the patients, to an increased amount of non-labelled substance, and potentially to detector saturation. Increasing scan time is also not always feasible due to patient comfort.

The other limiting factor is spatial resolution. PET in general is limited by spatial resolution and depending on the setup can additionally be influenced by relatively high noise levels. Thus, so-called partial-volume effects (PVEs) occur due to the limited spatial resolution and spill-in or spill-out of radioactivity from the surrounding tissue. PVE leads to an underestimation of radioactivity concentration and is becoming a prominent aspect when quantifying regions, which are smaller than 2–3 times the spatial resolution of the system. The PVE is especially crucial in structures that change their size during longitudinal monitoring, e.g. tumour size shrinks during therapy monitoring, as the PVE is dependent on the size of the structure.

Furthermore, the PVE is dependent on many factors such as the positron range of the used radioisotope, the size of the defined region, the SNR, the reconstruction algorithms and correction parameters, the position of the region within the FOV and the count rate of the acquisition. As no automatic PVE correction is available (both for clinical and preclinical PET), PVE related errors in quantification should either be avoided (defining large regions of interest) or at least discussed when presenting results.

4.3 Methodological aspect influencing image quantification

Methodological aspects that have to be considered when performing PET studies can be differentiated into pre- and post-study effectors. Pre-study aspects comprise the overall patient preparation and/or study design, when and how the radiotracer is administered and the actual PET acquisition parameters (energy window, scan time). For quantitative image analysis, PET images need to be corrected before obtaining and interpreting absolute values. Such corrections include dead time correction, radionuclide decay correction, detector normalisation, attenuation, random and scatter correction. Most of these corrections are automatically applied during reconstruction of the acquired data. Attenuation maps can be derived from the CT scan (when using a PET/CT scanner) or by using an external radioactive source rotating around the subject.

As described before, different image reconstruction algorithms are available both for clinical and preclinical PET. The choice of reconstruction algorithm and associated parameters strongly affects image quality and thus image quantification. FBP typically yields reproducible image quantification but provides images with lower spatial resolution and SNR. On the other hand, the iterative algorithms (MLEM) provide images with superior resolution, but image quantification might be biased. To our knowledge, to date there is no gold standard for image reconstruction available for all application areas.

Finally, image quantification depends on the method for regions of interest (ROI) delineation (e.g. manual, automatic, segmentation, region-growing) and the extracted parameters (e.g. mean, median, maximum value of the region).

4.4 Biological aspect influencing image quantification

Factors that influence the biological distribution *in vivo* highly affect PET image quality. For example, stress is a major impact factor in biological (preclinical) imaging studies, as it increases heart and breathing rates and activates and accelerates specific metabolic pathways. Consequently, stress-induced changes in metabolism may lead to a different distribution pattern of the same radiotracer. Hypoxia (lack of tissue oxygen) and hypothermia are additional aspects that directly influence the metabolic activity of the body. PET image quality and reproducibility is highly affected by biological parameters such as temperature, heart and breathing rate, fasting and anaesthesia (if used) but also the disease state, the animal model or physical and psychical stress. Depending on the biology under investigation these parameters will influence image quality and quantification.

By optimising data acquisition, reconstruction and data analysis the variability in image quantification can be reduced to <5% (as shown in phantom studies). However, biological variations are in the area of 20% (or more)—even between animals from the same age group and breeder. Thus, standardised subject handling requires special attention in order to obtain reproducible data.

5 Data processing

5.1 Image data analysis

In PET, image data analysis can be very complex. An easy method of image analysis is visual assessment of the images searching for regions with high radioactivity concentration. To obtain quantitative numbers from PET images, the image has to be calibrated in units of kBq/cc (or kBq mL^{-1} or kBq g^{-1}) of tissue. This calibration can be performed by scanning a tracer-filled cylinder of known activity and volume. Each voxel is assigned an intensity colour (also called radiation scale) according to a look-up table. Afterwards, several tools can be used for the definition of regions (or volumes) of interest so called ROIs/VOIs. A region of interest is usually defined by outlining a specific organ (whole organ or tumour) or part of the organ (e.g. cortex from brain). This can be done manually or by different image analysis tools, like region growing or segmentation or the use of a fixed-size region. Afterwards multiple parameters, e.g. the mean—median—max—five hottest voxels—peak value of the voxels in the region, can be calculated.

The percent injected dose per gram can be consequently calculated as:

$$\%ID/g = C_t \times \frac{V_t}{W_t} \times \frac{1}{A_{inj}}$$

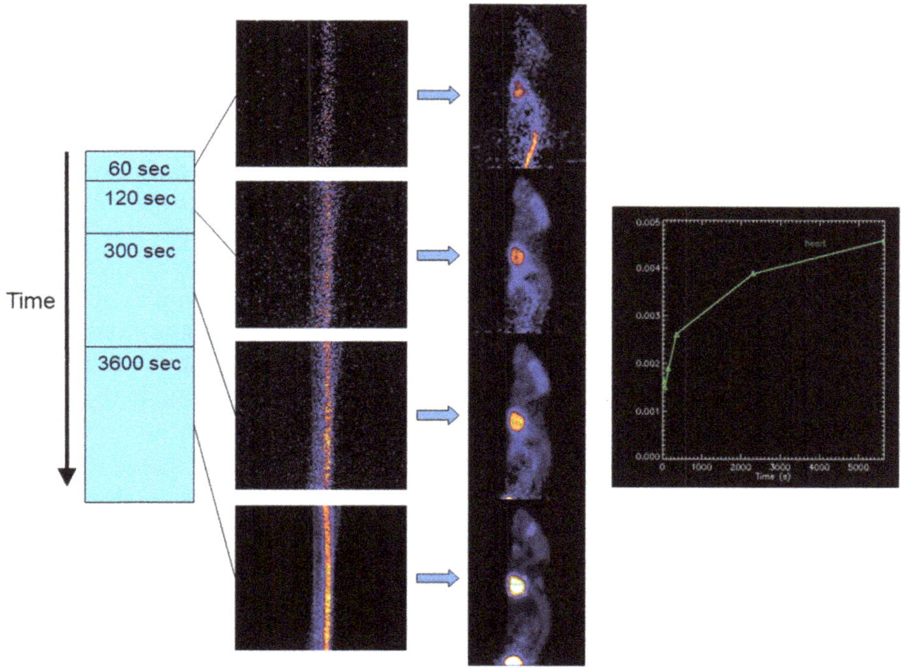

Figure 7. Illustration of the generation of a time-activity curve from a preclinical [^{18}F]FDG-PET scan. In this example an ROI over the heart was outlined.

where C_t is the tissue concentration (given in activity/volume), V_t is the tissue volume, W_t the tissue weight and A_{inj} the injected radioactivity. However, this only takes the total injected radioactivity into account.

Another method often used is the standardised uptake value (SUV):

$$SUV = (\%ID/g) \times W_P$$

Here the subject weight (W_P) is also taken into account.

In general, in PET imaging it is distinguished between static and dynamic scanning. Static imaging is applied when the pharmacokinetic of a radiotracer is already known and one is only interested at the radioactivity concentration of the tracer at a specific time point. In dynamic PET, the acquisition starts with the injection of the radiotracer. As such it is possible to follow the time course of radioactivity concentration in tissue over the defined measurement time. During image reconstruction, images are reconstructed for defined time intervals, e.g. the first image includes data acquired from 0–60 s after injection, the second image from 60–180 s after injection, etc. A so-called time-activity-curve (TAC) can be generated for each time frame (figure 7). Data from dynamic scans in combination with arterial blood (plasma) sampling allows for kinetic modelling. Kinetic modelling enables quantifying the interaction of radiotracer with tissue, however, as it is very complex and time consuming it is mostly used in clinical and preclinical research. Nevertheless, based on the kinetic modelling results target-to-non-target ratios can be established, which are then used in clinical routine. For [^{18}F]FDG, kinetic modelling enables quantification of the actual glucose metabolism and changes in [^{18}F]FDG transport rates.

5.2 Sample analysis

In preclinical and clinical research, it is common to also collect samples (e.g. blood) from the patient or laboratory animals to determine radioactive concentration in the blood or plasma. In addition, plasma might be analysed to obtain the fraction of unchanged radiotracer parent and radioactive metabolites. For pharmacokinetic studies this is very important as radiotracer metabolism can not only affect the amount of tracer that is able to bind to the target, but the generation of radioactive metabolites might interfere with image analysis. The PET detectors are not able to distinguish between intact radiotracer or radioactive metabolite as both substances are labelled with a PET radionuclide and thus emit photons with the same energy (511 keV).

As such samples are typically measured for radioactivity using so-called gamma-counters, which are based on a scintillation counter suitable for quantifying the radioactivity content in absolute values (e.g. kBq g^{-1}).

Radiotracer metabolism can be assessed by either high-performance liquid chromatography (HPLC) or thin-layer chromatography (TLC) and described in detail [17]. The obtained results (from the sample analysis and metabolites) are then incorporated into the kinetic modelling calculations.

6 Conclusions

PET is a functional imaging method for quantitatively measuring biochemical and physiological processes *in vivo* by using radiolabelled substances. Accordingly, the tissue concentration of the radiolabelled substance can be determined over time, and using appropriate mathematical models, the rate of specific biological processes can be extracted. Absolute quantification is a unique feature of PET.

6.1 Strength and limitations

PET is a nuclear medicine imaging modality used in clinical practice and research.

The strengths of PET are:

- It provides temporal and spatial resolved 3D images of the body;
- Versatile modality depending on the used radiolabelled substance;
- Needs low amount (mass) of radiolabelled substance (down to 10^{-12}–10^{-15} mol);
- Is a quantitative imaging modality. Biological processes can be mathematically described based on the image by applying kinetic modelling.

The limitations of PET are:

- Limited spatial resolution and sensitivity;
- Timing resolution in the region of some seconds;
- Limited axial FOV—except for preclinical and clinical total-body scanners;

- Cost intensive imaging modality (especially considering costs for radiotracer production);
- Requires highly trained personnel;
- Is based on the use of open radioactive material leading to radiation dose in personnel and subjects;
- Clinical reimbursement is an issue in many countries.

6.2 Future developments

Future developments in clinical PET technology will include new advanced detectors and electronics that enable higher SNR, improved spatial resolution and increased detection sensitivity such as faster TOF detectors, design of clinical total-body PET systems and also the introduction of organ-specific PET systems.

Development of dual-modality imaging probes is expected to further increase, e.g. radio- and fluorescent-labelled probes, as the combination of PET and optical imaging for integrated pre- and intraoperative targeted imaging via the same molecular receptor has shown to improve intraoperative cancer visualisation and maximise safer resection. Moreover, research in the development of PET/MR dual-modality imaging probes based on nanoparticles is also ongoing. As an example, iron oxide nanoparticles coated with a shell where positron emitters or small surface molecules are attached are under investigation. This approach aims to provide probes with high spatial resolution (MRI) and with high sensitivity (PET).

Finally, new reconstruction algorithms are expected to further improve image quality and resolution. These algorithms will include the possibility for motion correction and correction methods for PVE.

In preclinical imaging, standardisation will be an issue to enhance reproducibility of study results. It is expected that recommendations will be developed for certain protocols (acquisition, reconstruction, image analysis, animal handling) targeting specific radiotracer/animal model combinations.

References and further reading

[1] Phelps M E 2004 *PET: Molecular Imaging and Its Biological Applications* (Berlin: Springer)

[2] Wadsak W and Mitterhauser M 2010 *Eur. J. Radiol.* **73** 461–9

[3] Anderson C J, Ling X, Schlyer D J and Cutler C S 2019 A short history of nuclear medicine ed J S Lewis, A D Windhorst and B M Zeglis *Radiopharmaceutical Chemistry.* (Cham: Springer International Publishing) pp 11–26

[4] Antoni G 2019 The radiopharmaceutical chemistry of carbon-11: basic principles *Radiopharmaceutical Chemistry* ed J S Lewis, A D Windhorst and B M Zeglis (Cham: Springer International Publishing) pp 207–20

[5] Cherry S R, Sorenson J A and Phelps M E 2003 *Physics in Nuclear Medicine* 3rd edn (Philadelphia, PA: W. B. Saunders)

[6] Habib Z and Marie-Louise M 2006 *Curr. Med. Imaging* **2** 3–13

[7] Beyer T, Townsend D W, Czernin J and Freudenberg L S 2011 *Insights Imaging* **2** 225–34

[8] Vaquero J J and Kinahan P 2015 *Annu. Rev. Biomed. Eng.* **17** 385–414

[9] Cal-Gonzalez J, Rausch I, Shiyam Sundar L K, Lassen M L, Muzik O and Moser E *et al* 2018 *Front. Phys.* **6** 47

[10] Zhang X, Zhou J, Cherry S R, Badawi R D and Qi J 2017 *Phys. Med. Biol.* **62** 2465–85

[11] Defrise M, Kinahan P E and Michel C J 2005 Image reconstruction algorithms in PET *Positron Emission Tomography: Basic Sciences* ed D L Bailey, D W Townsend, P E Valk and M N Maisey (London: Springer) pp 63–91

[12] Surasi D S, Bhambhvani P, Baldwin J A, Almodovar S E and O'Malley J P 2014 *J. Nucl. Med. Technol.* **42** 5–13

[13] Herfert K, Mannheim J G, Kuebler L, Marciano S, Amend M and Parl C *et al* 2020 *Mol. Imag. Biol.* **22** 223–44

[14] Aide N, Lasnon C, Veit-Haibach P, Sera T, Sattler B and Boellaard R 2017 *Eur. J. Nucl. Med. Mol. Imaging* **44** 17–31

[15] Mannheim J G, Mamach M, Reder S, Traxl A, Mucha N and Disselhorst J A *et al* 2019 *J. Nucl. Med.* **60** 1483–91

[16] Stout D, Berr S S, LeBlanc A, Kalen J D, Osborne D and Price J *et al* 2013 *Mol. Imaging* **12** 1–15

[17] Rokka J, Gronroos T J, Viljanen T, Solin O and Haaparanta-Solin M 2017 *J. Chromatogr. B Analyt. Technol. Biomed. Life Sci.* **1048** 140–9

IOP Publishing

Imaging Modalities for Biological and Preclinical Research:
A Compendium, Volume 2
Parts II–IV: *In vivo* preclinical imaging, multimodality imaging and outlook
Andreas Walter, Julia G Mannheim and Carmel J Caruana

Chapter II.11

Single photon emission computed tomography

Karla Palma-Alejandro and Ludek Sefc

Single photon emission computed tomography (SPECT) is a tomographic imaging technique that visualises gamma-rays emitting radiopharmaceuticals introduced to the patient. SPECT along with positron emission tomography (PET) is the main molecular (functional) imaging method used for localisation of tumours, infections, cardiac or brain functions. In contrast to PET, SPECT detects single emitted photons. These photons need to pass through collimators before interaction with the detector. This process allows defining the direction to the source of gamma rays but also significantly decreases the sensitivity of the method. The proper collimator design enables a very good resolution. Compared to PET, SPECT can provide better resolution but worse sensitivity and poorer quantification. Recently, preclinical small-animal SPECT systems have become a major focus in nuclear medicine pushing the limits of SPECT into the submillimeter range.

This chapter describes the principles of SPECT, its relevance to biomedical applications, and the main acquisition parameters, which influence the image reconstruction.

1 Introduction

SPECT has a long history of development since its initial demonstration by Kuhl and Edwards in 1963 [1]. SPECT imaging allows detection of radiopharmaceuticals *in vivo* and has proven to be an excellent tool for visualisation of various physiological and pathophysiological processes. The main advances of SPECT in the last six decades have been: (1) improvements in SPECT collimators from simple parallel to pinhole and fan-beam collimators; (2) development from single-head to multi-head cameras; (3) improvements in detectors to use direct conversion of gamma rays into electrical signals and solid-state detectors improving the intrinsic

doi:10.1088/978-0-7503-3747-2ch15

resolution; (4) use of hybrid SPECT/CT systems that have improved image quality; (5) addition of new SPECT radiotracers that target disease mechanisms for more sensitive and specific detection; (6) development of organ-specific systems for cardiac, brain, or thyroid gland imaging [2].

Currently, new SPECT improvements are mainly focussed on detection, promising a significant enhance in the image quality and the sensitivity of the method.

2 Principles and setups

2.1 Principles of SPECT

SPECT is based on gamma-emitting radioisotopes to label biological compounds injected into the subject to acquire their distribution *in vivo*. Radioactive tracers consist of carrier molecules incorporating a radioactive atom. The carrier molecules vary depending on the purpose of the scan, e.g., some tracers employ molecules that interact with a specific protein or sugar in the subject. The radioactive tracer is administered to the subject of interest by intravenous injection, inhalation, oral ingestion, or by direct injection into the organ. The tracer administration mode will depend on the specific disease process that needs to be studied [2, 3]. The common radionuclides used for SPECT are shown in table 1.

A rotating gamma camera (also known as scintillation camera or Anger camera) is generally used to acquire a set of parallel contiguous sections, which serve for image reconstruction. This information effectively defines the three-dimensional distribution of the radiopharmaceuticals within the sample (see figure 1). The SPECT projection data results from the measurement of in-line integrals of the attenuation coefficients through the radiotracer distribution. The radiotracer distribution image can be reconstructed choosing the proper set of attenuation integrals when the emitted photons travel through attenuating media (transmission imaging), two processes are involved in the interaction: photoelectric absorption

Table 1. Decay properties of the most common radionuclides used in SPECT.

Radioisotope	Half-life	Energy (abundance)	Production method
^{99m}Tc	6.02 h	141 keV (89%)	Generator
^{123}I	13.22 h	159 keV (83%)	Cyclotron
^{125}I	59.49 days	35.49 keV (6.68%)	Reactor
^{131}I	8 days	364 keV (81%)	Reactor
^{111}In	2.80 days	171 keV (91%) and 245 keV (94%)	Cyclotron
^{201}Tl	3.04 days	69–83 keV (Hg X-rays), 135 keV (2.5%), 167 keV (10%)	Cyclotron
^{127}Xe	36 days	172 keV (26%), 203 keV (7%), 375 keV (17%)	Cyclotron
^{133}Xe	5.2 days	81 keV (37%)	Reactor
^{67}Ga	3.26 days	93 keV (37%), 185 keV (20%), 300 keV (17%), 395 keV (5%)	Cyclotron

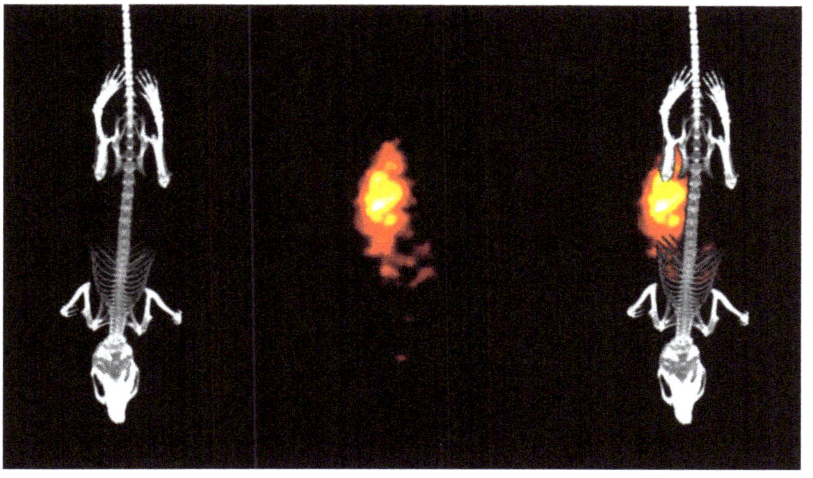

Figure 1. SPECT/CT tumour imaging via enhanced permeability and retention (EPR) effect after 3 days (HPMA (N-(2-hydroxypropyl) methacrylamide) polymer), ^{125}I, 30 MBq *i.v.*

and Compton scattering. The linear attenuation coefficient μ describes interaction of photons with the medium and is estimated by measuring the decrement in the intensity of a narrow beam of photons traversing an attenuating material [4]. The linear attenuation coefficient μ can be written in terms of the intensity observed at length l through the attenuator as a portion of the initial intensity I_0 [4].

$$I = I_0 exp(\mu l)$$

This equation does not take into account scattered photons, considering them to be removed from the narrow beam and not contributing to the detected signal and neither considers the distribution case of an extended source [4]. Scattered photons from an extended source, can only be removed from the detected signal using a detector with an infinitely high energy resolution or an infinitely high degree of collimation. However, the degree of collimation of a gamma camera is limited by sensitivity constraints and a finite value of 10% energy resolution [4].

2.2 Typical setups and state-of-the-art

At the beginning in the 1960s, the SPECT systems had limited image quality due to the imaging hardware and poor performance of the reconstruction software [4]. Improvements in the reconstruction algorithms and software development pushed forward by X-ray tomography, together with upgraded gamma camera hardware and performance, resulted in improved image quality of commercial SPECT systems in late 1970s. In the mid-1980s, the gamma camera performance improved significantly with microprocessor development [4].

Optimal performance of SPECT can be achieved with systems designed for a specific application. Some aspects of SPECT designs that are continuing and that may influence future development include solid-state readout, such as pin diodes (PD), avalanche photodiodes (APD) including position-sensitive devices (PSAPD), hybrid pixel detectors (Timepix), silicon photomultipliers (SiPM) and silicon drift detectors (SDD). To date, however, these efforts have largely been focused on potential preclinical and *in vivo* detectors for very low-energy emitters (e.g. ^{125}I) [2].

The collimator can be designed to detect more photons, but increased detection efficiency usually can be achieved only with a concurrent loss of spatial resolution [5]. The most crucial goal of SPECT instrumentation development is to increase the detection efficiency while simultaneously improving the imaging system spatial resolution; this can be pursued by adding more detectors around the subject.

The two most common methods for data acquisition are continuous acquisition and step-and-shoot. In continuous mode, the events are continuously recorded while the camera rotates around the subject. Thus, each projection view will be, to some extent, horizontally blurred or smeared along each row of pixels. On the other hand, in the step-and-shoot, the camera records the projection events in a stationary position diminishing the image quality [6].

3 Biomedical relevance

3.1 Application range and relevance

The demand for small animal imaging has significantly increased due to the recent advances in biomedical research. Imaging of internal biodistribution with SPECT needs an overlay of the molecular image with X-ray or another anatomical image to put the molecular information into the right anatomical context. For long-term drug biodistribution studies and treatment effect monitoring in a preclinical model, 125I is the most suitable radioisotope for SPECT imaging since it has relatively low energy (35 keV) and long half-life (60 days) [7].

SPECT is suitable for applications in neurosciences, oncology, cardiology, drug pharmacodynamics/pharmacokinetics, development of theranostic therapies, drug delivery systems, etc [3, 8].

For example, SPECT allows researchers to perform longitudinal studies to analyse drug cancer treatment effects [2, 3]. Whenever molecular oncologic imaging requires complex anatomy localisation, such as in gastric motility studies, or the target of imaging is tiny, such as small tumours and metastases, SPECT becomes a suitable imaging technique [9].

3.2 Sample preparation

3.2.1 Radiotracter preparation

The critical parts in radionuclide imaging such as SPECT are biomarkers (targets), radiopharmaceuticals (tracers), and the respective imaging device. The first step in preparing the radiopharmaceuticals is producing a radioisotope appropriate for imaging the disease in question. Usually, those used for SPECT imaging are readily available and have reasonable long half-lives—long enough to allow easy use but short enough to diminish the overall dose supplied to the imaging object (see table 1). The metastable nuclear isomer of technetium-99 (^{99m}Tc) is the most widely used radioisotope in SPECT and is generally produced in so-called technetium generators (figure 2). Other commonly used SPECT radioisotopes for diagnostic imaging are ^{201}Tl (myocardial perfusion studies), ^{111}In (prostate cancer), ^{67}Ga (infection or lymphoma), ^{123}I (thyroid gland) and ^{133}Xe (lung) [5].

Radioisotopes should be then incorporated into the tracers (chemical compounds with affinity to the molecular processes to be imagined). One or more atoms in the tracer molecule should be replaced by a corresponding radioisotope to visualise the site where tracers tend to accumulate. Such tracers are called radiotracers or radiopharmaceuticals.

3.2.2 Anesthesia

During manipulation/radiotracer injection and SPECT/CT imaging, the animals are immobilised by light anaesthesia such as isoflurane or sevoflurane. These volatile anaesthetics require a dedicated anaesthesia system, including vaporiser and flow meter. They are administered via a carrier gas inhaled by the animal (usually

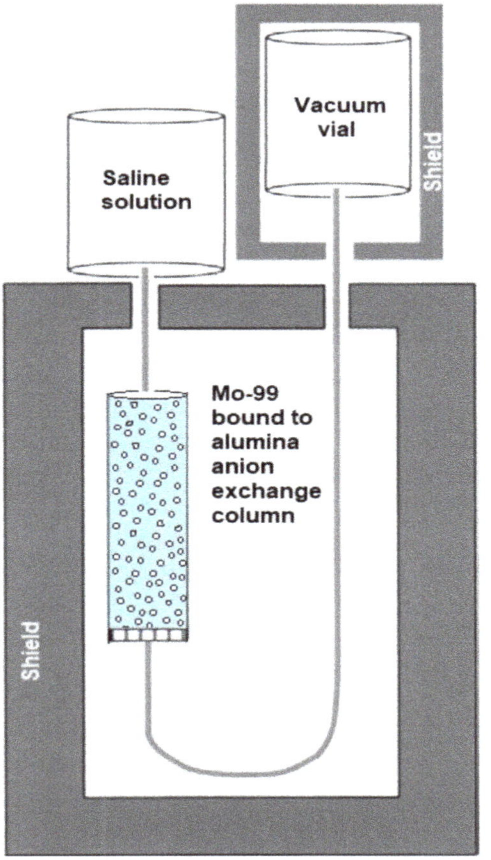

Figure 2. ^{99m}Tc radionuclide generator system filled with commercially produced molybdenum-99 (half-life 66 h) is the source of the ^{99}Mo decay product ^{99m}Tc (half-life 6 h) which is widely used in the SPECT medical imaging in hospitals.

oxygen or air). For long anaesthesia periods, the carrier gas may be humidified with water to avoid dehydration of the respiratory airways. For some applications, the injection anaesthesia with ketamine/xylazine is preferable.

Regardless of what type of anaesthetic agent is used, heating the animals is essential. Heating ideally starts before anaesthesia induction. Remarkably, mice with unfavourable mass-to-surface ratio are highly dependent on heat production through shivering thermogenesis. During anaesthesia, the mouse is immobilised; muscles do not produce enough heat and in ambient temperature, the mouse rapidly loses its body temperature, which can interfere with the experiment and even with the viability.

3.2.3 Application of tracers and drugs

The tracer is commonly injected intravenously via a tail vein or supraorbital sinus or inhaled/ingested by the animal. The blood volume of mice is about 58.5 mL kg^{-1} body weight [8], i.e. 1.46 mL for 25 g mouse. The recommended injection volume should not exceed 10%–15% of blood volume, i.e. approx. 150–200 μL. If using a syringe and cannula, the remaining radioactivity within them must be measured after injection to determine the exact amount of injected radioactivity. As the total tracer dose (usually 20–50 MBq) needs to be in a small volume, relatively high tracer concentrations are required.

The image quality in the area of interest can be increased by injecting a higher radiotracer activity; however, one should consider the safety limit of radiation exposure. Depending on the study type and the pharmacokinetics, the image acquisition should be performed at different time intervals after the radiotracer injection (minutes—hours—days).

4 Parameters of image quality

It could be essential to distinguish between parameters that can be seen or measured in reconstructed images versus parameters that are more ‘fundamental’ to the scanner. The latter type of parameters can also influence image quality, but it is not always directly deducible. SPECT image quality is mainly addressed to the spatial resolution as indicated by small structures in the image (small spheres or rods used to calculate recovery coefficients), image uniformity (signal-to-noise ratios, percentage standard deviation), the spillover of activity into cold regions and quantification (pixel values indicating Bq cm^{-3}). Some degrading factors include attenuation, scatter, and resolution effects, in addition to motion artefacts. The proper corrections and evaluations must be performed for a high-quality image. Assuming that spatial resolution can be directly affected by the collimator or pinhole diameter due to the scattering contribution, some corrections and measurements must be done. The intrinsic detector spatial resolution is directly proportional to the image quality [2, 3, 5].

4.1 Spatial resolution

Spatial resolution can be influenced by the collimator and the detector, thus is classified by (a) collimator resolution, (b) intrinsic detector resolution and (c) extrinsic resolution, which is the combination of collimator design, and intrinsic factors. This can be calculated as the quadratic sum of the full-width-half-maximum (FWHM) of both detector and collimator resolutions [11]. The resolution given by the collimator can be estimated as follows,

$$FWHM_c \sim \left[\frac{d}{l}\right](l + f + c),$$

where d is the hole diameter of the collimator, l is the length, t is the septal thickness of the collimator, f is the collimator to source distance, and c is the collimator to detector centre distance.

4.2 Sensitivity

This term can be defined as the ability to detect and record the percentage of the emitted events. In single-photon imaging (planar and SPECT), physical collimators are typically made from lead or tungsten. These materials reject photons that are not within a small angular range, thus defining the source direction. In SPECT, several approaches were investigated to increase sensitivity through the optimal design of the collimator. Novel types of collimators have been able to improve sensitivity without adversely affecting resolution [11]. Regarding diverse applications, it can be stated that high-resolution collimators may be desirable for small lesion detection in oncology. However, the low sensitivity of the high-resolution collimator requires a higher dose or longer acquisition times. Using high sensitivity collimators can improve the detection sensitivity at a cost of image resolution [12].

4.3 Noise

The noise in the reconstructed slices is a result of the back-projection process and does not follow simple nuclear counting statistics. A formulation for the noise has been given by Budinger *et al* [13]:

$$\%\text{rms noise} = \frac{120\,V^{3/4}}{\sqrt{N}},$$

where V *is* the total number of voxels covering the reconstructed object, based on a circular FOV and N is the total number of acquired events.

5 Data processing

Tomographic slices are reconstructed from the acquired projection data using either analytic or iterative algorithms. Analytic reconstructions represent an exact mathematical solution such as filtered back-projection based on the Radon transform, developed by Johann Radon in 1917. Although filtered back-projection is a relatively efficient operation, it does not always perform well on noisy projections and, as is the case with SPECT data, it generates artefacts when the projections are not line integrals of the internal activity. Iterative algorithms are a preferred alternative method for performing SPECT reconstruction. These types of algorithms are based on an estimate of the actual probability of a certain amount of radioactivity at a particular location of the detection by the imaging system at each particular point in each projection [14].

6 Conclusions

Since the commercial availability of the first rotating camera SPECT systems, there has been significant progress in the performance and reliability of SPECT systems, including a wide range of applications. SPECT continues to progress mainly in instrumentation and reconstruction, and it is possible to say that the limits of spatial resolution and sensitivity have not yet been reached. Furthermore, new biological targets are continuously being developed to study the disease progression and underlying mechanisms.

6.1 Strength and limitations

SPECT is a superior method for three-dimensional visualisation of radiolabelled molecules inside the body. Compared to PET, it offers the possibility to widen the observational time window (owing to the longer half-life of single-photon emitters), allowing longer time measurements of the fate of administered compounds. The resolution can also be superior because there is no need to consider the positron annihilation range that blurs localisation of tracer in PET imaging. The significant advantage of both SPECT and PET imaging for pharmacokinetic/pharmacodynamic studies is that they do not use large contrast agents to obtain the molecular signal such as optical, photoacoustic, or magnetic molecular imaging methods. The simple ion exchange for a radioisotope does not influence the transportation of the labelled compound.

The main disadvantage of SPECT compared to PET is significantly lower sensitivity (by approximately two to three orders of magnitude); i.e., the ability to detect and record a higher percentage of the emitted events. This disadvantage is due to collimator use. The collimators are needed to reject photons that are not within a small angular range; otherwise, the trajectory of detected photons will not be known. Collimators, therefore, exhibit very low geometric efficiencies (~0.01%). SPECT thus needs a higher radiation dose given to the object of imaging. Generally, longer half-lives of SPECT isotopes contribute to the high-dose exposition of the animal, which is somewhat compensated by their lower energies compared to 511 keV annihilation photons in PET imaging. Nevertheless, many of such γ-ray photons (due to their lower energy compared to PET) are absorbed and scattered by body structures before exiting the imagined subject, contributing to the image noise. Both SPECT and PET employ a significant radiation dose applied to the animal, favouring nonradioactive molecular imaging methods for many applications.

It is necessary to pay careful attention to compensate for image degrading factors, select appropriate techniques, and recognise the limitations of the SPECT, which are essential to achieve an accurate quantification. Noise, scattering, boundary accuracy, and attenuation are the main problems and must be solved with accurate and reliable corrections. Thus, it is possible to use SPECT to quantify tracer uptake in absolute terms of Becquerel per gram of tissue without having to make any assumptions or estimates of the background activity superimposed on the organ or area of interest [5].

The use of SPECT in animal research has become increasingly practical. In addition to having a wide choice of relatively inexpensive radiopharmaceuticals, long-lived radioisotopes with half-lives from a few hours to a few days are also available for antibody-binding studies and serial studies. SPECT has additional capabilities as well, including dual-isotope imaging and the imaging of radioisotopes used for tumour therapy where the dose is delivered at the same time as the image is produced (theranostics). New modalities, such as SPECT/MRI, could offer the advantage of additional anatomic and functional information, generating the next level of multimodality molecular imaging. In many aspects, contemporary preclinical SPECT imagers are still based on detectors rotating around the object. While this approach is relatively inexpensive, there are a number of shortcomings to rotational gamma camera designs that have led to the perception that PET, which uses coincidence detection with stationary detectors, is better than SPECT for preclinical imaging. However, in 2004, Beekman *et al* introduced ultra-high resolution 4D stationary multi-pinhole SPECT (4DSMP-SPECT) [15]. The advantage of this cutting-edge technology is that all image-blurring from motion imprecisions and slow acquisitions stemming from detector motion have been eliminated.

6.2 Future improvements

Improvements in detector technology [16, 17] and collimators will yield intrinsically better projection information. Compton cameras use the calculation of Compton scattering to determine the trajectory of incoming photons. In addition, regular two-layer Compton cameras, a novel and cheaper one-layer Compton camera with a broader field-of-view has been introduced [18]. Use of these cameras will allow increasing the sensitivity of SPECT from <0.1 to 80% for the widely used SPECT isotope ^{99m}Tc. SPECT imaging will require only a small portion of currently used radiation load to the patient, and the absence of collimator will allow combined CT/PET/SPECT devices with one common ring of detectors for simultaneous fast trimodal hybrid whole-body imaging with very low radiation load. Also, refinements in corrections for scattered radiation, attenuation, and spatial resolution losses will provide more accurate and quantitative reconstructed slices. The capability for combining this information with correlative imaging, both anatomical and functional, from other modalities is also expected to continue improving [4, 5, 16]. The MRI-compatible SPECT detectors will allow construction of SPECT/MRI devices and decrease further the radiation dose required for imaging [19].

References and further reading

[1] Kuhl D E and Edwards R Q 1963 *Radiology* **80** 653–62

[2] Hutton B F 2014 *Eur. J. Nucl. Med. Mol. Imaging* **41** 3–16

[3] National Research Council 1996 *Mathematics and Physics of Emerging Biomedical Imaging* (Washington DC: National Academies Press) pp 89–104

[4] Gillen G J 1990 Single photon emission computed tomography: performance assessment, development and clinical applications *PhD Thesis* University of Glasgow

[5] Tasneem R 2017 Design and evaluation of the newly proposed L-SPECT system based on light field imaging *PhD Thesis* University of New South Wales Canberra

[6] Groch M W and Erwin W D J 2000 *Nucl. Med. Tech.* **28** 233–44

[7] Kostiv U *et al* 2017 *Nanoscale* **9** 16680–8

[8] Pfau D, Thorn S L and Zhang J *et al* 2019 *Sci. Rep.* **9** 5791

[9] Franc B L, Seo Y, Flavell R and Aparici C M 2020 Preclinical SPECT and SPECT-CT in oncology *Molecular Imaging in Oncology* ed O Schober, F Kiessling and J Debus (Cham: Springer) pp 360–404

[10] Diehl K H, Hull R and Morton D *et al* 2001 *J. Appl. Toxicol.* **21** 15–23

[11] International Atomic Energy Agency Quality Assurance SPECT Systems 2009 *IAEA Human Health Series* 6 (Vienna: IAEA)

[12] Lau Y, Hutton B and Beekman F 2001 *Eur. J. Nucl. Med.* **28** 39–47

[13] Budinger T F 1980 *J. Nucl. Med.* **21** 579–92

[14] Madsen M T 2007 *J. Nucl. Med.* **48** 661–73

[15] Beekman F J, van der Have F and Vastenhouw B *et al* 2005 *J. Nucl. Med.* **46** 1194–200

[16] Kupinski M A and Barrett H H (ed) 2005 *Small-animal SPECT Imaging* (New York: Springer)

[17] Trojanova E, Jakubek J, Turecek D, Sykora V, Francova P, Kolarova V and Sefc L 2018 *J. Instrum.* **13** C01001

[18] Turecek D, Jakubek J, Trojanova E and Sefc L 2020 *J. Instrum.* **15** C01014–C

[19] Zajicek J, Burian M, Soukup P, Novak V, Macko M and Jakubek J 2017 *J. Instrum.* **12** P01015

Part III

Correlative multimodal imaging and image data fusion

Section 1

Biological correlative imaging

IOP Publishing

Imaging Modalities for Biological and Preclinical Research: A Compendium, Volume 2
Parts II–IV: *In vivo* preclinical imaging, multimodality imaging and outlook
Andreas Walter, Julia G Mannheim and Carmel J Caruana

Chapter III.1.a

Correlative light electron microscopy

Paul Verkade

Correlative light electron microscopy (CLEM) deals with the use of both light and electron microscopy in a single experiment. There are, however, many ways to do CLEM as there are many light microscopy techniques as well as several electron microscopy approaches. One can in principle combine any light microscopy and electron microscopy technique to fulfil the criterium of doing a CLEM experiment. It is therefore impossible to provide information on, for instance, resolution or the type of sample, as this will be dependent on the biological question as well as the technique used. The individual techniques are well described in the individual chapters in this book and in this chapter we will therefore deal with the general principles of CLEM, including brief discussions about probes, processing, and analysis.

1 Introduction

In this e-book almost all of the common imaging modalities used in the life sciences research area are highlighted. For each of these modalities it is described what can be achieved with the technique and what it is especially used for. But each of these techniques will also have its limitations. This can for instance be in the realm of the types of sample it can image, the type of information that can be provided, or the resolution that can be achieved. There is, however, no single imaging technique that can provide a complete picture of a sample.

Thus, in most research papers a combination of techniques will be used to draw conclusions on the biological sample in question. In some cases, however, it is essential to pull this information from one and the same sample. This can be because the object under study is very rare or the biological question itself requires the same sample to be investigated with multiple modalities. In this case we talk about correlated or multimodal imaging (CMI). The best established CMI technique is where fluorescence light microscopy is combined with transmission electron microscopy in so-called correlative light electron microscopy (CLEM).

2 Principles and setups

2.1 Principle of CLEM

The ability to use green fluorescent protein (GFP) together with technological advancements such as the development of the laser scanning confocal microscope in the mid-1990s transformed the way proteins of interested could be imaged in time and with higher resolution both in x and y but also in z-depth. One of the major drawbacks of fluorescence-based light microscopy (LM), however, is that only the labelled proteins of interest are lighting up and the exact location of the fluorescence has to be inferred from additional known fluorescence or shape recognition. In addition, the resolution of standard confocal microscopes is limited to approximately 200 nm, often well higher than the actual structure under investigation. This is where TEM comes into its own, providing both the resolution as well as the complete ultrastructure known as the 'reference space'. Except for a very few exceptions, TEM can, however, only analyse fixed samples. So, ideally one would want to combine the live imaging properties of fluorescence light microscopy with the contextual information of the electron microscope using CLEM [1].

2.2 Setup of a CLEM experiment

To illustrate how a CLEM experiment works we have designed a hypothetical experiment where a green fluorescent probe has been added to the outside of a model cell. For reference, the nucleus is depicted in blue, which can be done using a live nuclear dye but can also easily be seen using most bright-field LM techniques. A live imaging sequence would be acquired (figure 1(a), top row) to provide us with the history of a fluorescent structure (arrows) that has been internalised from the cell surface. Subsequently, the cell will be fixed and a higher magnification section of the

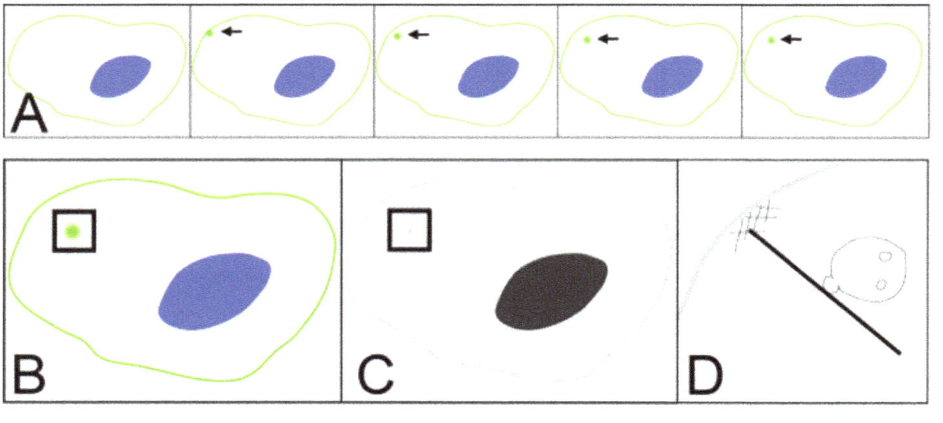

Figure 1. Concept of a CLEM experiment. This figure should be viewed in conjunction with section 2.2, which will explain the steps in a CLEM experiment.

last frame of the live sequence is shown in more detail in figure 1(b). This image needs to be matched with the overall structure of the same cell in the electron microscope (as in figure 1(c)). Once this same cell is retraced in the electron microscope (e.g. matched by its overall shape) one can zoom into the area of interest (boxed in figure 1(b and c)) where the fluorescent structure was observed. Now also the underlying ultrastructure is visualised including membranes and cytoskeleton (figure 1(d)). In this virtual experiment a fusion event between a membrane-bound endocytic structure that is taken up from the outside of the cell and a membrane-bound existing structure is captured.

3 Biomedical relevance

3.1 The power of CLEM

The example in figure 1 also provides us with the rationale why CLEM can be a powerful tool where the answer could not have been provided by a single standalone technique. The live LM alone would have given us information of the trajectory of fluorescent 'blobs', but the underlying structures would not have been visible. Similarly, the single high-resolution image alone (figure 1(d)) does not tell us whether the structure was fusing or actually moving away from the larger structure. Thus, the sum is more than its parts:

$$\text{CLEM: } 1 + 1 = 3$$

3.2 CLEM workflows

There are very many ways to do CLEM and the technique is not limited to just combining LM and EM but other modalities such as soft X-ray imaging and the combination with Mass Spec imaging are now included in correlative imaging (CI) workflows. All of the workflows do have some communalities;

Probes: The object of interest needs to be visualised in all modalities. Fluorescence is for instance not directly visible in EM.

Processing: Cells contain a lot of water but EM works under vacuum. One can go from live to fixed and then in EM.

Analysis: Finally, the data from both modalities needs to be combined and fitted together.

3.2.1 Probes

Whereas light microscopy is most commonly based around fluorescence, this fluorescence is not directly visible in an electron microscope. In EM it is the deviation of the electron beam by heavy atoms that generates contrast and we use this property to introduce clusters of heavy atoms into our sample to locate structures of interest. Most commonly these are gold particles but also quantum dots and electron dense precipitates can serve as good markers [1–3]. One of the earliest CLEM probes is a combination of fluorescence combined with an ultra-small gold particle called fluoronanogold that was developed in the mid-1990s. We have used this probe to visualise the specific concentration of a protein complex that is only present for approximately 90 s on a cell cycle of 22 h, or in other words less than 1 in 1000 cells would show this particular phenotype [4] at any given time. Without the aid of the fluorescence this would have been looking for the proverbial '**needle in the haystack**'. By using the fluoronanogold we were guided by the fluorescence to the area where we found silver-enhanced gold particles.

There can however be serious issues associated with using a gold particle in close proximity to a fluorophore, as the fluorescence can be absorbed by the gold particle and quench the fluorescence. It is unknown whether this is also the case for fluoronanogold. Similar to over-expression of fluorescently-tagged proteins that can lead to mis-localisation and misfunctioning, tagging a protein of interest with a CLEM probe can lead to a non- or mis-functional protein of interest [1]. One should therefore always be cautious and include the right controls when one starts using such probes.

It is not always essential to have a dual probe in a CLEM workflow. Researchers have been able to retain fluorescence (of GFP) after embedding in resin [5] and have even shown that GFP can resist the harsh conditions inside an EM [6].

3.2.2 Processing

One of the seminal papers using CLEM was published in 2000 by the group of Polishchuck, Luini, and Mironov [7]. They used the power of live GFP imaging to study biosynthetic carriers, being able to fix a moment in time and process this for study in the EM. They employed a relatively simple gridded coverslip (see figure 2) as their main tool and started the wave of interest in CLEM that is still growing to this day. From that year the interest in CLEM took off and the number of papers using CLEM published each year is still increasing. With it, more and more different CLEM workflows were developed as each biological question demanded its own specific workflow, highlighting another aspect of a good CLEM experiment: The biological question drives the CLEM technique that is to be used. Over the years we have compiled a large number of these CLEM and CI approaches in a series of books and one can refer to those for an overview of the different techniques [8].

3.2.3 Analysis

How does one first image a sample in a light microscope and then following the whole processing procedure for EM find again that one particular point of interest? Well, it needs a series of landmarks that are recognisable in both modalities. As highlighted, Polishchuck used a very simple finder pattern that was embossed in the glass. This pattern is easily recognised in the bright field of the light microscope and because the pattern stands out, it is also embossed in the resin block in which the sample is embedded (figure 2(a)). Using a stereo microscope the block can be trimmed down to a very small size containing very few cells for which the pattern itself can be recognised. If the live imaging aspect of the CLEM workflow is not essential one can also collect sections on a finder grid (figure 2(b)). This pattern is visible in both LM and EM.

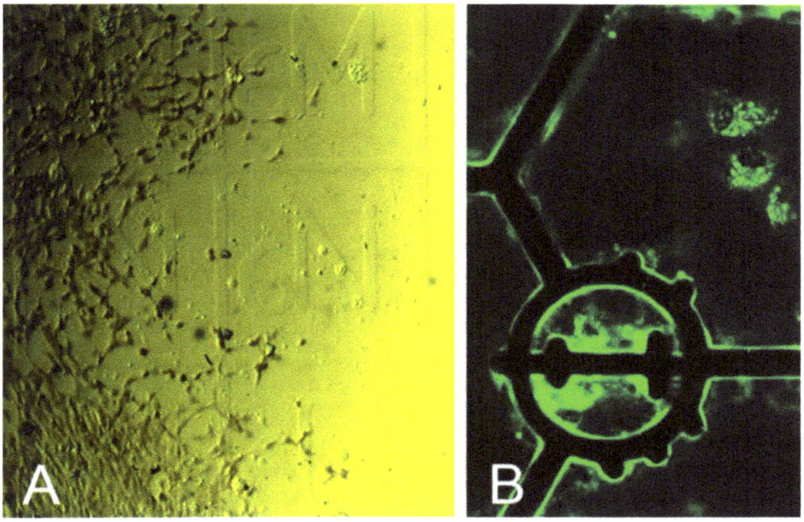

Figure 2. Means for retracing specific tools are available to aid in finding again the same object in the electron microscope that was previously imaged in the light microscope. In (a), an imaging dish with a special coverslip glued into the bottom has been used (Mattek gridded coverslip dishes). The finder pattern is embossed in the glass and is transferred when the sample is embedded in resin. After lift-off of the coverslip the finder pattern is still visible ((a), 6N). In addition, the stained cells are also very well visible. It is important to note that the cell density on the bottom left of the block is too dense to trace back individual cells. Only the part in the middle and top left of image (a) is suitable to recognise specific patterns of cells and individual cell shapes. In (b), GFP-expressing cells were grown on a coated finder grid. The cells were cryo-fixed and the cells were imaged using a cryo-fluorescence microscope in a cryo-CLEM workflow, not discussed in detail but see [9].

Alternatively, fiducial markers like FluoSpheres [5] can be added to the sections and they can be used to align both data sets.

Finally, at the end of a CLEM process the data from the imaging modalities need to be merged and the accuracy of the overlay can be crucial for the outcome of an experiment. The addition of super resolution fluorescence microscopy to the CLEM workflow will aid in that respect. There are now commercial and freely available software packages such as ecCLEM available that help align both data sets.

3.3 A real CLEM example

So far we have only discussed the theory of doing a CLEM experiment. In this last part I would like to highlight a real experiment as has been done in my lab. We are interested in the uptake, trafficking, and sorting of Transferrin (Tf) and epidermal growth factor (EGF) in cells [1]. Simply put; EGF binds to its receptor, EGFr, is endocytosed into the cell and signals that the cell needs to divide. As this signal needs to be shut down after signalling, the EGFr is routed to the degradative pathway. Tf on the other hand takes care of iron transport into the cells. It also binds to its receptor, Tfr, and is endocytosed. The iron is released from the Tf-Tfr complex and Tf and Tfr can be re-used and thus they recycle back to the plasma membrane. While both receptors are internalised together there must be a segregation event inside the

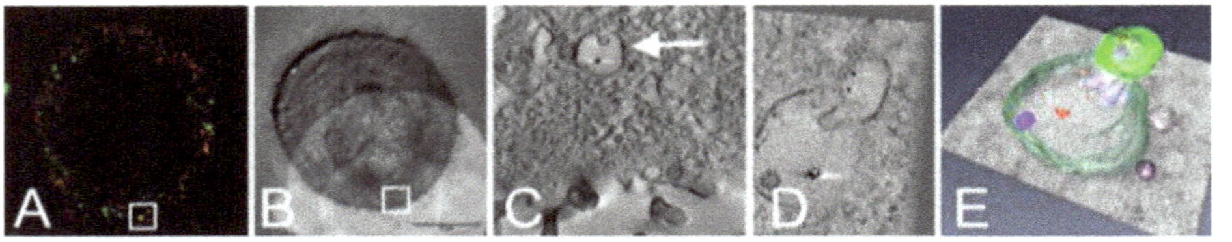

Figure 3. A real CLEM experiment. HeLa cells were grown on a gridded finder dish as shown in figure 2(a) and Tf-Alexa488-5nmgold and EGF-Qdots655 were added to the cells. Cells were analysed using a confocal laser scanning microscope (a, last frame from a live sequence). There is very little overlap between the green Tf signal with the red EGF signal. Only in the bottom of the cell is there an area where both signals overlap and a yellow 'blob' is visible (boxed). The sample was chemically fixed and processed for EM, where 300 nm thick sections were cut for electron tomography (ET). The same cell is retraced in the EM (b) and the area of interest is retraced (box in (b), zoom in (c), arrow pointing to the yellow fluorescent structure of interest from (a)). The markers are still barely visible because of the thickness of the section. Only after ET and reconstruction of the volume are individual markers visible in individual slices from the tomogram (d). (e) By manually drawing the outlines of the endosomes and identifying the probes (EGF as yellow dots, Tf as red dots) we identify one smaller endosome (light green) which contains both probes that is connected with a larger endosome (dark green) almost solely containing Tf. Scalebar in (b) is 1 μm for reference.

cell. From the above it is clear that only a fraction of the receptors will be in the process of segregation at any point in time and will require fluorescence microscopy to visualise that point. The ultrastructure of the membrane-bound endosomes on the other hand needs EM to visualise them. This is why CLEM is applied to this question.

We have used the gridded coverslips [7] to grow the HeLa cells on and then added EGF coupled to quantum dots 655 (Qdots655) together with Tf coupled to Alexa488 and 5 nm gold to the cells. The cells were imaged under the confocal microscope (figure 3(a)) and then fixed and processed for EM. The cell was retraced using the grid pattern (figure 3(b), grid pattern not shown). We can now zoom into an area of interest (figure 3(c)) and analyse whether both probes are present and where they are in relation to each other. This was done using ET (figure 3(d)). To better understand the possible connections between structures the data from the tomogram can be presented in a 3D model (figure 3(e)).

4 Conclusions

4.1 Strengths and limitations

The correlative microscopy field, with CLEM as its major representative is still evolving [3]. Higher resolution in the light microscope will result in better precision in the alignment of the data. Retention of fluorescence in the electron microscope will also aid in this process. Faster and better fixation technologies are still being developed further to fill the gap between the live data and what is being analysed in EM.

4.2 Future developments

The CLEM field is now spreading very wide, with one area looking at the very highest resolution using cryo CLEM. Another area is going in the other direction, zooming out and using large volumes to correlate with each other in VolumeEM and VolumeCLEM approaches. All of these technologies are generating massive amounts of data and in the recent book *Correlative Imaging; Focusing on the Future* [10] it was highlighted by all the experts in the field that DATA will be the biggest challenge for the future.

Acknowledgements

I would like to thank all the members of my lab and the CLEM community that have provided insights into CLEM. A special thanks to Benjamin Hicks, a 3rd year University of Bristol Biochemistry student who acquired the data for figure 3 with the help of Dr Lorna Hodgson and Judith Mantell.

References and further reading

[1] Van Weering J, Brown E, Mantell J, Sharp T and Verkade P 2010 *Methods in Cell Biology: Electron Microscopy of Model Systems* **96** 619–48

[2] de Boer P, Hoogenboom J P and Giepmans B N 2015 *Nat. Methods* **12** 503–13

[3] Ando T *et al* 2018 *J. Physics D: Appl. Phys.* **51** 443001

[4] Olmos Y, Hodgson L, Mantell J, Verkade P and Carlton J G 2015 *Nature* **522** 236–9

[5] Kukulski W, Schorb M, Welsch S, Picco A, Kaksonen M and Briggs J A 2011 *J. Cell Biol.* **192** 111–9

[6] Peddie C J, Domart M C, Snetkov X, O'Toole P, Larijani B, Way M, Cox S and Collinson L M 2017 *J. Struct. Biol.* **199** 120–31

[7] Polishchuk R S, Polishchuk E V, Marra P, Alberti S, Buccione R, Luini A and Mironov A A 2000 *J. Cell Biol.* **148** 45–58

[8] Mueller-Reichert T and Verkade P 2014 *Methods in Cell Biology: Correlative Light Electron Microscopy II* 124 (Elsevier: Amsterdam)
Mueller-Reichert T and Verkade P 2017 *Methods in Cell Biology: Correlative Light Electron Microscopy III* 140 (Elsevier: Amsterdam)
Mueller-Reichert T and Verkade P 2021 *Methods in Cell Biology: Correlative Light Electron Microscopy IV* 162 (Elsevier: Amsterdam)

[9] Sartori A, Gatz R, Beck F, Rigort A, Baumeister W and Plitzko J M 2007 *J. Struct. Biol.* **160** 135–45

[10] Verkade P and Collison L 2019 *Correlative Imaging; Focusing on the Future* (New York: Wiley)

IOP Publishing

Imaging Modalities for Biological and Preclinical Research: A Compendium, Volume 2

Parts II–IV: *In vivo* preclinical imaging, multimodality imaging and outlook

Andreas Walter, Julia G Mannheim and Carmel J Caruana

Chapter III.1.b

Correlative atomic force microscopy

Adelaide Miranda, Ana I Gómez-Varela, Catarina C Moura, Pieter A A De Beule, Michal Sarna, Michal M Dykas and Andreas Stylianou

Atomic force microscopy (AFM) is a powerful technique that can provide a wide range of quantitative and qualitative information concerning biological samples. AFM delivers multifunctional data related to topography as well as nanomechanics and obtains nanoscopic spatial resolution and piconewton force sensitivity. However, only near-surface biological interactions can be studied, and sample preparation protocols require immobilisation because of limited temporal resolution. Furthermore, the vertical and lateral imaging range is limited to a few μm. Some of these limitations can be overcome through correlation of AFM with other techniques. In this chapter, we present the general principles and setups that can be used for the correlation of AFM with other modalities. As these correlations are numerous, we subsequently focus on details of two system implementations, i.e. AFM and differential spinning disk (DSD) fluorescence optical sectioning microscopy and AFM and super-resolution structured illumination microscopy (SR-SIM). Last but not least, we present a novel method (correscopy) that can be used to combine AFM with a wide range of other modalities by using an easy-to-use commercially available setup.

1 Introduction

AFM can be expanded to become part of a multi-functional microscope in order to provide a plethora of quantitative and qualitative information concerning biological samples [1]. AFM provides high-resolution imaging on the nanoscale and nanomechanical biophysical properties characterisation of samples with high accuracy. These properties define AFM as a unique tool for basic research and preclinical studies for a wide range of pathological conditions [1], from single molecules up to tissue level studies [2]. However, AFM cannot provide detailed information for the

doi:10.1088/978-0-7503-3747-2ch17

biochemical characterisation of samples or sensitivity to specific components of the specimen. Another limitation of AFM is the screening of large surface areas covering a substantial number of cells. Several of these limitations can be overcome by combining AFM with other fluorescence microscopy techniques. Here we present system integrations of AFM with differential spinning disk (DSD) and super-resolution structured illumination microscopy (SR-SIM). We conclude by the presentation of correscopy, a commercially available tool that uses a unique setup to enable the combination of AFM with other modalities.

2 Principles and setups

2.1 Physical principles

AFM does not require any special sample preparation, which is quite unique, and the sample can be observed under different environmental conditions such as air, liquid and vacuum at different temperatures. The combination with other modalities is only restricted by the fact that the AFM probe, i.e. the tip extending from the end point of a ~100 μm long cantilever, scans the sample's surface.

AFM is frequently combined with optical techniques, like traditional bright-field optical microscopy, fluorescent microscopy, laser scanning confocal microscope (CLSM), infrared spectroscopy, Raman spectroscopy, and super-resolution fluorescence microscopy; optical techniques offer complementary information to AFM studies. For instance, fluorophores can identify specific cellular or tissue components as well as exhibit sensitivity to environmental parameters such as, e.g., alterations of calcium and hydrogen ions [3].

The application of an inverted optical microscope is most widespread for the benefit of system integration between AFM and fluorescence microscopy. This implementation is due to the typical AFM head design that approaches samples from the top. Hence, the AFM probe does not interrupt the downward optical path of bright-field imaging as well as phase contrast imaging and allows for epifluorescence imaging. This geometry is the most optimal for thin and transparent specimens like cells, films and gels. When the specimen is more bulky and non-transparent such as tissue samples, an upright optical microscopy configuration is more suitable, but this approach is more complex and challenging. A number of different solutions have been proposed in the literature: (i) both AFM head and optical components including objective lens, camera, fluorescence illumination optics and excitation are located above the specimen; (ii) the excitation light is below the sample requiring the sample to be transparent and (iii) the optical components have access to the sample surface from an angle, the side-view approach. Furthermore, novel methods and techniques are emerging in order to avoid complicated setups where the imaging and characterisation of the sample is not performed simultaneously but sequentially, i.e. first AFM data acquisition followed by optical microscopy or vice versa. This approach does not require sophisticated AFM setups, but issues with the correlation of the same areas arise, especially with non-stationary living cellular samples. Concluding, the approach for joining AFM and fluorescence microscopy depends to a large extent on the available equipment and the specific needs of the experiment and specimen.

AFM can also be correlated with other more complex imaging systems, like scanning electron microscopy (SEM). The combination of AFM with SEM provides a very powerful system since these techniques are two of the most frequently used high-resolution non-optical scanning microscopes. SEM has great depth of field, can measure chemical composition of the surface specimen and offers high-resolution in two-dimensions. Conversely, AFM has limited depth of field but excellent resolution on the surface topology of flat samples, it can provide measurements of physical and chemical properties, i.e., mechanical characterisation and interaction forces. AFM provides sample height information encoded in a 3D map of the sample surface.

In the next sections, we introduce two system integrations that allow simultaneous operation of AFM with fluorescence microscopy. At first, we introduce a system integration of AFM with a system implementing the aperture correlation approach to confocal microscopy in the form a DSD. Then we discuss the benefits offered by a system integration of AFM with SR-SIM. We conclude with a presentation of correscopy, a commercially available tool that can be applied to facilitate the combination of AFM with optical microscopy and SEM alike.

2.2 Typical setups and state-of-the-art

Adapted in part from https://aip.scitation.org/doi/10.1063/1.4931064.

2.2.1 Combining DSD fluorescence optical sectioning microscopy with AFM for simultaneous image acquisition

2.2.1.1 Introduction to DSD–AFM

Here we present a combined microscopy platform merging AFM with fluorescence optical sectioning microscopy through aperture correlation microscopy (ACM) implemented through a DSD approach [4]. The illumination scheme of a DSD unit provides a time-independent illumination of the AFM cantilever enabling one to integrate simultaneously recorded fluorescence microscopy data with AFM.

2.2.1.2 Physical principles

Optical sectioning in fluorescence microscopy with DSD technology is obtained with incoherent illumination through the application of rapidly fluctuating structured illumination patterns. The latter are generated by placing an illumination mask consisting of a rectangular grid of alternating transparent and reflective areas on a circular disk spinning at a high constant speed of 100 Hz (figure 1). Hence, the disk rotates many times during the fluorescence camera integration time creating a conjugate (C) and non-conjugate (NC) image following the transparent and reflective areas of the disk, respectively. Confocal-like imaging can be achieved by subtracting the NC image from the C image [5]. The disk integrated in the DSD

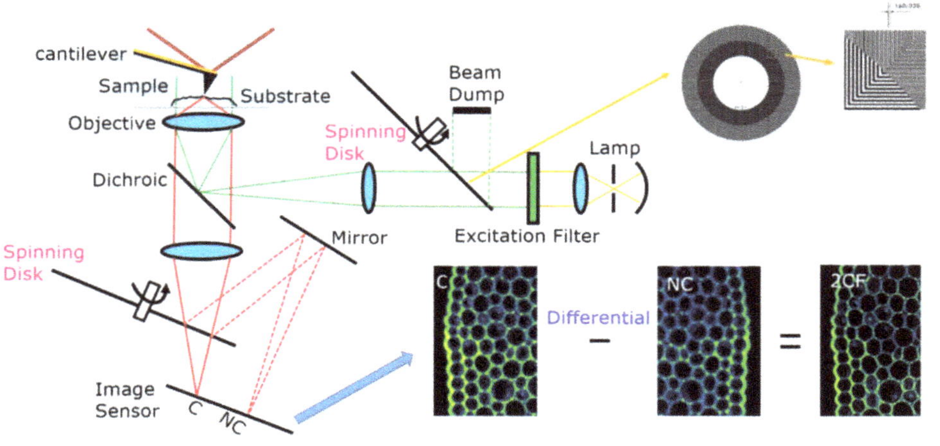

Figure 1. AFM–DSD setup with an illustration of optical pathways for fluorescence excitation and emission, respectively.

set-up comes with two spacings of the rectangular grid in order to provide different levels of optical sectioning with a focus on high confocality in cellular sample or high contrast in tissues samples, respectively.

Combining the DSD with AFM imaging in contact or intermittent contact modes allows for simultaneous imaging due to the time-independent illumination of the AFM cantilever. Also, force-distance curves can be recorded at extremely high speeds in a pixel-by-pixel approach while performing fluorescence imaging similar to force volume imaging but an order of magnitude faster [6]. The trace curve contains the contact point between sample and tip and thus the topographic information can be derived; the mechanical properties of the sample can be derived from the curve after contact occurs. Furthermore, the retraction curve reveals information on chemical binding, be it simple adhesion or specific binding forces between tip and sample.

2.2.1.3 Typical setups and state-of-the-art

Reprinted in part from https://aip.scitation.org/doi/10.1063/1.4931064.

A general experimental setup integrating a DSD imaging platform and an advanced bioscience AFM system is presented in figure 2. The entire system is mounted on an ultra-low noise floor constructed to comply with the NIST-A vibration specifications [7]. This platform is loaded with an inverted optical microscope equipped with an AFM. A dedicated condenser lens with an extra-long working distance of 75 mm is installed to adjust to the height of the AFM unit. Fast automated objective focus control and scanning is provided by a piezo unit. For optimal fluorescence detection, a high numerical aperture objective is used. A metal halide light source with built-in fluorescence excitation filter wheel provides excitation light for the DSD unit. A high-dynamic range camera is necessary to cope with the differences in intensity of both C and NC images. For this purpose, an interline CCD camera is the preferred camera of choice for DSD image capture.

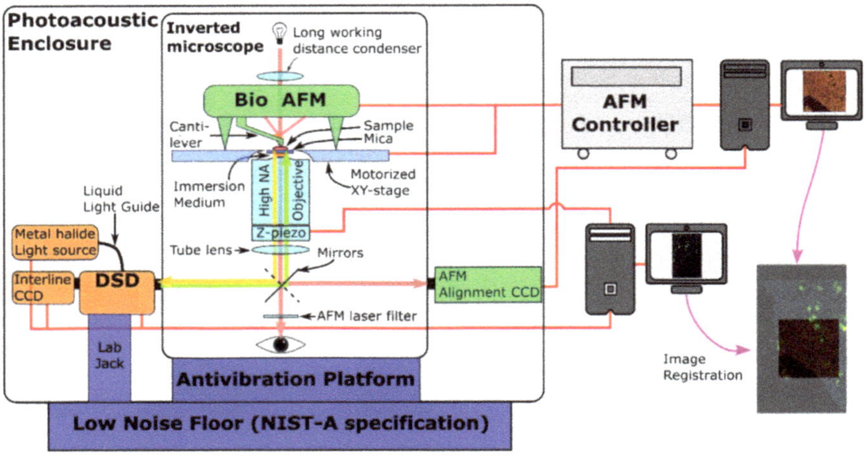

Figure 2. Experimental setup for combined differential spinning disk fluorescence optical sectioning and atomic force microscopy. Adapted from reference [4] with permission of AIP Publishing.

A standard monochrome CCD camera is coupled to a microscope port for AFM cantilever laser spot alignment for better performance.

2.2.1.4 Application range and relevance

The combined microscopy approach can be used to derive mechanical properties, such as Young's modulus, in tissues or cells using the AFM and at the same time provide information on the localisation and dynamics of several molecular interactions at the cellular level using the DSD. In contrast to system integrations of AFM with total internal reflectance fluorescence microscopy (TIRFM), events away from the coverslip can also be studied.

2.2.1.5 Sample preparation

For DSD–AFM simultaneous operation, one should consider choosing fluorophores (a) with an excitation spectra away from the 850 nm AFM laser to avoid bleaching and (b) resistant to photobleaching and non-blinking. All biological water or oil immersion objectives are designed to be used with 0.17 mm glass coverslips. A strong variation on the thickness of coverslips has a significant influence on resolution. Ideally, the sample should be embedded in a medium matching the refractive index of the immersion oil. For imaging cells the cantilever should have a low spring constant, a typical tip radius of curvature about 30 nm. Sometimes the cantilevers can act as an unwanted source of autofluorescence. However, due to the confocal-like imaging peformed with the DSD, this background signal gets filtered out.

2.2.1.6 Parameters of image quality

Reprinted in part from https://aip.scitation.org/doi/10.1063/1.4931064.

The combined system presents two sources of noise: AFM cantilever distortion induced by fluorescence light excitation and DSD mechanical noise transfer to AFM [4].

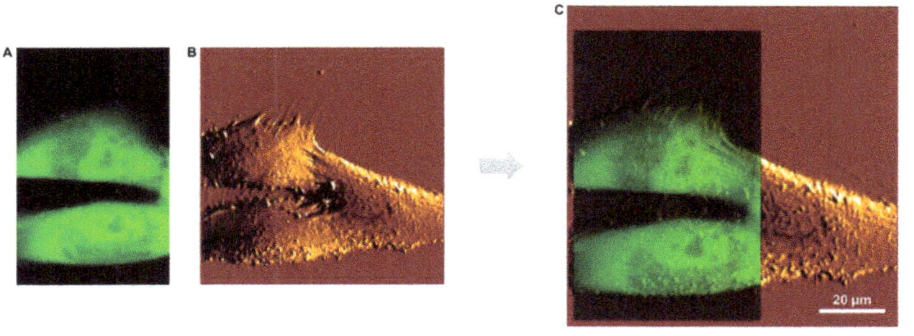

Figure 3. Osteosarcoma cells: (a) DSD image where cells were labelled with green fluorescent protein (GFP), (b) AFM topographical image with an edge detector algorithm, and (c) combined DSD and AFM images.

Concerning the noise transfer, DSD spinning disk motion can induce cantilever distortions at 100 Hz with an amplitude of 5.1 nm using a Pyrex-Nitride Probe—TRiangular (PNP-TR) type cantilever. This noise can be removed by using a mechanical vibration absorbing C-mount for the DSD or controlling the mounting of the XY sample stage by limiting the screw locking torque. Using the latter approach, the noise amplitude has been reduced to 75 pm [4]. Regarding the cantilever distortion, the fluorescence excitation light is capable of introducing cantilever bending disturbing AFM operation. This bending can be linked to light absorption by the gold coating on the cantilevers, which leads to heat absorption, to promote AFM laser reflection. For example, the DSD fluorescence excitation source in combination with a 60× optical microscope imaging objective is capable of inducing a maximal cantilever deflection of 9 nm for a PNP-TR cantilever type A. Application of a heat model linking cantilever deflection to temperature rises links this bending to a 0.33 °C temperature change, indicating that fluorescence excitation sources can induce significant temperature gradients in the vicinity of AFM cantilevers, that can change the behavior of cell membrane receptors [8].

2.2.1.7 Data processing

One of the principal challenges of combined optical and AFM microscopy is posed by the need for accurate nanoscale image registration. Shear and stretch are commonly encountered distortions in the optical image. This has to be compared to the real-space image created with highly linear piezo-elements of an AFM. As a general rule, a coordination transformation matrix between the AFM and the fluorescent optical images from the DSD should be obtained so that the exact position of the AFM probe is known. Figure 3 depicts an example of image registration between the two instruments using osteosarcoma cells as specimen [9]. The images of both techniques were combined using a software module available with the commercial AFM to calibrate the optical image. The AFM cantilever was displaced to a set of predefined coordinates to register an optical image at each position. From these images a transform function between the optical and the AFM image is determined and the overlaying of the two images is made possible.

2.2.1.8 Strength and limitations

The integration of DSD and AFM technology allows for the simultaneous image acquisition without suffering from time-dependent AFM cantilever disruption and heating induced by the fluorescence excitation nlight. Specifically, this feature is envisioned to be of advantage for the observation of signaling processes in live-cell investigations in cellular biology. Integration of this combined microscopy relies on avoiding DSD spinning disk mechanical noise transfer to AFM cantilever motion.

The overall light efficiency of DSD image formation compared to state-of-the-art confocal technology is relatively low, due to the presence of many extra optical components to fold the optical pathway and the necessity for a folded optical axis design of the image formation process. Hence, relatively strong and stable fluorophores are required to avoid excessive photobleaching of the sample which has the possibility of compromising image formation of dynamical processes.

2.2.1.9 Future developments

DSD technology has undergone several recent advances to improve optical losses in the optical layout (figure 1). While it is not yet demonstrated that these recent developments will not hamper system integration with AFM, they represent an opportunity to close the performance gap with state-of-the-art confocal microscopy. A further development would be the integration of DSD technology with high speed AFM (HS-AFM). Currently, HS-AFM allows simultaneous assessment of structure and dynamics of single molecules, with a typical spatial resolution of 2–3 nm and ~0.15 nm in the lateral and vertical directions, respectively, at 10–20 frames per second (fps). [10] The combination of an HS-AFM with DSD second generation would shed light on questions related to protein molecules such as interaction with small ligands and encapsulated substrate proteins and subtle changes that are only detectable with fluorescence energy transfer.

2.2.2 Combining SR-SIM with AFM for simultaneous image acquisition

2.2.2.1 Introduction to SR-SIM/AFM

Several microscopy schemes have been developed taking advantage of structured illumination microscopy (SIM) principles, whereby the sample is illuminated by a spatially modulated light pattern [11]. Super-resolution (SR) SIM approaches make use of an interference pattern with a periodicity close to the Abbe resolution limit. This can enhance lateral resolution (2D-SIM) [12] or both lateral and axial resolution (3D-SIM) [13] in fluorescence imaging by up to a factor of two in each spatial direction in comparison to a wide-field image, corresponding to ~140 nm lateral and ~300 nm axial resolution, respectively, for a high numerical aperture objective. SR-SIM illumination is particularly favorable when combined with AFM since it avoids cantilever bending due to fluorescence excitation light, a significant problem when integrating AFM with other SR fluorescence microscopy techniques.

2.2.2.2 Physical principles

The SIM and AFM physical principles have been thoroughly explained in previous chapters of the book. Figure 4 depicts the schematic setup of the SR-SIM/AFM platform first presented by Gómez *et al* [14]. For SIM illumination, only the 0 and ±1 orders diffracted by a grating are allowed into the light path. The diffraction orders are focused on the back focal plane of the objective lens and the grid pattern generated excites the sample. Fluorescence light collection is done with a high aperture objective, a dichromatic mirror (DM), and a tube lens (TL). Structured illumination is provided by the N-SIM illuminator module, while alignment of the AFM laser spot and fluorescence detection are done with a CCD and a scientific CMOS (sCMOS) camera, respectively.

2.2.2.3 Typical setups and state-of-the-art

Figure 4(a) depicts an experimental setup integrating an SIM imaging platform and an advanced bioscience AFM system. The AFM scan head (XYZ) and sample positioning stage (XY) are mounted on the inverted optical microscope. Laser spot alignment of the AFM is accomplished using a standard monochrome CCD camera while fluorescence signal is acquired with an sCMOS camera. The N-SIM E illuminator module provides the structured illumination.

2.2.2.4 Application range and relevance

SR-SIM can reveal dynamic processes in live biological samples in 3D due to its resolution providing information on the localisation and dynamics of several

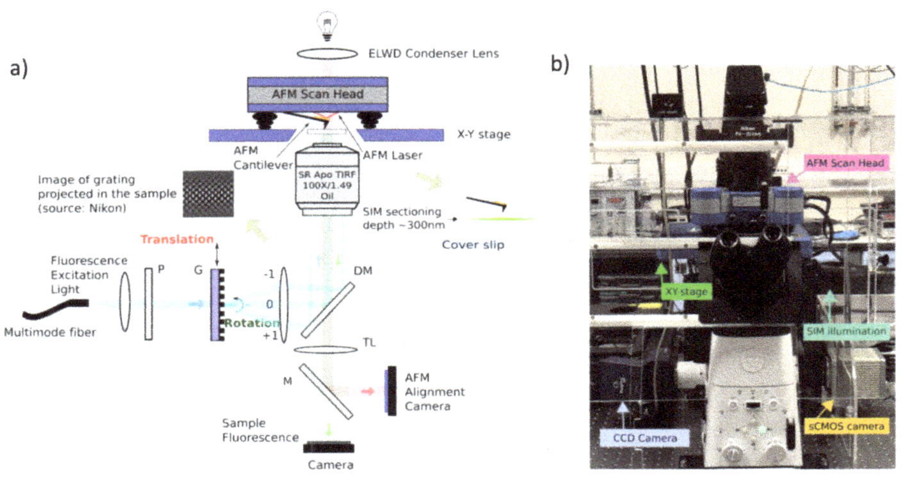

Figure 4. SR-SIM/AFM microscope: (a) schematic setup using a grating for SIM illumination. In 3D-SIM the interference of three coherent beams produces a pattern projected onto the sample that fades away into a blur a short distance away from focus. This is favorable for providing uniform illumination of the AFM cantilever and allowing synchronous operation of both microscopes; (b) image of the combined NanoWizard®3 AFM from JPK and N-SIM E Nikon microscope capable of simultaneous operation. Adapted from [14] with permission of Springer, under a Creative Commons Attribution 4.0 International License.

molecular interactions in cells while simultaneously the AFM determines cellular nano-mechanical properties such as elasticity or stiffness. A modality of SR-SIM combined with AFM promises a more detailed understanding of cells and tissues, which is absolutely crucial for research and development of novel biomedical applications.

2.2.2.5 Sample preparation

For SR-SIM/AFM simultaneous operation, fluorophores with an excitation spectra away from the 850 nm AFM laser should be considered to avoid bleaching. Also, probes should be resilient, i.e. exhibit low photobleaching and nonblinking behavior. For instance, the use of fluorescein is not advisable as it bleaches rapidly. SIM presents advantages over other super-resolution techniques as it does not require the use of specialised fluorophores or laborious sample preparations. For either conventional fluorescence or super-resolution imaging, a strong variation on thickness of the coverslips has a significant effect on resolution and it is recommended to use high performance cover glasses for great accuracy. Autofluorescence originating from the AFM cantilevers [4, 14] does not affect the fluorescence image formation process as the images obtained after processing are optically sectioned, i.e. reject out-of-focus fluorescence contributions. Also, for imaging living cells, cantilevers should have low spring constants and be unsharpened (tip radius of 30 nm).

2.2.2.6 Parameters of image quality

SIM image resolution can be evaluated by performing an initial sample calibration using nanometer-scale samples such as calibration standards or beads. Nonetheless, the resolution may differ for the sample of interest as it greatly depends on specimen properties such as refractive index, light scattering, choice of fluorophore, and acquisition parameters. Thus, it is advisable to determine the resolution through the full-width at half-maximum (FWHM) of a line profile over a nanometer-sized part of the sample (for instance in filopodia-like structures). The performance of the simultaneous imaging of the system can be assessed by testing the noise response with the SIM illumination in place and without it. According to the results shown in [14], non-significant effects are observed on AFM surface parameters when the SIM illumination is on, corroborating the fact that it does not disrupt the cantilever operation.

2.2.2.7 Data processing

AFM is able to acquire 'real-space' images [4], while optical images can suffer from aberrations introduced by the lenses. In the first combination presented [14], the authors have addressed this problem by combining the images from both techniques with a software module available with the commercial AFM to calibrate the optical image. This calibration is performed by displacing the AFM cantilever to a set of predefined coordinates to register an optical image at each position. From these images a mathematical function is derived allowing for spatial registration of AFM and the fluorescence image data. Figure 5 shows an example of a simultaneous SR-SIM/AFM acquisition carried out on fixed U2OS cells.

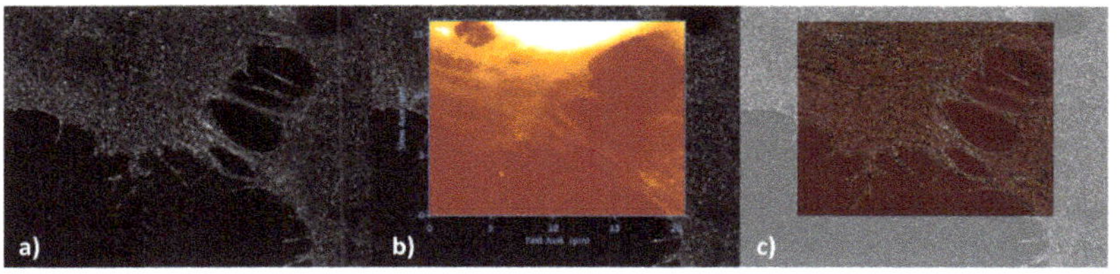

Figure 5. Correlation of the SR-SIM and AFM data. The SR-SIM images of human bone osteosarcoma epithelial cells (a) are superimposed with the height/topography channel of the AFM data (b). The correlated SR-SIM/AFM channels on cells in a semi-transparent mode (c), include a mean-curvature enhanced AFM-topographic information. XY-scales of the AFM image in B are 20.74 µm and 16.12 µm, respectively, recorded at a resolution of 256 × 199 pixels. Reprint from [14] with permission of Springer, under a Creative Commons Attribution 4.0 International License.

2.2.2.8 Strengths and limitations

SIM microscopy holds the most promising of all super-resolution schemes to unravel dynamics processes of biological samples, as it provides a good compromise regarding resolution, phototoxicity, compatibility with conventional fluorophores, and higher acquisition speeds for longer periods of time. SR-SIM illumination characteristics also make this technique very suitable for integration with AFM because it does not interfere with AFM cantilever operation, requiring rather low light exposure and therefore allowing simultaneous measurements. Most of the work using SR fluorescence microscopy requires high intensity lasers and/or long acquisition times, making them incompatible with observation of molecular processes in living cells. In addition, it is important to note that SIM is capable of multicolour fluorescence detection and 3D imaging.

However, some limitations of SR-SIM affecting the combined system are: (i) the moderate resolution improvement being approximately a factor of two, compared to other SR techniques, (ii) the sensitivity to artefacts from complex image reconstruction typically introduced by sample movement [15] and (iii) in principle, not appropriate for thick samples (>15 µm), although this limitation was addressed by combining a single plane reconstruction algorithm with hardware for high-speed switching between illumination patterns and rapid acquisition of fluorescence images [16].

2.2.2.9 Future developments

The capability of the SR-SIM/AFM system for synchronous imaging is very promising for dynamic processes visualisation during AFM data acquisition. Further resolution enhancement could be obtained combining the HS-AFM with a non-linear SIM (NL-SIM) system. This is an alternative SIM approach that uses the non-linear response of a fluorophore to generate an effective illumination pattern that contains higher spatial frequencies [17]. However, the system needs to be tested to verify there is no disruption on the AFM cantilever operation as NL-SIM requires higher light exposure values than linear SIM. Another pathway to explore is combined HS-AFM and video-rate SIM [18] on living cells, which could potentially harmonise AFM and fluorescence scan rates.

2.2.3 Combined AFM and high-resolution fluorescence confocal microscopy using correscopy

Correscopy is a correlative imaging solution that enables the use of a standard standalone microscope for correlative imaging. Thus, 'standard' AFM and high-resolution fluorescence confocal microscopes can be used to correlate the mechanical properties of live cells with its subcellular structure. The correscopy solution is based on serial imaging (one microscope after the other), which is in contrast to parallel imaging when using a standard correlative microscope. The correscopy setup for AFM and high-resolution correlative microscopy consists of a universal sample holder to mount the sample throughout the imaging, microscope adapters dedicated to AFM and light microscopy, and correscopy software that allows one to obtain the co-localised data for correlative analysis. Such a serial approach allows for the preparation of the sample specifically for a dedicated microscope, just before the analysis. This allows performing first the AFM live-cell imaging (which is one of its key benefits), and then fluorescence imaging of the fixed cell, as shown in figure 6. Correscopy is easy and a straightforward solution for correlative imaging that simplifies the correlative imaging process and sample preparation, and eliminates many artefacts. Importantly, the use of this method enables the correlation of AFM with almost any imaging technique as it is microscope and modality independent. The applications of correscopy are not limited to any particular topic and they depend on the research question that needs to be addressed.

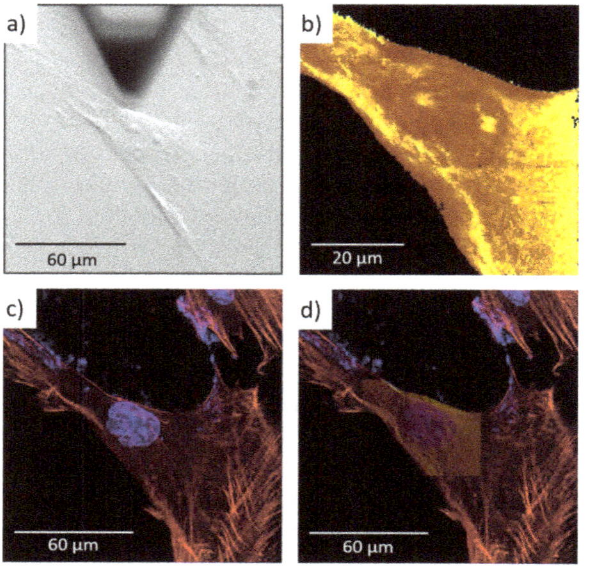

Figure 6. HeLa cell imaged using (a) a simple light microscope installed on an AFM; (b) an AFM microscope (Bruker Catalyst) image; (c) high-resolution fluorescence confocal microscopy (Carl Zeiss LSM 880 with Airyscan). (d) Shows an overlay of the images. (Movies are included in the online version of the chapter, available at https://iopscience.iop.org/book/978-0-7503-3747-2).

3 Conclusions

3.1 Strength and limitations

The combination of AFM with other modalities has been demonstrated by many researchers through in-house developments but also using several commercial solutions. The unique properties of AFM include nanoscale high-resolution imaging and nanomechanical characterisation of specimens. Through a correlation with different imaging modalities, it becomes possible to combine biochemical characterisation of the sample, optical sectioning, acquisition of information on larger areas or large cell populations and specificity to different specimen components. However, correlation of AFM with other modalities faces a multitude of challenges to achieve smooth integration. The real-time correlation of AFM with fluorescence microscopy reveals challenges related to asynchronous imaging since AFM data acquisition is usually two orders of magnitude slower than fluorescence imaging. The emergence of novel AFM techniques, like high-speed AFM, facilitates correlative modes and helps to overcome the limitation of the low speed of AFM. Another significant drawback is the matching of the resolution of the AFM with that of other modalities. In the case of correlative AFM with optical techniques, this is even more crucial due to the relative low resolution of the conventional optical modalities. However, this limitation is overcome by the development of novel systems combining AFM with super-resolution optical techniques, such as SR-SIM, scanning near-field optical microscopy (SNOM), stimulated emission depletion (STED) microscopy and photoactivated localisation microscopy (PALM). Furthermore, a challenge of using AFM with other modalities is the development of appropriate protocols for sample preparation so as the endogenous properties of the samples are preserved for all the involved modalities. For instance, for fluorescence microscopy, staining protocols might be performed that can demand the fixation of the biological samples (i.e., cells and tissues), but these protocols alter the nanomechanical properties of the specimens that can be assessed by AFM, or there is the risk of altering the structure of the specimen.

3.2 Future developments

For the correlation of AFM with other modalities different developments are expected for the various modalities. Generally, future developments of the current systems and techniques include the optimisation of the imaging speed, the sample/image alignment between the different techniques, and the sample preparation protocols. Besides, the development of new fluorescence labelling strategies or small molecule markers appropriate for such correlative imaging systems is required.

References and further reading

[1] Stylianou A, Kontomaris S V, Alexandratou E and Grant C 2019 *Scanning* **2019** 845285

[2] Stylianou A, Lekka M and Stylianopoulos T 2018 *Nanoscale* **10** 20930–45

[3] Krautbauer R, Rief M and Gaub H E 2003 *Nano Lett.* **3** 493–6

[4] Miranda A, Martins M and De Beule P A A 2015 *Rev. Sci. Instrum.* **86** 093705

[5] Wilson T, Juškaitis R, Neil M A A and Kozubek M 1996 *Opt. Lett.* **21** 1879–81
[6] Stamov D R, Kaemmer S B, Hermsdörfer A, Barner J, Jähnke T and Haschke H 2015 *Micros. Today* **23** 18–25
[7] Amick H, Gendreau M, Busch T and Gordon C 2005 Evolving criteria for research facilities: vibration *Proc. SPIE* **5933** 593303
[8] Timoshenko S 1925 *J. Opt. Soc. Am.* **11** 233–55
[9] Moura C C, Miranda A, Oreffo R O C and De Beule P A A *et al* 2020 *Biochem. Biophys. Res. Commun.* **20** 392–97
[10] Ando T, Bhamidimarri S P and Brending N *et al* 2018 *J. Phys. D: Appl. Phys.* **51** 443001
[11] Neil M A A, Juškaitis R and Wilson T 1997 *Opt. Lett.* **22** 1905–7
[12] Gustafsson M G L 2000 *J. Microsc.* **198** 82–7
[13] Gustafsson M G L, Shao L, Carlton P M, Wang C J R, Golubovskaya I N, Cande W Z, Agard D A and Sedat J W 2008 *Biophys. J.* **94** 4957–70
[14] Gómez-Varela A I, Stamov D R, Miranda A, Alves R, Barata-Antunes C, Dambournet D, Drubin D G, Paiva S and De Beule P A A 2020 *Sci. Rep.* **10** 1122
[15] Förster R, Müller W, Richter R and Heintzmann R 2018 *Opt. Express* **26** 20680–94
[16] Shaw M, Zajiczek L and O'Holleran K 2015 *Methods* **88** 11–9
[17] Rego E H, Shao L, Macklin J J, Winoto L, Johansson G A, Kamps-Hughes N, Davidson M W and Gustafsson M G L 2012 *Proc. Natl Acad. Sci.* **109** E135–43
[18] Markwirth A, Lachetta M, Mönkemöller V, Heintzmann R, Hübner W, Huser T and Müller M 2019 *Nat. Comm.* **10** 4315

Section 2

(Pre)clinical hybrid imaging

IOP Publishing

Imaging Modalities for Biological and Preclinical Research: A Compendium, Volume 2

Parts II–IV: *In vivo* preclinical imaging, multimodality imaging and outlook

Andreas Walter, Julia G Mannheim and Carmel J Caruana

Chapter III.2.a

PET/CT

Wendy McDougald

Positron emission tomography/computed tomography (PET/CT) is a key non-invasive imaging tool used for studying disease development and progression as well as in the preclinical field for the development of novel radiotracers and pharmaceuticals. It supports the investigations and evaluations of underlying biological mechanisms, physiological processes in healthy, as well as in diseased subjects/models. Using target-specific positron emitting radiotracers in conjunction with X-ray, PET/CT allows for the acquisition of anatomical and functional information in one imaging data set. Figure 1 shows a PET/CT rodent acquisition, displaying the CT, PET and fused PET/CT image.

1 Introduction

The idea of combining PET and CT into one scanner started with the desire to co-register functional PET images with high spatial resolution anatomical CT images without moving clinical patients, as one scanning system would avoid the need to image patients on two separate systems and eliminate the process of correlating images from two scanners. The integration of PET with CT in one system allowed for simplified co-registration of imaging data sets. Thus, PET/CT provides a precise identification of the distinct anatomical regions of PET uptake. First PET/CT images and systems where introduced in 2000 by Beyer *et al* [1].

Combining the two modalities also introduced the concept of using CT data for generating attenuation correction maps for PET. Previously, PET attenuation corrections were done by acquiring a 'blank' (no object in the bore) transmission scan using a radioactive source (usually caesium-137 (^{137}Cs) or germanium-68 (^{68}Ge) of a known activity), followed by a transmission scan acquired with the patient in the bore. Attenuation factors were calculated based on the subtraction of both scans [2]. Using the patient's own CT scan information for attenuation

doi:10.1088/978-0-7503-3747-2ch18

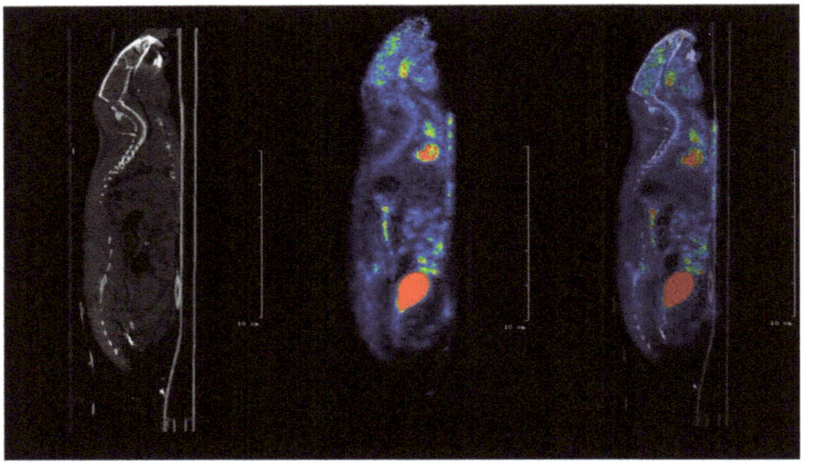

Figure 1. PET/CT sequence of images (CT, PET and PET/CT) of a rat following injection of [^{18}F]FDG. Panel (a) displays the CT (X-ray) anatomical information. Panel (b) shows the PET image (functional information). Panel (c) displays the PET/CT fused images.

correction removed the need for a separate PET transmission scan being applied to the patient's scan and/or a separate CT scan on a CT-only system.

However, converting CT data to correlate with PET posed several problems. The main concern revolved around the different photon energies. PET is mono-energetic with 511 keV and CT is a polychromatic X-ray range usually from 35 to 160 keV. Furthermore, energy dependent attenuation (absorption/scatter) processes for PET and CT differ, e.g., Compton scattering versus photoelectric absorption in denser materials (bone) and Compton scattering in less dense material (soft tissue).

Feasibility work generating CT attenuation correction maps started during the late 1990s. By 2002, Burger *et al* presented results showing the process for converting CT Hounsfield Units (HU) to PET attenuation coefficients to create accurate attenuation maps. Their segmented bilinear algorithm consigning CT values to PET for human tissue is shown below.

$$\mu^{\mathrm{PET}} = \begin{cases} \mu_w^{PET}(CT + 1000/1000) & \mathrm{CT} \leqslant 0\mathrm{HU} \\ \mu_w^{PET} + CT\dfrac{\mu_w^{PET}\left(\mu_\beta^{PET} - \mu_w^{PET}\right)}{1000\left(\mu_\beta^{CT} - \mu_w^{CT}\right)} & \mathrm{CT} > 0\mathrm{HU} \end{cases} \qquad \begin{aligned} \mu_w^{PET} &= 0.096\ \mathrm{cm}^{-1} \\ \mu_\beta^{PET} &= 0.172\ \mathrm{cm}^{-1} \\ \mu_w^{CT} &= 0.184\ \mathrm{cm}^{-1} \\ \mu_\beta^{CT} &= 0.428\ \mathrm{cm}^{-1} \end{aligned} \tag{1}$$

where in equation (1) w and β represent the attenuation coefficients of water and bone, respectively. Attenuation coefficients (μ, cm^{-1}) represent the linear attenuation coefficients of PET 511 keV energy and the effective CT energy (80 keV) for water and bone [3].

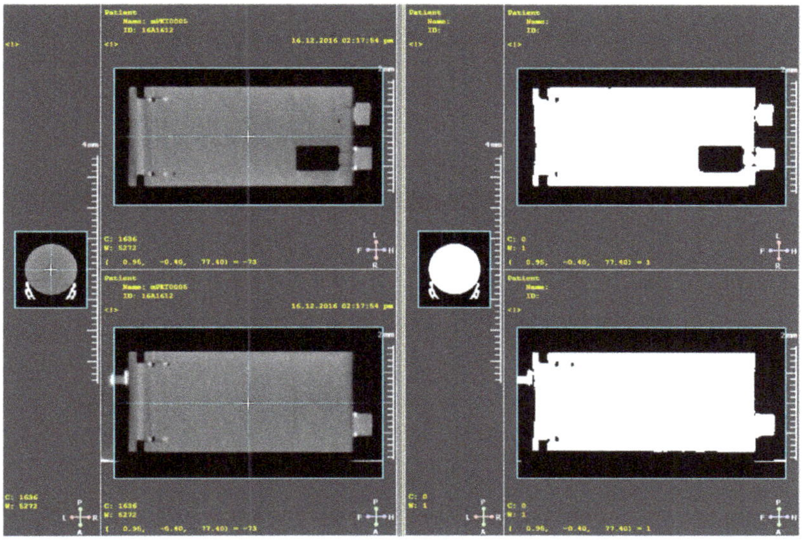

Figure 2. Left panel shows the CT image of a phantom. Image is displayed in grey scale (multiple shades of grey). Right panel shows the attenuation correction map to be applied to the PET image. In this case the attenuation correction maps are binary (white and black).

An example of CT attenuation correction maps for a PET scan are shown in figure 2 for a phantom scan.

2 Principles and setups

PET/CT systems function in the exact same fashion and contain the same physical components as in a stand-alone PET or CT (please refer to chapters II.10 and II.9). In brief, the PET design is based on a ring-design of scintillations crystals attached to the front end of detectors (typically photomultiplier tubes (PMTs), avalanche photodiode (APDs) or silicon photomultiplier (SiPMs)) with the read-out electronics connected to the back of the detectors. The CT consists of the X-ray tube (anode and cathode) fixed to a rotating ring with the detector system (scintillation crystals mounted on photo diode arrays) fixed opposite the X-ray tube. Greater details of PET and CT physical principles and setups can be found in chapters II.10 and II.9, respectively.

2.1 Physical principles

The physical principles of PET and CT are discussed in detail in chapters II.10 and II.9, respectively.

2.2 Typical setups and state-of-the-art

The integration of PET with CT consists of four main components: rotating gantry (CT), stationary gantry (PET), patient table and work station console, figure 3 below.

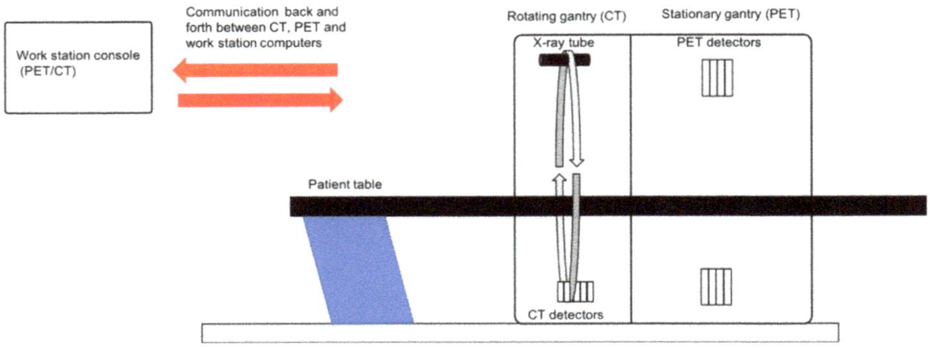

Figure 3. Diagram outlining a PET/CT system. Note, CT gantry rotates around the table for the X-ray scan. The X-ray tube and detectors are situated on the circular rotating ring, whereas the PET is a full circular stationary ring around the table for the PET emission scan. Diagram does not show full ring of PET detectors. Table moves up/down and inside each gantry.

The work station console contains the interface for separate communications with the table, CT, PET and intercom for patient communication when on the bed. This allows for setting up the imaging parameters and the table positions for each modality. The work station computers serve as the collection and depository for all the raw imaging data, except PET list mode, which is typically temporarily left on the PET gantry storage if requested. List mode files can be used to create sinograms of different durations and binning modes. Software for displaying and reconstructing imaging datasets also resides on the work station. Typically, the PET/CT imaging data are stored in the digital imaging and communication in medicine (DICOM) format and reconstructed prior to being moved off the computers for display or further analysis.

3 Biomedical relevance

3.1 Application range and relevance

All clinical/preclinical imaging techniques have important biomedical applications and relevance, with some of the imaging modalities overlapping. For example, ultra sound, CT, and MRI provide information covering anatomy and physiology. PET, MRI and optical imaging will provide information on physiology and metabolism. The combination of PET with CT expands the information compared to a single imaging acquisition. PET/CT's relevance covers areas of anatomy, physiology, metabolism and molecular information.

There are several radiotracers which will 'map' the physiological or biochemical process applicable for use in PET/CT. Hundreds of compounds have been developed for labelling with positron emitting radioisotopes, such as ^{11}C, ^{13}N, ^{15}O, ^{82}Rb, ^{18}F, ^{64}Cu ^{89}Zr. Radioisotopes are labelled to small organic molecules, antibodies and/or peptides. The most popular analog used for labelling is glucose. Generally, the common and widely used tracer is 2-deoxy-2-[^{18}F]fluoro-D-glucose ([^{18}F]FDG). Some of the research areas/physiological processes studied using PET radiotracers

Table 1. Provides a limited list of radiotracers used in PET/CT and their application.

Biological process	Radiotracer
Blood flow/perfusion	
	$H_2{}^{15}O$
	$^{82}RbCl$
	$^{13}NH_3$
	^{11}CO
Metabolism	
	$^{15}O_2$
	^{11}C-acetate
	^{18}F-fluorodeoxyglucose
	^{11}C-palmitic acid
Hypoxia	
	^{18}F-fluoromisonidazole
Proliferation	
	^{18}F-fluorothymidine
Protein synthesis	
	^{11}C-leucine
	^{11}C-methionine
Receptor systems	
	^{18}F-fluoro-L-DOPA
	^{18}F-fluoroethylspiperone
	^{18}F-flumazenil
	^{18}F-fluoroestradiol
	^{11}C-raclopride
	^{11}C-CFT
	^{11}C-carfentanil
	^{11}C-altanserin
	^{64}Cu-DOTATATE

DOPA: 3,4-dihydroxypenyalalanine CFT: [N-methyl-^{11}C]–2-β-carbomethoxy-3-β-(4-fluorophenyl)-tropane. DOTATATE: dodecanetetraacetic acid (DOTA), tyrosine-3-octreotate (Tyr3-octreotate or TATE)
Table reference [4]

are listed in table 1. Table 1 is not all inclusive, as research using PET/CT ranges greatly. Though the major applications for PET/CT cover oncology, neurology and cardiovascular diseases. Here, the choice of radiotracer is dependent on the research question and the experimental design. Refer to chapter II.10 for greater detail.

3.2 Subject preparation

When using the combined PET/CT, whether clinical or preclinical, the subject preparation remains essentially the same as in the dedicated PET or CT (please refer to chapters II.10 and II.9). The subject is prepped for both at the same time.

Literature, recommendations and guidelines outlining subject preparation is available in both clinical and preclinical cases. For example, the European Association of Nuclear Medicine, Society of Nuclear Medicine and Molecular Imaging and the American College of Radiology all offer subject guidelines (mainly clinical) [5, 6]. These guidelines provide assistance ranging from patient diet, medications, environmental conditions, injections to time between multiple imaging. Preclinical guidelines for animal handling, husbandry and welfare can mainly be found through the National Centre for the Replacement Refinement and Reduction of Animals in Research (NC3Rs) as well as from currently literature. Such literature also includes animal care regarding anesthesia and injection procedures and preparation [7, 8].

4 Parameters of image quality

Image quality is dependent not only on scanner characteristics performance but also on the calibration and maintenance of the scanner. In other words, scanner calibrations and maintenance are different from scanner performance testing, such as National Electrical Manufacturers Association (NEMA) evaluations in PET.

In this section, firstly, section 4.1 briefly reiterates the basic PET and CT factors that degrade image quality (artefacts), whilst section 4.2 outlines the main artefacts seen in PET/CT. Secondly, sections 4.3 and 4.4 briefly cover clinical PET/CT scanner calibrations and section 4.5 describes annual testing. In regard to preclinical calibrations, maintenance and annual testing are similar to clinical systems though not carried out as often.

4.1 Factors degrading image quality

Combining PET with CT in one imaging system did not eliminate any scanner artefacts seen in stand-alone PET or CT. A more detailed explanation regarding PET or CT artefacts can be found in chapters II.10 and II.9. However, using CT to generate attenuation correction maps with the integration of PET and CT introduced new potential artefacts.

4.1.1 PET artefacts

PET artefacts include; patient motion, partial volume effects, scatter effects, detector failures and calibration errors.

4.1.2 CT artefacts

CT artefacts include: patient motion, partial volume effects, beam hardening, photon starvation, undersampling, centre-of-rotation, detector failures, patient prostheses implants, truncation and calibration errors

4.2 PET/CT artefacts

The advantage of combined PET/CT is that it provides mechanically aligned anatomical and functional images in the same bed position. With that said, the alignment of separately acquired PET and CT still remains a potential source of

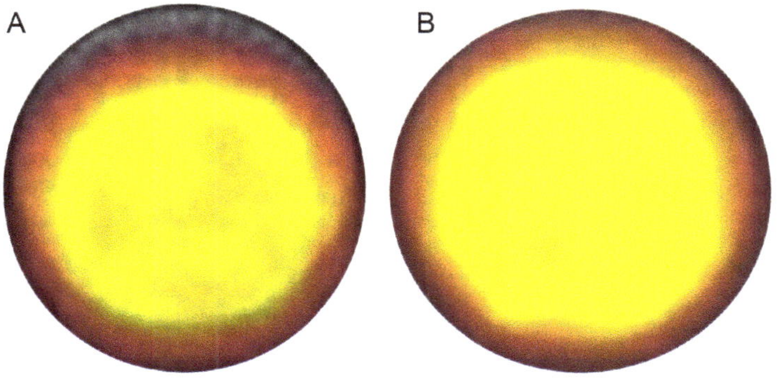

Figure 4. Panel (a) shows the misalignment between the PET image (colour) and the CT image (light grey ring). This is due to scanner hardware being misaligned. Panel (b) shows the automatic corrected image.

artefacts, as shown in figure 4. However, the common artefacts in PET/CT are related to metallic implants, respiratory motion, truncation and contrast errors [4, 9].

Patient metallic prostheses implants (i.e. hip prosthetics, chemotherapy ports) can create a higher CT HU value, being a denser material than bone or tissue. This will generate streaking artefacts and potentially influence the attenuation correction applied to the PET image data. Reviewing a PET image with no attenuation correction as well as with the correction will assist in correctly assessing the image data.

Patients can move between scans causing respiratory motion artefacts resulting in discrepancy in the chest area between the CT and PET images. Additionally, respiratory mismatches between PET and CT can be created from the displacement of the diaphragm due to breathing. Furthermore, a PET image acquisition is sustainably longer than that of a CT. During the PET acquisition the patient's breathing pattern is normal, lasting over many breathing cycles. A CT acquisition is short, which covers only a specific time point (seconds) in the breathing cycle. This difference in the respiratory motion causes a curvilinear artefact along the diaphragm region and/or liver. Not only does this lead to the underestimation of activity in the PET data, lesions in the liver might be misplaced. Most clinical CT protocols are done as a 'breath-hold' in order to eliminate this artefact. Unfortunately, this is not always the case. Some patients will not be able to follow the breath-hold for the required time. Other possible solutions involve gated imaging (not covered here).

In PET/CT, the field of view (FOV) for the CT is generally smaller than that of the PET FOV. Therefore, CT truncation artefacts can occur. The most common area for truncation errors is around the shoulders and/or if the patient's arms are not moved over the head instead of lying on each side of the patient in the field of view. CT truncation errors correspond to CT attenuation correction errors being applied to the PET image. This will cause higher attenuation correction factors (area thought to be more dense) and will produce artefactually higher PET values in that region.

The use of CT contrast agents, which typically contain iodine or barium, for enhancing the CT image can potentially cause streaking artefacts. These contrast agents will increase the CT (X-ray) attenuation and thus the CT HU values. This sort of artefact is not as common and is also dependent on the type of contrast agent administered, the injection routine (intravenous or oral), the distribution, clearance, and time between administration and acquisition.

4.3 PET calibration and quality control

Regular maintenance and clinical scanner calibrations are done quarterly, semi-annually or as suggested by the manufacturer. This time frame may vary for preclinical scanners. The first steps in calibrating a PET scanner require using the system diagnostics software. The software program allows for running analytical evaluations on the detector system (scintillations crystals, detectors and read-out electronics). The resulting software assessments provide information on the detector gains and energy (i.e. PMTs output), crystal positions/read-out (crystal maps) and coincidence timing evaluation. Corrections, updates of gains, energy, position and timing are done and replacements of detectors or circuit boards are amended accordingly. Blank scans using 2D and 3D mode are also acquired to evaluate sinograms. Along with diagnostic evaluations, visual inspections of internal components, laser light alignments and filter cleaning are carried out. Once any modification or corrections are finalised, a 12 h transmission scan is acquired for normalising the detectors, using a ^{68}Ge rod source. For preclinical scanners the normalisation scan is usually around 4 h. The time of the normalisation scans will slightly vary based on the activity of the ^{68}Ge source.

Second, clinical PET scanner calibration entails using a 20 cm cylindrical water phantom filled with measured amounts of fluorine-18 (^{18}F) or a solid source germanium-68/gallium-68 (^{68}Ge/^{68}Ga) phantom. The preclinical scanners' cylindrical phantom is sustainably smaller (generally 3.5 cm). Generally, the scan times are 20 min but may vary depending on the system. Acquisition of the phantom provides the scanner's global activity correction factor, which corrects for internal scanner variations. The calibration of a scanner also sets a 'new' baseline for the expected scanner performance. Sites typically run daily quality checks, which verify that the scanner is performing up to the standard expectation of the set baseline performance.

4.4 CT calibration and quality control

As with PET, regular CT maintenance and clinical scanner calibrations are done quarterly, semi-annually or as suggested by the manufacturer. First, visual inspections of internal components, laser light alignments and filter cleaning are carried out. Second, using the system diagnostics software along with a CT performance phantom placed inside the bore, assessment of the X-ray tube and detectors is carried out. A CT performance phantom contains several separate sections, each designed to test a specific performance. Mainly composed of acryl, sections consist of different sized lines or strips, circles and an area with different density rods. This phantom allows for the testing of CT spatial resolution (high and low contrast scale), uniformity noise and slice thickness. Using the scanner software, a diagnostic test is also carried out on CT numbers (HUs) for each tube voltage.

Processed calibration files remain on the PET/CT work station and the raw calibration files are typically stored on the clinical picture archiving and communication system (PACS) subsystem or other comparable systems.

4.5 PET/CT annual testing

Annual testing provides an additional level of assessing scanner consistency and stability in performance. The current gold standard guidelines on annual testing for PET can be found in the NEMA NU 4–2008 preclinical and NEMA NU 2–2018 clinical evaluation protocols. Clinical guidelines and procedures for testing and reporting are also found in NEMA Standards Publication NU 3–2004 for PET and under the American College of Radiology (ACR) for PET and CT.

Clinically and preclinically, PET annual testing mainly focuses on evaluating scanner performance measurements for sensitivity and spatial resolution in air and in scattered medium. Annual tests currently cover: volume sensitivity, count rate, scatter fraction, coincidence timing winding, count losses and random events, accuracy of corrections (attenuation and scatter), time-of-flight (if applicable) resolution, and co-registration accuracy with CT.

This is accomplished using a variety of PET image quality (IQ) phantoms filled with water and a specific amount of ^{18}F (with areas for activity and no activity measurements) and a phantom designed as a line source filled with a relatively high activity of ^{18}F. The basic PET IQ phantom is comprised of sections with rods, an area for uniformity and chambers for no activity to determine recovery coefficients, percentage standard deviation and spill-over effects, respectively.

The annual testing of CT focuses on measurements of the X-ray tube beam energies, ionising radiation, absorbed dose and detector performance. Annual testing on CT also assists in monitoring correct absorbed doses, thus, potentially avoiding overexposure due to incorrect X-ray beam energies. This test uses a similar CT phantom to the one used in CT calibrations (most popular: Gammex 464 phantom, Sun Nuclear, FL, USA) with an ion chamber and radiochromic film. This phantom consists of four sections to evaluate high contrast, uniformity/noise, low contrast and alignment, with each section containing the necessary density rods and lines/circles. Similar to calibration, these parameters are measured along with the tube current (mAs) linearity and slice thickness. Annual CT testing also covers measuring the CT dose index using an ion chamber, for several tube voltages and/or tube current. The X-ray beam width is measured using the radiochromic film for each slice thickness using the routine CT protocols.

5 Data processing

The processing of PET/CT data sets contains the same initial methods for the reconstruction of a stand-alone PET or CT (please refer to chapters II.10 and II.9). The basic reconstruction methods used by clinical and preclinical sites are defined by two mathematical methodologies; analytical (inverse Radon transform) and iterative (statistical estimation/Poisson likelihood function) methods. The filtered back projection (FBP) algorithm is an analytical approach, whereas ordered subsets expectation maximisation (OSEM) or maximum likelihood expectation maximisation (MLEM) algorithms are iterative approaches [10]. Currently, FBP and OSEM remain the most commonly used methods in PET and FBP in CT. OSEM is also now becoming a popular reconstruction method for CT. However, not all currently available clinical and preclinical CT systems have an OSEM reconstruction implemented. For more details on reconstruction algorithms and data processing refer to chapters II.10 and II.9.

The difference between a stand-alone PET or CT to PET/CT in regard to data processing are the additional steps involved to generate CT attenuation correction maps for PET and to co-register the images. As noted in the introduction, CT data is scaled to fit to the PET energy level. The CT pixel values (HU) are thereby converted into linear attenuation coefficients to match the PET photon energy (511 keV). The two main issues of scaling CT to PET energies are the different photon energies itself (due to scattering and attenuating) and the attenuation of different materials

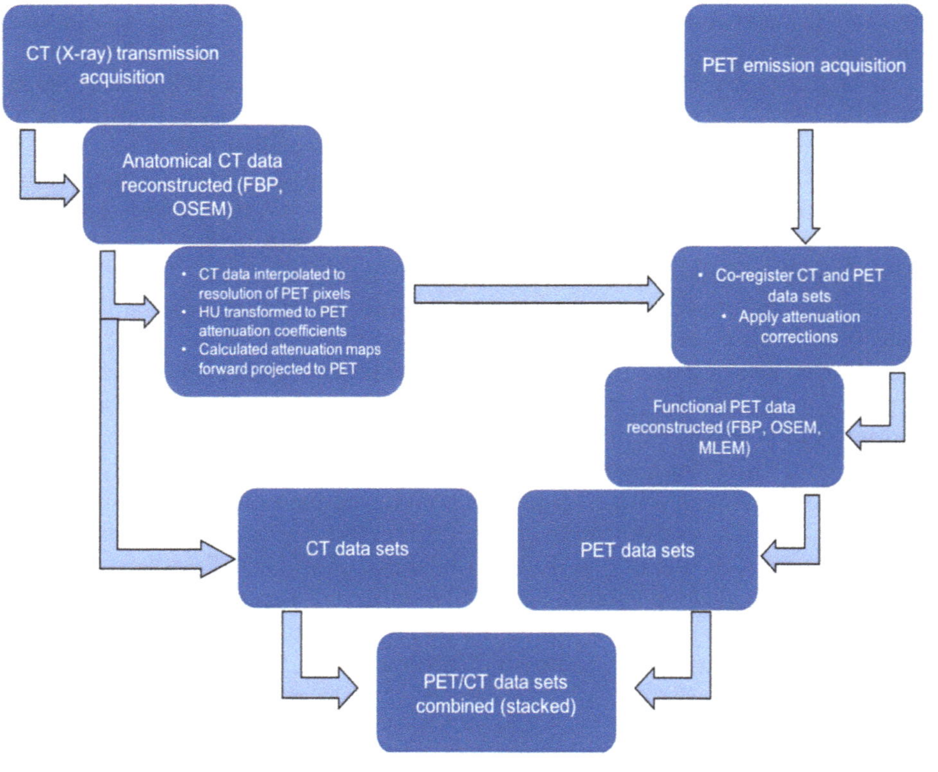

Figure 5. The flow chart displays the general steps from acquisition, reconstruction, generating the attenuation correction maps for PET, imaging data sets to the final stacking of PET and CT data sets. PET and CT imaging data can be displayed separately or as a PET/CT image, as shown in figure 1.

(e.g., soft tissue versus bone). To compensate for the various differences in densities, pixels (CT) are generally scaled using a bilinear fitting function, broken into at least three categories. Thus, pixel values are placed into air, soft tissue or bone using a different linear fit. The known linear attenuation coefficient for each category/material at the 511 keV energy is then substituted, generating an attenuation coefficient map. Figure 5 shows a basic outline of the data processing steps. It should be noted, not all preclinical PET/CT scanners apply CT generated attenuation correction maps, but it's a routine step in clinical PET/CT imaging.

6 Conclusions

Preclinical and clinical facilities and research sites have taken advantage of the benefits of PET/CT multimodality imaging. Besides the advanced co-registration, PET/CT also offers the efficiency of acquiring two imaging data sets in a single imaging study.

6.1 Strength and limitations

As noted above in section 4.3, the primary advantage of PET/CT is that it provides mechanically aligned anatomical and functional images. Another benefit is the use of CT for the attenuation correction of the PET data. As with any imaging system a major limitation are the artefacts introduced by the lack of proper scanner calibration. PET/CT misalignment and potential false CT correction factors might hinder correct PET quantitative analysis.

6.2 Future developments

Integrating multiple imaging modalities continues to show great promise, though not without difficulties. Further developments in hardware and software are expected, e.g., improved detector configurations and materials in PET and CT as well as enhancement of reconstruction algorithms and scatter corrections. Additionally, and importantly, research continues on the development and/or repurposing of PET radiotracers and biomarkers. For example, the use of ^{64}Cu or ^{89}Zr in oncology studies, which have been showing promise with more preclinical and clinical trials emerging. Other examples of newer PET and/or CT technologies can be found in chapters II.10 and II.9.

References and further reading

[1] Beyer T, Townsend D W and Brun T *et al* 2000 *J. Nucl. Med.* **41** 1369–79

[2] Beyer T, Kinahan P E, Townsend D W and Sashin D 1995 *Proc. of 1994 IEEE Nuclear Science Symp. - NSS'94 (Norfolk, VA)* **4** 1573–7

[3] Burger C, Goerres G, Schoenes S, Buck A, Lonn A and Von Schulthess G 2002 *Eur. J. Nucl. Med.* **29** 922–7

[4] Cherry S, Sorenson J and Phelps M 2012 *Physics in Nuclear Medicine* 4th edn (Amsterdam: Elsevier)

[5] Boellaard R, Doherty M J O and Weber W A *et al* 2009 FDG PET and PET / CT : EANM procedure guidelines for tumour PET imaging *Eur. J. Nucl. Med. Mol. Imaging* **37** 181–200

[6] Surasi D S, Bhambhvani P, Baldwin J A, Almodovar S E and O'Malley J P 2014 *J. Nucl. Med. Technol.* **42** 5–13
[7] Vanhove C, Bankstahl J P, Krämer S D, Visser E, Belcari N and Vandenberghe S 2015 *EJNMMI Phys.* **2** 1–25
[8] Kagadis G, Ford N L, Karnabatidis D and Loudos G K 2016 *Handbook of Small Animal Imaging: Preclinical Imaging, Therapy, and Applications* (Boca Raton, FL: CRC Press)
[9] Blodgett T M, Mehta A, Mehta A, Layman C, Carney J and Townsend D W 2011 *Clin. Imaging* **35** 49–63
[10] Alessio A and Kinahan P 1995 PET image reconstruction *Nuclear Medicine* vol 1 (Philadelphia, PA: Mosby)

IOP Publishing

Imaging Modalities for Biological and Preclinical Research: A Compendium, Volume 2
Parts II–IV: *In vivo* preclinical imaging, multimodality imaging and outlook
Andreas Walter, Julia G Mannheim and Carmel J Caruana

Chapter III.2.b

PET/SPECT/CT

Adriana A S Tavares, Christian Vanhove, Joost C Haeck, Sandra T van Tiel and Monique R Bernsen

Positron emission tomography (PET) and single photon emission computed tomography (SPECT) are important imaging tools used to study molecular and physiological processes in living organisms non-invasively. Applying molecular imaging to laboratory animal research is extremely important, because it can accelerate the development of new imaging biomarkers and drugs, it can increase our understanding of pathophysiologic processes in many areas of medicine, and it can advance technological developments that can be directly translated to the clinic. Unfortunately, PET and SPECT have relatively poor anatomical context and therefore are often combined with computed tomography (CT), which produces three-dimensional data from a series of X-ray images taken at different angles around the subject's body. CT images provide good delineation of internal organs and structures, namely bones, lungs and soft tissue. When co-registered with PET and SPECT, CT images add anatomical context to the molecular dimension of PET and SPECT. As hybrid-imaging technology continues to develop, studies looking to measure multiple molecular pathways simultaneously become possible. This is particularly interesting as medical research evolves towards a 'theragnostic' approach, where patient therapy is guided by diagnostic imaging and total-body imaging.

1 Introduction

PET and SPECT are versatile imaging techniques that can be used to study physiological and molecular processes in living organisms [1]. As reviewed in previous chapters, PET and SPECT have enormous potential in the areas of diagnostics, treatment monitoring, and personalised medicine. For example, PET or SPECT are able to provide molecular/cellular information that is unattainable

doi:10.1088/978-0-7503-3747-2ch19

with other imaging technologies or that would require more invasive procedures, such as biopsy or surgery. They can also identify diseases in their earliest stages often before symptoms occur or abnormalities can be detected with other diagnostic tests. Despite the exceptional characteristics of PET and SPECT imaging in the context of molecular medicine, these imaging modalities have limited spatial resolution and poor anatomical context compared to other imaging techniques, such as CT, which is less sensitive and conveys less specific molecular information but has higher spatial resolution [2].

To ally the best of both worlds, hybrid-imaging systems combining molecular imaging techniques (e.g. PET and SPECT) and anatomical imaging techniques (e.g. CT) have become increasingly popular. The use of hybrid systems in small animal research has become particularly important, because it can accelerate the development of new theragnostic applications in many areas of medicine. Importantly, because PET, SPECT and CT can be used in both clinical and preclinical studies, they convey huge translational potential and can rapidly impact biomedical research.

2 Principles and setups

2.1 Physical principles

The fundamental principles of PET, SPECT and CT imaging were presented in detail in chapters II.10, II.11 and II.9, respectively. Below we provide a short review of the key principles of each modality for ease of readability of the rest of this current chapter.

PET and SPECT imaging rely on the administration of relevant molecules labelled with radionuclides that emit gamma radiation. Gamma radiation can efficiently pass through tissue and can be detected outside the body with PET scanners or gamma cameras (figure 1(a) and (b)). Examples of frequently used PET radioactive isotopes include fluoride-18 (^{18}F), carbon-11 (^{11}C) and gallium-68 (^{68}Ga). An excess of protons in their nucleus, which requires positron-decay to achieve stability, characterises positron-emitting radioisotopes. A proton rich nucleus (X) can be converted into a daughter nucleus (Y), which has the same

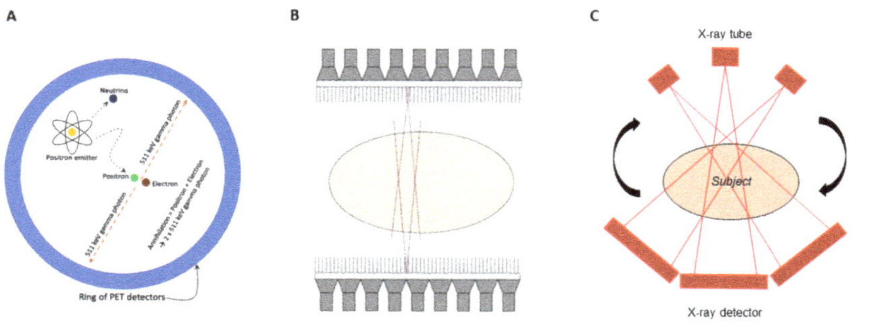

Figure 1. Schematic representation of main physical principles of PET (A), SPECT (B) and CT (C) imaging.

mass number (A) but lower atomic number (Z), with the emission of a positron (β^+) and a neutrino (ν):

$$ {}^{A}_{Z}\mathrm{X} \rightarrow {}_{Z-1}{}^{A}\mathrm{Y} + \beta^{+} + \nu \qquad (1) $$

As the name implies, SPECT examinations make use of radionuclides that emit a single gamma photon per decay, for instance technetium-99 (^{99m}Tc). There are, however, radionuclides which emit multiple gamma photons, e.g. indium-111 (^{111}In), or co-emit gamma photons with alpha or beta(-) particles, e.g. bismuth-213 (^{213}Bi) and lutetium-177 (^{177}Lu). Radionuclides which co-emit alpha or beta(-) particles with gamma photons allow for diagnostic therapy monitoring through SPECT images, as the localisation and number of the therapeutic alpha or beta(-) radionuclides can be derived from the number of emitted gamma photons, detected with SPECT systems.

Unlike emission techniques, such as PET and SPECT, which rely on the administration of radiolabelled molecules, the transmission technique CT uses external X-ray sources to generate images of the subject's internal organs and structures. CT images are obtained by generating X-rays, firing those around the subject's body at selected angles and collecting the X-rays that traveled through matter in the CT detector (figure 1(c)).

2.2 Typical setups and state-of-the-art

In preclinical PET systems, a small detector ring diameter (±10 cm compared to ±80 cm in the clinic) is combined with an axial bore length that is mostly equivalent to the size of a mouse (±10 cm compared to ±20 cm in the clinic). This more compact detector ring results in larger solid-angle coverage, compared to the clinical situation, resulting in a larger number of gamma photons that can be recorded and therefore in high system sensitivities (approximately five times larger than clinical systems). The majority of current preclinical PET systems is based on detectors that contain arrays of individual scintillation crystals, often coupled to silicon photomultipliers or photomultipliers (figure 2(a)) [3, 4]. A recent breakthrough in preclinical PET is the development of benchtop systems that combine monolithic scintillation crystals with silicon photomultipliers (figure 2(b)). These novel

Figure 2. Pixelated scintillation crystal (a) versus monolithic scintillation crystals (b). Example of a last generation PET system (Molecubes β-Cube system, c).

PET systems for laboratory animals offer sub-millimeter spatial resolution and sensitivities >10% (figure 2(c)).

Unlike PET dual-gamma emission, the information from a single gamma photon does not allow for spatial localisation of the source of decay by SPECT. Consequently, collimators are used to allow gamma rays to reach the detector only if emitted from a specific angle (figure 1(b), parallel hole-collimator). Unsurprisingly, collimators are an important factor in SPECT imaging. Small openings in SPECT collimators limit the angles from which a gamma photon can reach the detector giving higher spatial resolution, but at a cost of the sensitivity of the measurement due to the lower number of gamma photons passing through and reaching the detector (figure 1(b)). In the clinical setting, most SPECT collimators are based on the parallel-hole design, which suffers from limited spatial resolution (6–10 mm). Conversely, in the preclinical setting, the use of pinhole collimators is widespread (figure 3). Pinhole collimators allow for magnification of the object onto the detector, enabling sub-mm imaging resolution, which is far higher than preclinical PET (1–2 mm). Sensitivity is increased by utilising multiple pinhole collimator designs.

Using novel pinhole designs, certain preclinical imaging systems can collimate photons with higher energies than conventional systems. This opens up the opportunity to create images from PET radionuclides and other gamma-emitting radionuclides, such as the aforementioned bismuth-213 (gamma of 440 keV) and iodine-131 (gamma of 606 keV). By separating the energies from different radionuclides used in a single imaging experiment, it is possible to image two or more radionuclides simultaneously, and reconstruct the images from the different radionuclides separately by choosing the corresponding energy peaks from the list mode data (figure 4).

The use of CT alongside PET and/or SPECT serves two main purposes: (1) improve anatomical localisation of radioactive areas detected with PET and/or SPECT; and (2) generate attenuation maps. The importance of the first point has

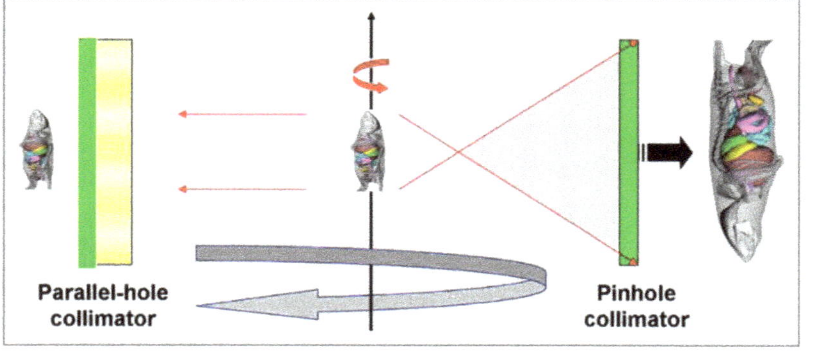

Figure 3. Comparison between parallel-hole and pinhole collimation. Through magnification of smaller objects at short distance onto large detector area, pinhole collimation geometry offers substantial increase in detection efficiency over that offered by parallel-hole collimation geometry, and spatial resolution is determined by size of pinhole aperture. Reprinted from [2] with permission from Hindawi.

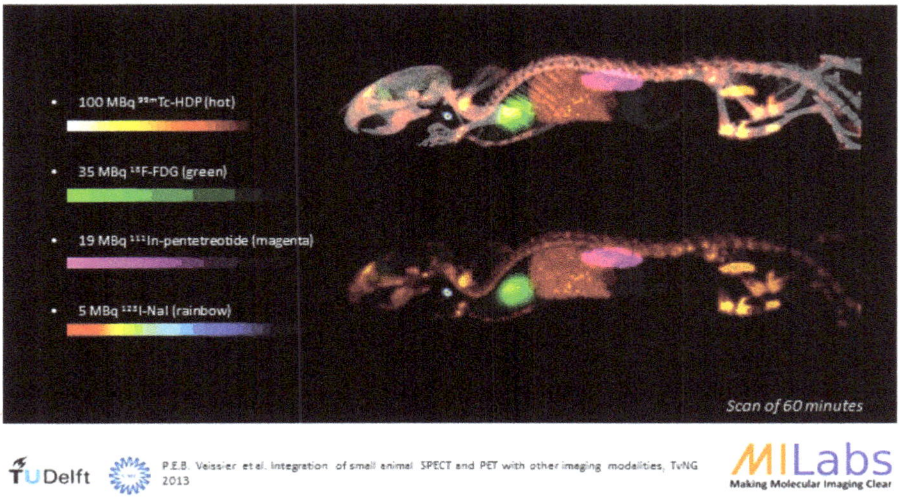

Figure 4. Multi-isotope PET/SPECT study using a high-energy collimator and the MiLabs VECTor system (taken from vendor brochure with permission). Different targeting molecules are labelled with different radionuclides to examine the *in vivo* biodistribution. Using listmode data, the radionuclides are reconstructed in separate images and overlaid to show the different uptake patterns: ^{99m}Tc-HDP targets bone, ^{18}F-FDG is taken up through glucose transporters, ^{111}In-pentreotide is excreted through the kidneys and ^{123}I-NaI is metabolised in the thyroid.

been elucidated above and the second point relates with the need to obtain quantitatively accurate PET and SPECT images (details below under 'Data processing' section).

Two or more imaging modalities can be combined in the preclinical setting either in a single platform, where the bed translates across different system bores (i.e. in line imaging or 'clip on' modality, figure 5(a–c)), or in separate imaging systems where the subject translates between imaging systems using a multi-modal bed (figure 5(d)).

3 Biomedical relevance

3.1 Application range and relevance

PET, SPECT and CT have a wide range of applications in the field of biomedical research. Preclinical PET and SPECT can be used to visualise expression of novel biological targets by using radiolabelled compounds such as antibodies, antibody fragments and peptides that have high affinity for these targets [1 and 3]. Generally, pharmacokinetics and biodistribution of these novel imaging probes are evaluated in suitable animal models. In addition, preclinical PET, SPECT and CT studies can be used to monitor the effectiveness of new therapies, namely by using probes or biomarkers that are routinely used in the clinic, e.g. ^{18}F-FDG (glucose metabolism), ^{99m}Tc-HDP (bone calcification) and CT Hounsfield units (characterisation of atherosclerotic plaque or lung fibrosis). Preclinical PET, SPECT and CT imaging techniques are increasingly used as platforms for translational research and drug

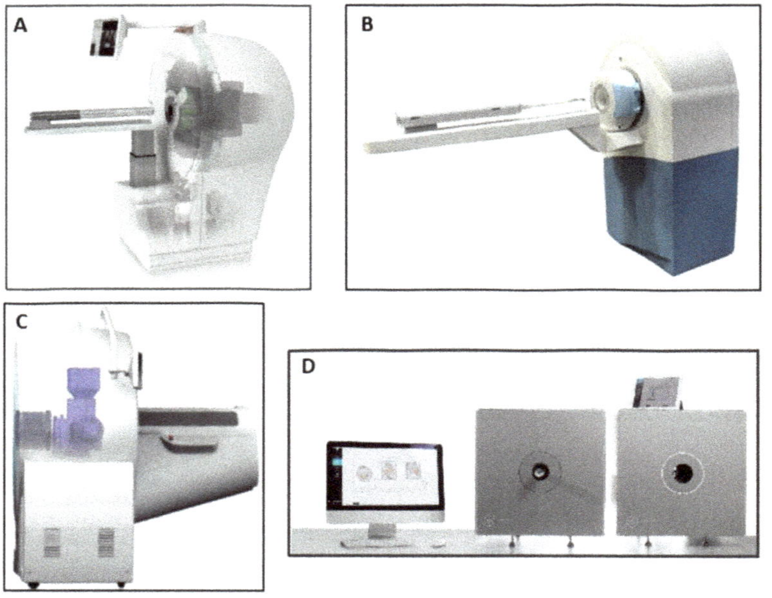

Figure 5. Different scanner arrangements for multi-modal preclinical PET/SPECT/CT imaging. (a) Imaging system where PET and CT components are integrated (Mediso NanoScan system, taken from vendor webpage with permission); (b) imaging system where the PET module was clipped onto the CT scanner (MR Solutions system, taken from vendor webpage with permission); (c) imaging system where the SPECT, CT and PET components are integrated inline from right to left (Siemens Inveon system, taken from [5] with permission from Hindawi; and (d) PET and CT systems where both components are separate (Molecubes X-Cube β-Cube combination, taken from vendor webpage with permission).

development. The ability to extrapolate from laboratory animals to human studies makes these imaging techniques ideal for both the validation of new imaging probes that might be relevant to the evaluation of human diseases and preclinical testing of new therapeutic drugs.

Furthermore, preclinical imaging has a particularly pertinent role these days in the assessment of total-body multi-organ responses to disease, as well as, understanding kinetics of radiotracers (and drugs) in the whole organism as opposed to a single organ. This emerging field is leveraged by the recent development of the clinical counterpart, the first total-body human scanner. Ultimately, preclinical PET/SPECT/CT imaging can play an important role in widening our understanding in pathophysiology by truly delivering multi-parametric and multi-dimensional imaging datasets, thus aiding with realising the potential of personalised medicine. Importantly, through the improved capabilities and generation of hybrid multimodal imaging, the number of animals used in biomedical research can be substantially reduced [3] while increasing statistical validity due to longitudinal studies of the same animal, and, at minimum, by a factor correspondent to the number of imaging techniques fully integrated into a single platform.

3.2 Sample preparation

Generally, preclinical PET/SPECT/CT imaging is suitable for imaging mice and rats. Similar to the clinical situation, PET and SPECT imaging is contingent on the administration of radiotracers, typically via intravenous (or intraperitoneal) injection. Depending on the radiotracer properties and the biological process to be studied, animals can be scanned immediately after injection of the radiotracer or after a certain period of uptake time that can vary from a few minutes to a few days post-radiotracer administration. Also similar to the clinical situation, CT contrast agents can also be used to enhance tissue attenuation and consequently organ contrast. The smaller blood volume of laboratory animals, compared to humans, limits the volume of the PET and/or SPECT radiotracer, and/or CT contrast agent, which can be administered [6].

In most cases, laboratory animals are anaesthetised during the scanning session. A volatile anesthetic, such as isoflurane, is preferred because the duration and depth of anesthesia can be easily monitored and controlled. However, one should be aware that anesthesia might significantly affect imaging outcomes. Most anesthetics not only interfere with neurotransmission in the brain, but might also affect cardiovascular, respiratory parameters and other physiological functions [6]. On its own, a PET, and SPECT or CT scan can last about 10–60 min, 20–60 min and 2–10 min, respectively. However, collectively they accumulate to longer scanning sessions, which is an important limitation of hybrid multi-modal imaging, given that anesthesia is known to have effects on an animal's physiology.

4 Parameters of image quality

PET, SPECT and CT imaging require a sufficient signal-to-noise ratio for optimal image quality. All these modalities require good uniformity (typically SD <10%), good recovery coefficients (ideally as close to 1 as possible) and good spill-over ratios (ideally as close to zero as possible). Possible solutions to increase signal-to-noise ratio in PET/SPECT/CT imaging include: the use of more sensitive collimators with larger pinholes, at the cost of resolution (SPECT); more pinholes; modified reconstruction settings (PET/SPECT/CT) or longer scanning experiments to collect more counts (PET/SPECT/CT).

Critically important, PET and SPECT image quality can be influenced by the energy of the detected gamma photons, in particular scattered and attenuated photons. In fact, if photon scattering is overlooked, the scattered radiation results in a loss of resolution and an apparent migration of radioactivity from areas of high concentration to areas of low concentration in the images. Therefore, it is important to use scatter correction methods during reconstruction of radionuclide imaging datasets, in order to assure maximum image quality [7].

Furthermore, disregarding photon attenuation will always lead to an underestimation of the measured radioactivity concentration. For PET, the underestimation can range from 10%–20% for mice and from 20%–35% for rats. Consequently, multi-modal solutions (PET/CT or SPECT/CT or PET/SPECT/CT) are particularly advantageous to maximise potential of each imaging modality by combining their symbiotic nature. For example, gamma photon attenuation (PET or SPECT) is often performed by co-registering the map of attenuation coefficients to a relevant photon energy (e.g. 511 keV for PET). CT images can be pivotal to generate attenuation maps. This is done by segmenting the CT scan of the subject into different structures, typically bone, soft tissue and air; then applying a correction factor per image voxel based on the calculated attenuation coefficients of the measured CT signals. Attenuation correction of PET and SPECT data, via this CT method, results in sharp images and quantitatively accurate datasets.

Generally, SPECT/CT or PET/CT are used as multimodal imaging modalities to combine functional (SPECT or PET) with anatomical (CT) information. However, the simultaneous use of a SPECT and a PET tracer, as a dual-isotope technique, in the same animal is much more challenging. This is caused by down-scatter of the high energy PET gamma photons that penetrate the SPECT collimator and scatter into the detector, potentially within the SPECT photopeak window. A potential solution here is to administer approximately 10 times less activity for the PET-tracer than for the SPECT-tracer.

5 Data processing

Once the PET/SPECT/CT data is acquired, it needs to be reconstructed to generate images for subsequent processing and analysis. Usually an iterative reconstruction technique based on the maximum-likelihood expectation-maximisation (MLEM) algorithm is used for reconstruction of PET data [8], which usually results in a quantitative image where the value of each voxel is expressed as amount of radioactivity per unit of volume or activity concentration (i.e. Bq/mL). MLEM is also routinely used for reconstruction of SPECT data because it can better handle a complex pinhole geometry (e.g. multiple detectors using multiple-pinhole collimators). Quantification of the PET/SPECT signal requires activity calibration. This calibration is needed to convert the detected counts to the amount of activity (MBq per unit of volume). This is performed by imaging a source with a known amount of activity of the radionuclide of interest used for a specific experiment. The counts from the known source are used to calculate the calibration factor (CF):

$$CF = \frac{\text{Activity}}{\text{Unit of volume} \times \text{Counts}} \tag{2}$$

In regards to CT, typically, preclinical CT images are reconstructed using filtered-back projection. Sir Godfrey Hounsfield derived a CT scaling factor for quantifying the distribution and attenuation of X-ray transmission through tissue. Numerical values were ascribed, relating to the attenuation calculations (attenuation coefficients) to tissue densities. These values are known as Hounsfield Units (HU, equation 3), and are used to differentiate and compare tissue densities (e.g. water was set to be 0, and air set at −1000 HU). Each CT system will generate slightly varied HU numbers, which can be calculated as:

$$HU = 1000\frac{(\mu_m - \mu_w)}{\mu_w} \tag{3}$$

where μ_m represents the soft tissue attenuation coefficient and μ_w represents the water attenuation coefficient [9]. Refer to chapters II.9, II.10 and II.11 for a more detailed overview of reconstruction algorithms for each modality.

Once the PET/SPECT/CT images are reconstructed, they can be processed for visualisation and/or quantification depending on the biological research question. Typically, volumes of interest (VOIs) are drawn around a target organ to measure the radioactive concentration (PET and SPECT) or tissue density (CT HU). In general, PET data is reported using standard uptake values (SUVs), which are calculated as follows:

$$SUV = \frac{C_T}{D/W} \tag{4}$$

where C_T = radioactive concentration in tissue, W = weight of the animal and D = injected dose of radiotracer.

SPECT data is often reported as percent injected dose ($\%ID$) per mL that can be calculated by dividing the measured activity concentration in the VOI (A_{VOI}) with the injected amount of radioactivity (A_{inj}):

$$\%ID = \frac{A_{VOI}}{A_{inj}} \tag{5}$$

Because radiotracer experiments (PET and SPECT) are inherently time-dependent processes, the (semi-)quantitative metrics (%ID and SUV) plotted over time (so-called time-activity curve, TAC) can describe the radiotracer pharmacokinetics *in vivo*. Therefore, the full potential of the quantitative nature of PET and SPECT can be exploited when a suitable mathematical description of the tracer kinetics (i.e. kinetic modelling) is applied to the dynamic data to obtain quantitative values of relevant physiological or biochemical transport constants [10]. CT image analysis often rely on drawing VOIs to report regional HU and, when scanner calibrations are correctly performed, this can be an important outcome measure to investigate, for example, longitudinal changes in bone density or cardiovascular calcification (calcium score analysis). The HU can also be used to characterise lung fibrosis in rodent models of infection or cystic fibrosis. CT image analysis could also include quantification and comparison of organ dimensions in genetically engineered rodent models, or even measurement of spinal curvature angles in the context of scoliosis.

6 Conclusions

6.1 Strength and limitations

Multi-modality imaging, namely PET/SPECT/CT platforms, can be extremely powerful tools in preclinical research by allying molecular information (from the PET and SPECT modules) with anatomical and structural information (from the CT module). Molecular and structural information provide complementary insights into physiological and pathological processes in living organisms over time, thus widening biomedical knowledge. Because these imaging modalities also exist in the clinical setting, their translational value is immediate. Furthermore, by allowing for longitudinal imaging of animals, these multi-modal platforms can substantially reduce the number of animals used in biomedical research. However, by combining multiple modalities into a single scanning session, there is an inevitable increase of an animal's exposure to long periods of anesthesia, which can affect the animal's physiology and consequently molecular imaging findings.

6.2 Future developments

Current technological developments, including hardware and software, will increase sensitivity and resolution of preclinical PET/SPECT/CT systems. The use of hybrid systems in small animal research is likely to play a particularly important role in the development of new theragnostic applications and the understanding of systems biology by permitting total-body imaging of a full mammalian organism.

Furthermore, machine learning and other artificial intelligence techniques might further improve image quality (e.g. noise reduction) and simplify the segmentation of, for example, CT images.

References and further reading

[1] James M L and Gambhir S S 2012 *Physiol. Res.* **92** 897–985

[2] Kemp R, Epstein F, Catana C, Tsui B and Ritman E 2010 *J. Nucl. Med.* **51** 18–32

[3] Kiessling F and Pichler B J 2011 *Small Animal Imaging: Basics and Practical Guide* 1st edn (Berlin, Heidelberg: Springer)

[4] Zaidi H 2014 *Molecular Imaging of Small Animals: Instrumentation and Applications* 1st edn (New York: Springer Science+Business Media)

[5] Khalil M M, Tremoleda J L, Bayomy TB and Gsell W 2011 *Int. J. Mol. Imaging.* **2011** 796025

[6] Vanhove C, Bankstah J P, Krämer S D, Visser E, Belcari N and Vandenberghe S 2015 *EJNMMI Phys.* **2** 31

[7] Prasad R and Zaidi H 2014 *Mol. Imaging Biol.* **16** 199–209

[8] Lange K and Carson R 1984 *J. Comput. Assist. Tomogr.* **8** 306–16

[9] Hounsfield G 1973 *Br. J. Radiol.* **46** 1016–22

[10] Dupont P and Warwick W 2009 *Methods* **48** 98–103

IOP Publishing

Imaging Modalities for Biological and Preclinical Research: A Compendium, Volume 2
Parts II–IV: *In vivo* preclinical imaging, multimodality imaging and outlook
Andreas Walter, Julia G Mannheim and Carmel J Caruana

Chapter III.2.c

PET/MR

Andreas M Schmid, Prateek Katiyar and Julia G Mannheim

The combination of positron emission tomography (PET) and magnetic resonance imaging (MRI) is a noninvasive *in vivo* imaging approach to study both metabolic and morphological processes. PET is based on the detection of annihilated photons of radiolabelled targets and can provide a variety of metabolic imaging parameters, such as glucose uptake, proliferation, and hypoxia. Compared to PET, MRI utilises the spin of nuclei to detect their distribution by using magnetic fields and dedicated radio-frequency (RF) coils to excite the nuclei. The detection of the resulting signal enables the acquisition of both morphological and metabolic information, such as tissue diffusion or perfusion.

The combination of both standalone techniques into one simultaneous imaging approach offers the unique potential of studying disease mechanisms on a multi-parametric level with the patient/subject in the same physiological status compared to sequential PET and MR acquisitions.

1 Introduction

The combination of PET and MRI into one system has revolutionised the imaging field over the past decade. Compared to routinely used PET/CT systems that enable the sequential acquisition of PET and CT data to coregister the metabolic data provided by PET with the structural information measured by CT, the combination of PET/MRI enables the simultaneous acquisition of metabolic information provided by PET and morphological as well as metabolic information by MRI at the same physiological state of the subject (e.g. perfusion, body temperature, anesthesia). The correlation of multiple parameters measured by PET and MRI has facilitated new insights in biological research and diagnosis [1].

doi:10.1088/978-0-7503-3747-2ch20

2 Principles and setups

2.1 Physical principles

The basic physical principles of PET and MRI are explained in detail in chapters II.10, II.7.a and II.7.b, respectively. Therefore, only the technical aspects that are specific to simultaneous PET/MRI systems will be discussed in this chapter (subsections 2.1.1 and 2.1.2).

2.1.1 PET

To enable simultaneous PET/MRI acquisitions, specific PET detectors are required that operate in magnetic fields. Commonly used photomultiplier tubes (PMTs) are nonfunctional in strong magnetic fields, as the electron path within the vacuum tube is affected by the magnetic field; thus these detectors cannot be utilised for simultaneous PET/MRI (see figure 1(a)). The transition to using avalanche photodiodes (APDs, see figure 1(b)) and more recently to using silicon PMs (SiPMs, see figure 2), both of which are functional in magnetic fields, has been crucial for the development of the fully integrated systems [2].

Although PMTs are still superior in terms of timing resolution compared to APDs, the development of SiPMs as the latest generation of MR-compatible PET detectors has revealed similar, if not improved, performance characteristics compared to PMTs. SiPMs are much smaller in size, making them the ideal choice for simultaneous PET/MRI system designs. Furthermore, SiPMs, due to their fast timing properties, enable the use of time-of-flight information to be used for

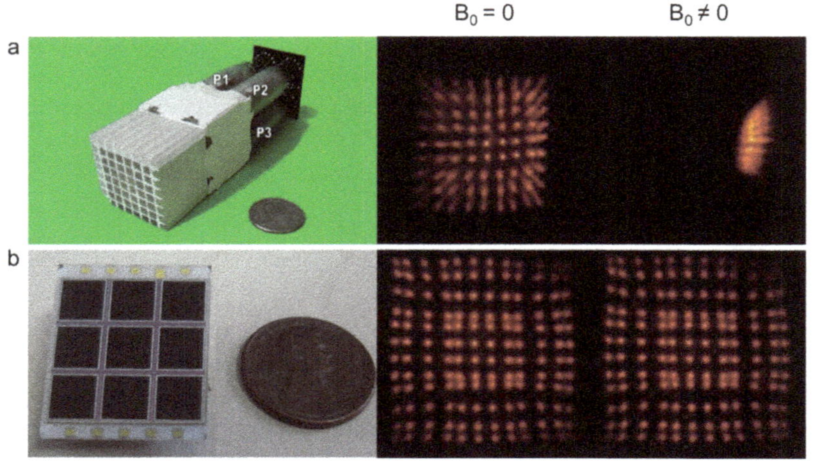

Figure 1. Comparison of different detector designs. (a) The photomultiplier tubes (PMTs) P1–3 are much larger in size than modern avalanche photodiodes (APDs) (b). While detectors based on regular PMTs, coupled to a crystal array, lose all spatial information of the crystal profile inside a magnetic field, detectors based on APDs preserve the crystal profile even at high magnetic fields, e.g. inside MRI scanners. Parts of the figure are reprinted and amended with permissions from the authors [3]. Parts of the figure are adapted from [4] and reprinted after purchasing reprint permission.

Figure 2. An array of a 4 × 4 silicon photomultiplier with 3 × 3 mm^2 pixels, in which each individual pixel is comprised of 3600 Geiger mode avalanche photodiodes. The figure is adapted from [5], copyright (2008) with permission from Elsevier.

localisation of photon annihilation. This information can be utilised for image reconstructions to significantly improve the signal-to-noise ratio (SNR).

Furthermore, in addition to the detector choice and while all considerations for the design of standalone PET and MR systems are still valid, additional issues arise for simultaneous PET/MRI system designs. The space is strongly limited inside an MRI scanner, and additional material for RF shielding for the PET detectors must be included, while ensuring at the same time that the shielding does not impact the MR performance. Especially for preclinical systems, space within the magnet bore is strongly limited, as preclinical magnet bore diameters are typically in a range of 30 cm or less. Integrating a PET insert within such a small bore size automatically limits the active PET field-of-view (FOV) size. Currently, most available preclinical PET inserts can only cover whole body mouse or rat brain imaging.

Furthermore, the temperature stabilisation of PET inserts is specifically important [2], as the quantification accuracy of the acquired data depends on the detector temperature and can be significantly biased by an unstable temperature performance.

Compared to PET/CT systems, in which the CT is used as the gold-standard tool to acquire (tissue-)attenuation data to correct attenuated PET photons, tremendous research efforts have been conducted over the last decade to develop MR-based attenuation correction approaches (see figure 3) [6, 7]. MR contrast is based on the number of protons in the underlying tissue and the corresponding relaxation properties but does not provide direct information on the potential attenuation of PET photons in tissue.

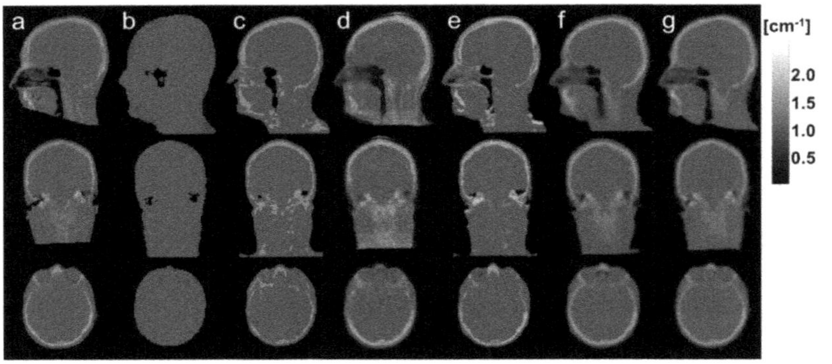

Figure 3. Comparison of different attenuation correction approaches. μ-maps based on the gold-standard CT (a), or MRI- and atlas-based methods such as (b) Dixon, (c) dual ultrashort echo time (UTE), (d) a network-based artificial neural atlas, (e) dual UTE-based R2, (f) an atlas based on magnetisation-prepared rapid acquisition gradient echo (MPRAGE), and (g) an MPRAGE-based template. All of the MR methods have inherent strengths and drawbacks when identifying the different tissues (soft-tissue, bone, air) and quantifying the respective attenuation factor. The figure is adapted from [6], copyright (2016) with permission from Elsevier.

Current state-of-the-art clinical PET/MRI systems correct for attenuation based on a segmentation approach [7], in which the MR data acquired using special MR sequences (e.g. zero echo time or ultrashort echo time sequences, see figure 3) are divided into different tissue classes, such as lungs, air, soft tissue and bone. In sharp contrast to clinical PET/MRI systems with established MR-based attenuation correction methods, preclinical PET/MRI attenuation correction approaches are not as standardised or broadly available. Furthermore, preclinical PET inserts are often not stationary built-in into the MR bore, and the position of the RF coils within the PET FOV might change (e.g. flexible head coils); hence, attenuation coefficients might change. Especially in the preclinical field, attenuation correction is still an active field of research to enhance the quantification accuracy of the acquired PET data, in which standardised coil positions would enhance the comparability of data.

2.1.2 MRI

As discussed in earlier chapters (see chapters II.7.a and II.7.b), MRI is a very comprehensive tool providing multi-parametric readouts like anatomy, metabolism (MRS), perfusion (DCE-MRI), tissue connectivity (diffusion MRI), function (fMRI), and more; however, these options are very sensitive to disturbances by objects placed inside the scanner. Furthermore, to achieve the highest B_0 homogeneity, the magnet's bore size is usually kept to a minimum. Both paradigms complicate the installation of a PET insert for simultaneous PET/MR imaging, as the design of the MRI scanner has to accommodate at least the PET detectors while still keeping the minimum free bore size needed to conduct MRI acquisitions. To optimise the SNR, the entire signal detection chain of both, PET and MRI, should be accounted for. Specifically, designing RF coils with minimal attenuation to the PET signal, invariant to translation and rotation of the coils and applying shielding to all PET components, is of high interest to prevent any interaction with the RF signal with only causing negligible distortions of the B_0 field and also not generating RF noise, hence affecting the B_1 field. This may not always be feasible, but in order to achieve high-quality PET data throughout the study, the overall setup must not change within a study. With increasing complexity of the study design, e.g. the assessment of functional parameters (see chapters II.7.a, II.7.b and II.7.c (MRI, fMRI and hyperpolarised MRI)) in stimulation settings, additional equipment (electrodes, cables, intubation systems, etc) may be necessary, and the need for a high SNR may require dedicated surface coils. In this case, it may not be possible to design translationally invariant setups, but MRI-detectable equipment may allow for these items to be included into PET-attenuation correction approaches and restore PET quantification accuracy as much as possible.

2.2 Typical setups

Combined PET/MRI systems are available in two configurations to acquire PET and MRI data, namely, sequential or simultaneous PET/MRI (see figure 4). In a sequential setup, standalone PET and MRI scanners are utilised to acquire the data, and the images are fused in postprocessing steps. Both systems can be physically located in different rooms or close to each other with adequate shielding, thus enabling the same bed components to be used to move the subject from one scanner to the next.

A sequential setup (see figure 4(b)) offers a highly flexible study design in terms of subject preparation, access during the scan (for injections) and the RF coil choice. Furthermore, both scanners can be used independently at the same time, enabling potentially a higher throughput. Another advantage of a sequential setup is the fact that the PET detectors do not have to be optimised to work in magnetic fields, and the potential impacts of PET components on the MR image quality (and vice versa) are prevented. In sum, a sequential setup allows the use of all options that are available on either system, including in a preclinical setup more sophisticated animal preparation and animal beds. However, transporting the animal between the scanners or adding specific RF coils during the sessions may change the animal's position, which in turn hampers an accurate subsequent fusion of image data. Furthermore, one of the clear benefits of simultaneous PET/MRI, namely, acquiring functional data during the same physiological state of the subject, is not facilitated, as well as the possibility to compensate and correct for motion of the PET data to enhance SNR.

In contrast to this, in a simultaneous setup, an MR-compatible PET insert is either permanently installed inside the bore of the MRI scanner (see figure 4, bottom) or as an insert, that can be removed and installed within the MR bore (see figure 4 top). This requires a sufficient magnet bore size and also puts further constraints on the study design, as RF coils may cause notable attenuation of the PET signal. Therefore, in a preclinical setting, two approaches to ensure an accurate

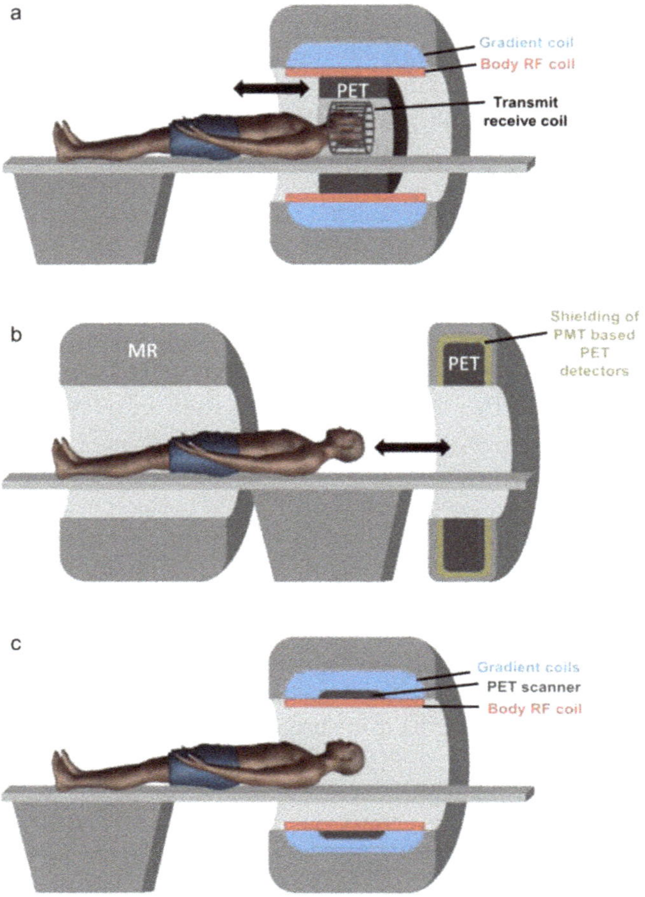

Figure 4. Concepts for the integration of PET and MRI. (a) A PET insert installed in a whole body MRI for a simultaneous acquisition; (b) a sequential PET/MRI scanner; (c) a fully integrated simultaneous PET/MRI scanner, in which the PET system is permanently built into the MR scanner. The figure is reproduced and adapted from [8] under the terms of the Creative Commons Attribution 3.0 license (https://creativecommons.org/licenses/by/3.0/).

quantification of PET data are feasible—either reducing all potential attenuation contributors to a minimum (e.g. thin animal beds, only one RF coil, only the most necessary equipment) or having a setup that is completely reproducible, meaning the same coils in the same positions, fixed animal positions on the bed (potentially including stereotactic holders), perfect localisation within the FOV between animals, etc. As sufficient attenuation correction approaches are implemented in a clinical setup, including the attenuation maps of coils, the reproducibility of scans is not as affected as it is in the preclinical setting. However, simultaneous PET/MRI offers the unique opportunity to address biological mechanisms with two independent modalities within the same physiological state, allowing a direct correlation between the imaging biomarkers/readout. Furthermore, the correlation of multiparametric data will allow new insights into the pathways and progression of multiple diseases.

3 Biomedical relevance

3.1 Applications in biological/preclinical research

Over the past decade, simultaneous PET/MR imaging has become part of preclinical and clinical research, as well as clinical diagnostic routines. Multiple applications have benefited from the simultaneous acquisition of multiple PET and MRI parameters.

Directly upon the first installation of a simultaneous PET/MRI system, comparisons to the gold-standard PET/CT system were performed to demonstrate the benefits and usefulness of simultaneous PET/MR imaging in various applications. Especially for pediatric applications, as well as for longitudinal studies, simultaneous PET/MR imaging has been proven to be very beneficial as the CT radiation can be avoided [2].

The correlation of simultaneous multimodal PET/MR imaging contributed significantly to the understanding of multiple diseases and their pathways. To name a few examples, PET/MRI is utilised for therapy stratification and responses in oncological applications, in which the metabolic information provided by PET (e.g. [^{18}F]Fluorodeoxyglucose ([^{18}F]FDG) uptake as a marker of energy consumption) is correlated with the underlying anatomical tissue as determined by MRI, as well as diffusion-weighted sequences to determine water-mobility in tissue [2].

Furthermore, structural and functional MR imaging to determine neuronal degeneration, as well as [^{18}F]FDG or tau-specific PET tracers to detect amyloid and tau plaques, are used to improve dementia diagnosis [9].

3.2 Sample preparation and requirements

In general, the same preparation routine as that for standalone MR or PET imaging is required (see chapters II.7.a, II.7.b and II.10 for MR and PET preparation, respectively).

For human PET/MRI systems, the patients need to be screened for any magnetic implants and dependent on the application, for the tolerance of MR contrast agent injection.

For animal research, MR-compatible animal physiological equipment for temperature regulation, breathing rate monitoring, etc is required. Furthermore, separate catheters for contrast agent and activity injections might be required if the protocol encompasses constant infusion of a tracer or contrast agent. In a simultaneous approach, the dead volumes of the catheters need to be considered, as the length of catheters is increased. All animal beds must be fully MR-compatible. The MR RF coil choice needs to be carefully considered to provide a good SNR for MRI and not to affect the PET image quality due to signal attenuation. Ideally, the RF coil is included in the PET attenuation correction process.

4 Parameters of image quality

The same image quality parameters are to be determined for simultaneous PET/MRI, as for standalone PET or MRI (refer to chapters II.7.a, II.7.b and II.10 for a detailed description of standalone PET and MR image quality parameters). However, when combining PET and MRI in a simultaneous approach, it is important to fully evaluate potential influences on the image quality from one modality to the other to ensure that no signal degradation has occurred due to the simultaneous setup.

Regarding PET, a full national electrical manufacturing association (NEMA) evaluation is being performed to determine scanner-specific parameters, such as spatial resolution, sensitivity, scatter fraction, etc. In the specific case of a PET insert being used (often the case for preclinical systems), which can be installed within the MRI, the evaluation is performed with and without the insert being installed inside the MR scanner, so that a direct comparison of the potential impacts of the insert on the MR performance or vice versa can be determined. For a permanently installed PET system within the MRI (often the case for clinical systems), influences need to be determined in comparison to PET- or MR-only systems, as well as if they meet the criteria of the respective NEMA specifications.

Importantly, when performing tests to determine the potential impact of MR sequences on the PET performance, multiple MR sequences that are routinely used should be tested, including more gradient-demanding sequences, such as EPI. In this regard, temperature stabilisation of the PET system is a crucial point—as the MR gradient might heat up the detectors—and needs to be monitored, as changes in temperature will impact the image quality and consequently affect quantification. Furthermore, the shielding of the detector modules from the RF-pulses needs to be investigated to avoid inducing RF-noise during PET acquisition.

In terms of MRI data quality, similar approaches to those used with PET are valid. The MRI performance should not be affected by the presence or absence of the other scanner. While this is obvious for sequential setups, it should be carefully evaluated in simultaneous setups. Deviations introduced by the PET insert should not affect the SNR, B_0 homogeneity after shimming and general peak-linewidths or other measures of system performance such that they have a significant impact on the data.

5 Data processing and visualisation

Combined PET/MRI allows for the recording of a plethora of metabolic and anatomical information. The scale and complexity of the multimodality data may, however, make manual data analyses non-practical and in some cases unfeasible. Whereas the direct benefits of the complementary multiparametric imaging data can be obtained with simple regionwise quantification techniques, advanced data processing methods are required to exploit the holistic benefits of combined PET/MRI.

Very well-known cases for using cross-modality improvements in PET/MRI examinations are MRI-based motion estimation and the correction of the PET data.

Subject motion severely degrades PET quantification metrics, which are already limited due to the intrinsically poor spatial-resolution of PET. In contrast to that, the high spatial contrast and SNR of MR images simultaneously acquired with PET can be used to estimate the subject's motion (e.g. respiratory motion) and subsequently correct the corrupted PET data. The standardised uptake value (SUV) measures derived from the motion-corrected PET images have been clinically shown to improve lesion localisation and detectability, particularly in PET-avid tumours (see figure 5). Similarly, the variance of compartmental modelling rate constants derived from PET time-activity curves in dynamic studies is highly reliant on the accuracy of the arterial input function. Image-derived arterial input functions are an appealing noninvasive alternative, but are prone to activity spill-in, spill-out and partial volume effects. In this context, high-resolution MRI angiography data are extremely useful in defining the major arteries needed to correct the adverse effects of the aforementioned factors and for determining the PET input function [10, 11].

Beyond data correction and enhancement, the biggest advantage of simultaneous PET/MRI is its ability to capture differing biological mechanisms of the disease. The complementary information allows machine learning algorithms to be leveraged to build tools of high predictive value that can be used for disease characterisation, monitoring and therapy stratification. For example, basic clustering techniques applied to the combination of the PET tracer uptake and MRI apparent diffusion coefficient (ADC) parameters acquired in tumours have been shown to precisely characterise intratumoural regions of phenotypic heterogeneity, which differ in their cellularity and the degree of glucose consumption. These two complementary properties cannot be quantified by using either one of the modalities alone. Likewise, [^{18}F]FET PET/MRI has enabled amino acid imaging, which has shown promise in improving the distinction between recurring gliomas and post-treatment effects compared to MRI alone [12].

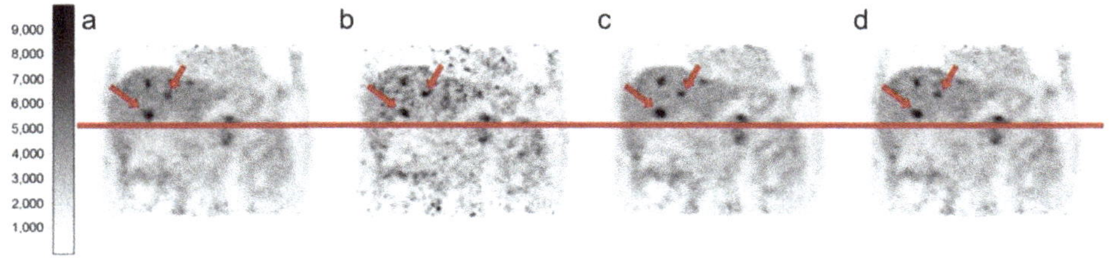

Figure 5. Representative PET slices of uncorrected static scans (a), gated scans (b), motion correction based on MRI (c) or on PET scans (d). Liver lesion detections is significantly impaired due to motion blurring (a). Gating (b), as well as motion correction based on MR and PET (c and d), increase lesion detection, with noise parameters of the MR and PET motion correction comparable to the uncorrected image, while gating revealed increased noise patterns due to the reduced number of counts. This research was originally published in JNM. Amended from [13] by the Society of Nuclear Medicine and Molecular Imaging, Inc.

6 Conclusions

The combination of PET and MRI in a simultaneous hybrid approach has been one of the major developments in medical imaging within the last decade. However, to date, the full potential of simultaneous PET/MRI has not yet been fully exploited. Among the many examples of potential applications, projects that include a joint data analysis indicate the overall potential of simultaneous imaging. Multidimensional correlations between imaging parameters are now being researched and have the potential to elucidate (and quantify) biological mechanisms at new levels. Especially in combination with an increasing parameter space, further developments of advanced analysis methods to characterise certain diseases on a multiparametric level are essential and bear tremendous opportunities. While the transition of these diagnostic tools from research into clinical diagnosis is certainly challenging, the benefits already demonstrated by PET/MRI are—without a doubt—of great merit.

6.1 Strength and limitations

To date, the combination of PET and MRI has already shown great benefits in several research and medical questions. Major limitations arise from the extended initial and running costs, the need for more advanced imaging protocols, and, especially in preclinical research, unsolved issues in PET attenuation correction; however, these limitations are outweighed by several, unique opportunities. Due to the inherent qualities of non-invasive imaging, longitudinal studies are possible, enhancing the statistical quality of the data and reducing the total number of animals, which is greatly appreciated with respect to the 3R-principle (reduction, refinement and replacement). Furthermore, simultaneous acquisitions generate a live view of biological mechanisms at the same physiological state, accompanied by an overall reduction of the total anesthesia time. In addition, tissues (lesions) with otherwise indifferent uptake can now be characterised on the basis of detailed anatomical and functional landmarks. Due to the similarities among scanners and data, preclinically established workflows, including data acquisition and analysis, may directly be transferred to the clinics. Furthermore, simultaneous PET/MRI allows the cross-calibration of the same target with different approaches and the clear identification of mismatches, leading to a better understanding of the tools themselves. Several advanced MRI-related options are currently explored in basic and clinical research, e.g. PET/MRI using hyperpolarised compounds or other X-nuclei applications of MRI. In conclusion, PET/MRI is a unique tool that has already shown great power, but there is still more to come.

6.2 Future developments

Regarding technical developments for PET, continuous developments and research have been performed over the last decade to improve the timing and spatial resolution of detectors. Improving the timing resolution is especially important to facilitate the time-of-flight (TOF) information for reconstructions to enhance the SNR. To date, one commercially available clinical PET/MRI scanner already incorporates TOF information for reconstructions; however, in the preclinical setting, due to the smaller FOV, the detector timing resolution needs to be further improved to enable the benefits of TOF in preclinical imaging. Improvements in spatial resolution will impact quantification accuracy, as the partial volume effect scales with object size and spatial resolution.

Future MRI technical developments are necessary to optimise the coil design for simultaneous PET/MR imaging, especially to minimise the potential attenuation of the PET signal.

In addition to technical challenges and developments, advanced image analysis techniques that correlate the variety of multiparametric data are needed to prove the benefits of simultaneous PET/MR imaging also in a routine clinical setting. MR-based attenuation correction of preclinical data, including the respective animal bed and RF coils (without prior knowledge), is a mandatory development to be implemented in the future to ensure an accurate quantification. Furthermore, (automated) MR-based segmentation of the PET data is essential for partial volume correction and should be implemented in the preclinical settings as well. Using MR-based gating independent of the specific MR sequence as prior for the PET data could significantly improve the SNR and is an active research field.

References for further reading

For a more detailed list, please also consult the recommendations in chapters II.7.a, II.7.b and II.10.

1. Phelps M E 2004 *PET Molecular Imaging and Its Biological Applications* (Berlin: Springer)
2. Tofts P 2003 *Quantitative MRI of the Brain: Measuring Changes Caused by Disease* (Chichester: Wiley)
3. Catana C 2017 *Magn. Reson. Imaging Clin. N. Am.* **25** 231–43.

References

[1] Bailey D L, Pichler B J, Guckel B, Antoch G, Barthel H and Bhujwalla Z M *et al* 2018 Combined PET/MRI: global warming—Summary report of the 6th International Workshop on PET/MRI, March 27-29, 2017, Tübingen, Germany *Mol. Imaging Biol.* **20** 4–20

[2] Wehrl H F, Sauter A W, Judenhofer M S and Pichler B J 2010 *Technol. Cancer Res. Treat.* **9** 5–20

[3] Pichler B J *et al* 2008 *J. Nucl. Med.* **49** Suppl. 2 5s–23s

[4] Pichler B J *et al* 2008 *Semin. Nucl. Med.* **38** 199–208

[5] Kato T *et al* 2011 *Nucl. Instrum. Methods Phys. A* **638** 83–91

[6] Cabello J, Lukas M, Rota Kops E, Ribeiro A, Shah N J and Yakushev I *et al* 2016 *Eur. J. Nucl. Med. Mol. Imaging* **43** 2190–200

[7] Bezrukov I, Mantlik F, Schmidt H, Scholkopf B and Pichler B J 2013 *Semin. Nucl. Med.* **43** 45–59

[8] Vandenberghe S *et al* 2015 *Phys. Med. Biol.* **21** R115–54

[9] Barthel H, Schroeter M L, Hoffmann K T and Sabri O 2015 *Semin. Nucl. Med.* **45** 224–33

[10] Broski S M, Goenka A H, Kemp B J and Johnson G B 2018 *AJR Am. J. Roentgenol.* **211** 295–313

[11] Hope T A, Fayad Z A, Fowler K J, Holley D, Iagaru A and McMillan A B *et al* 2019 *J. Nucl. Med.* **60** 1340–6
[12] Lohmeier J, Bohner G, Siebert E, Brenner W, Hamm B and Makowski M 2019 *Sci. Rep.* **9** 14603
[13] Fürst S, Grimm R and Hong I *et al* 2015 *J. Nucl. Med.* **56** 261–9

IOP Publishing

Imaging Modalities for Biological and Preclinical Research: A Compendium, Volume 2

Parts II–IV: *In vivo* preclinical imaging, multimodality imaging and outlook

Andreas Walter, Julia G Mannheim and Carmel J Caruana

Chapter III.2.d

Fluorescence molecular tomography/CT

Fatih Alioglu, Julia Scholda, Karla-Nikita Singeorzan, Aileen Prosche, Manfred Ogris and Haider Sami

Fluorescence molecular tomography (FMT) allows investigation of spatiotemporal biodistribution of macromolecules or drug carriers labelled with fluorophores emitting in the near infrared (NIR). When combined with computed tomography (CT), precise organ allocation of the signal can be achieved. Here we describe the use of an imaging device allowing simultaneous FMT and CT measurements. As an example, we describe the quantitative measurement of the biodistribution of small interfering RNA (siRNA) covalently labelled with a NIR emitting dye (AF750) after intravenous injection into mice. A detailed protocol is provided for animal pretreatment and handling, imaging procedure, data acquisition, image analysis and data processing. In principle, this approach is applicable for all NIR labelled macromolecules and drug carriers and represents a versatile tool for convenient and cost effective pharmacokinetic and pharmacodynamic studies in preclinical research and development.

1 Introduction

Optical imaging represents a versatile technique for functional and molecular *in vivo* imaging. Bioluminescence imaging allows signal detection at a very high signal/noise ratio, but requires the expression of an enzyme (usually a luciferase enzyme) and the application of a suitable substrate. *In vivo* fluorescence imaging (FLI) is achieved by exciting appropriate fluorophores and measuring their emitted light. While the detection of fluorophores emitting in the visible range of light is limited to superficial signals, far-red and near infrared (NIR) excitable and emitting fluorophores allow the detection of deep seated signals by harnessing the ‘optical window’ in *in vivo* imaging ranging from approximately 650–900 nm. In a clinical setting (e.g. for image guided surgery), FLI allows real-time tracking of NIR-emitting probes, whereas in

doi:10.1088/978-0-7503-3747-2ch21

preclinical research different biological processes, biomarkers or the distribution of macromolecules and drug carriers can be studied. In addition to two-dimensional epifluorescence imaging, FLI can be employed for fluorescence molecular tomography (FMT) [1]. FMT enables the reconstruction of a fluorescence source within 3D volumetric space combining FLI data with surface topography of the imaging subject. This combination provides anatomical context to the molecular/functional information arising from fluorescence imaging. Longitudinal studies on expression of disease specific molecular biomarkers or disease models can be performed with FMT with a low animal burden. FMT based imaging is also very relevant in preclinical biomedical research, especially for pharmaceutical applications.

2 Principles and setups

2.1 Physical principles

Within epifluorescence *in vivo* imaging, excitation is applied from and emission measured at the same position relative to the surface of an object. Hence, fluorescence signals are mainly surface weighted, whereas deeper seated signals will be attenuated due to absorption and scattering both during excitation and emission. While light absorption by tissue is less pronounced in the NIR range, scattering within the tissue precludes a precise allocation of deep seated signals. FMT can allocate deeper seated signals and account for signal loss due to scattering and absorption enabling quantitative measurement. For FMT signal reconstruction, the signal has to be recorded sequentially with the light source illuminating the object from different positions or angles relative to the object. Fluorescence can be recorded either in reflectance or in transillumination mode, where the latter mode is advantageous for detecting deeper seated signals. With these signals, a mathematical model is developed within which scattering and absorption are defined as a function of the signal position within the object. In the case of contact free measurement, i.e. with a camera, the knowledge of surface geometry of the object is crucial to properly model the light propagation within the tissue. Such a surface can be generated by structured light illumination using laser galvanometry or by applying morphological imaging techniques, i.e. CT or magnetic resonance imaging (MRI).

2.2 Typical setups and state-of-the-art

While for NIR epifluorescence measurements there are several systems available for preclinical and also clinical imaging, only a few devices are commercially available for FMT imaging. PerkinElmer offers the FMT imager series with surface generation by structured light illumination using laser galvanometry. Transfer cassettes (to avoid movement of the animal between measurements) are available for the FMT devices to allow a successive measurement in a separate μCT device. With the FMT-XCT technology, correlative imaging of NIR and CT is enabled in one machine by multi-angle acquisition with laser excitation [2], although this system is not commercialised.

The *in vivo* FMT/CT imaging system described in this chapter is an IVIS Spectrum CT Imaging device (PerkinElmer® Inc., Waltham, MA, USA), which allows correlative imaging of FMT, bioluminescence and CT. Besides FMT, this machine can in principle also record bioluminescence in tomographic view combined with CT. For correlative FMT/CT measurement, the anaesthetised mouse is placed on a grid plate on a heated stage, and fluorescence is recorded in trans-illumination mode using a tungsten lamp with extended near infrared range as light source (figure 1). A suitable excitation filter can be chosen from a filter wheel placed above the lamp, whereas an emission filter can be similarly chosen placed in front of the camera (a highly sensitive back-thinned and thermo-electrically cooled CCD). The movable light source is consecutively placed at selected openings of the grid plate to cover the whole region within the animal intended for imaging and a set of 2D fluorescent surface radiance pictures is generated corresponding to the location of the fluorescent source [3]. Of note, the excitation light source is placed close to the animal maximising tissue penetration and minimising background signal. An X-ray computed tomography source is used to generate the surface of the imaging subject to reconstruct a three-dimensional or tomographic view. X-ray images are acquired while the imaging subject rotates on the stage between the X-ray source and the detectors. The fluorescence images along with the positional information and the surface topography obtained via CT serve as an input for reconstructing the localisation of the fluorescence signal within the three-dimensional space of the imaging subject [3]. Post FMT/CT data acquisition, imaging data are reconstructed and evaluated by the image analysis software Living Image® Software (Version 4.5.2., PerkinElmer Inc., Waltham, MA, USA).

3 Biomedical relevance

3.1 Application range and relevance

FMT/CT is applicable for a broad range of fluorophores which can be excited in the far red and near infrared. This comprises commercially available dyes, like Alexa Fluor

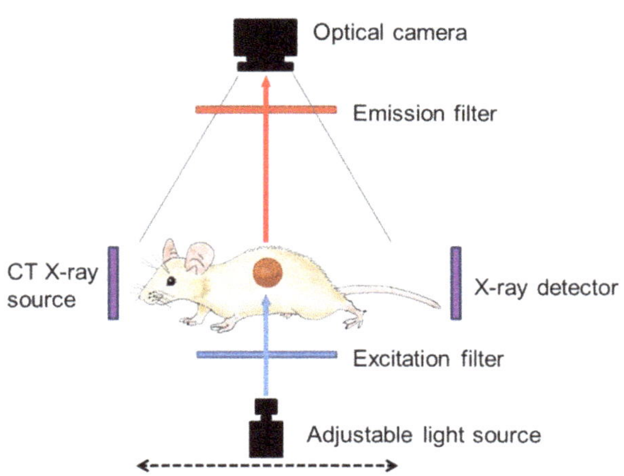

Figure 1. Principle and setup of fluorescence imaging tomography (figure adapted from ref [3] copyright 2015–2020, PerkinElmer, Inc. All rights reserved. Printed with permission.).

750 [4], but also NIR-fluorescent quantum dots or far red/NIR excitable reporter proteins [5]. After covalent attachment of fluorophores, the method is suitable to study the biodistribution of almost any macromolecule. In principle, also transgenic reporter mice expressing NIR excitable reporter proteins can be analysed and also the growth and therapy response of gene-marked tumours can be followed [5]. Signal allocation to anatomical structures is significantly improved when using CT-contrast agents.

3.2 Sample preparation

Mice C57BL/6-Tyrc albino strain (B6N-Tyrc/BrdCrCrl) were purchased from Charles River Laboratories (Sulzfeld, Germany) and housed in individually ventilated cages (Type 2L; Tecniplast, Hohenspeißenberg, Germany) under specified pathogen-free conditions with a 12 h/12 h dark/light circle. Animals received autoclaved water and a special low fluorescent food (EF AIN 76A rodent diet, Purified Diet, ssniff Spezialdiäten GmbH, D-59494 Soest, Germany) for at least three to four weeks prior to fluorescence imaging. Mice were transferred individually in a special custom-built anesthesia induction chamber (Combi-Vet® anesthesia System, Rothacher Medical, Heitenried, Switzerland) to the experimental room (figure 2(a)). Anesthesia was initiated with 5% isoflurane (Isoflurane CP 1 ml ml^{-1}, CP-Pharma, Burgdorf, Germany) in oxygen (flow rate 5 liter min^{-1}) and maintained at 2%–3% (flow rate 2 liter min^{-1}). Animals were placed on a heated small animal anesthesia table (Combi-Vet®, Rothacher Medical) and fur removed on the ventral and lateral side of the thorax and abdomen by using a hair clipper (ISIS Aesculap® GT420, B.Braun, Buchbach, Germany) (figure 2(b)). Eyes were protected by applying eye ointment (Vit.A POS® 5g; Ursapharm, Saarbrücken, Germany). Mice were injected intravenously with siRNA-AF750 diluted in buffer (sterile filtered 5% glucose (w v^{-1}), 20 mM HEPES pH 7.4) into the dilated lateral tail veins by an insulin syringe (Omnican® 100, B.Braun, Melsungen, Germany) using an in-house built restrainer (50 ml conical tube, bottom partially removed and lid with opening for mouse tail). Immediately prior to FMT/CT imaging session, CT contrast agent (120 μl Iopamidol (Scanlux, Sanochemia, Vienna; 300 mg iodine ml^{-1}) and 180 μl of 5% (w v^{-1}) glucose) was injected intraperitoneally to enhance organ contrast. Mice were transferred onto the heated imaging platform using the transillumination grid plate for FMT/CT within the IVIS Spectrum CT Imaging System, connected with in- and outlets for anesthesia. Mice were placed either in supine or prone position with the nose inserted in the cone and fixed with non-reflecting black tapes on paws and limbs. Vital parameters (e.g. breathing state) were continuously checked. Following this, the imaging chamber was closed and the imaging procedure started (figures 2(c) and 2(d)).

4 Parameters of image quality

4.1 Factors affecting sensitivity of fluorescence measurement

Several factors can affect the fluorescence signal measurement during fluorescence imaging tomography [3]. A major source of background fluorescence is chlorophyll containing ingredient within the standard rodent diet (in particular alfalfa), which

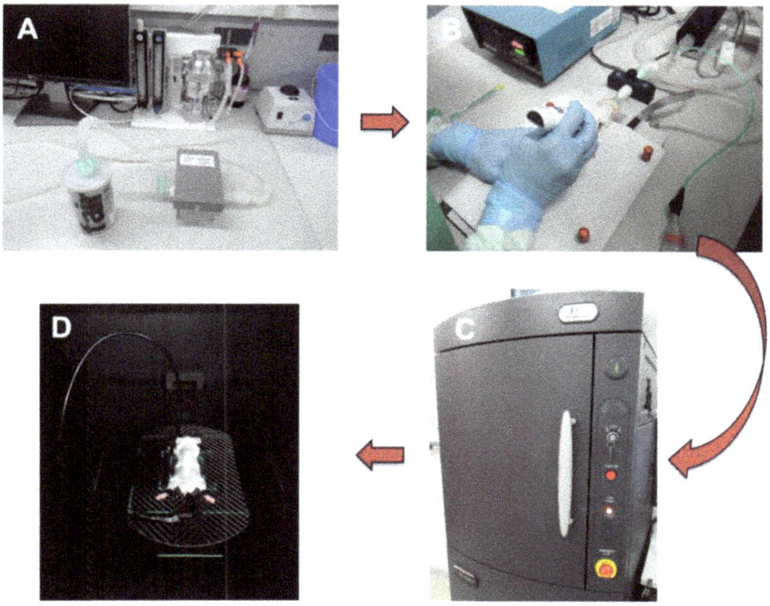

Figure 2. Preparation and imaging workflow for FMT/CT imaging of animals. For details see text in section 3.2.

shows a strong fluorescence signal in the red and far-red region (650–750 nm). To reduce this signal in the gut area, animals receive alfalfa free, low-fluorescent diet for at least three to four weeks before starting the experiment. Another important factor for consideration is skin pigmentation of the imaging subject, as pigmented mice cannot be used for FMT measurements due to the presence of light absorbing melanin in the fur and skin. Here, albino mouse strains (e.g. C57BL/6-Tyrc albino strain instead of C57/BL6) are recommended for such measurements. The presence of fur results in poor image quality due to light scattering and absorption. Either hairless mice should be used, or fur must be removed both from the abdominal and dorsal side without injuring the healthy skin. Another important factor to obtain consistent results is the positioning of the mice within the imaging device. For longitudinal studies, it is recommended to ensure that animals are always in the same position at each measurement of a time series (either prone OR supine). In the example shown here with intravenously applied siRNA-AF750, consistent results were obtained by positioning mice either in prone (ventral) or in supine (dorsal) position and by placing the animal in the center of the imaging field.

4.2 Imaging parameters ensuring desirable signal acquisition

Prior to imaging session, the Living Image software was setup for 3D fluorescence imaging tomography (FMT) as per the instructions from the operating manual. The parameters listed in table 1 were used in our example of siRNA-AF750 injected

Table 1. Overview of parameters for FMT/CT imaging of animals systemically treated with siRNA-AF750.

Field of view:	B
Fluorophore	AF750
Transillumination points:	12–14
Subject height:	1.5 cm
Exposure time:	Auto
Counts	Min. 600
	Max. 60 000
Binning	8
F-Stop:	2

animals. These parameters have to be optimised depending on the experimental setup and fluorophore used.

First, a 2D-epifluorescence image and a reflected light image of the mouse were recorded to know where to set the transillumination points. Transillumination points are projected onto the reflected light image by the software and have to be marked individually as per the instructions from the operating manual. It is recommended to select additional points surrounding the field intended for scanning to achieve accurate results. Transillumination pictures are recorded using the normalised transmission fluorescence (NTF) method. NTF subtracts nonspecific light leakage at areas where tissue is thinner by using an additional picture recorded with a neutral density filter. This method allows imaging of deep seated signals with fewer artefacts. After the scanning procedure (CT and transillumination imaging), mice were transferred to a cage placed on a heated table to recover from anesthesia. Ensure that individual treatment groups and control groups are kept in separate cages, as fluorophores can otherwise be transferred from one animal to another due to ingestion of faeces, leading to fluorescence signals also in otherwise untreated animals. At the end of the study, mice were euthanised and organs explanted and placed on a non-fluorescent, black plastic plate to perform *ex vivo* imaging.

4.3 Absolute signal quantification, fluorescence database

FMT/CT offers the option of absolute fluorescence signal quantification within imaging subjects. For this, a fluorescence database was generated by 2D-FLI imaging of a multi-well plate with known amounts of fluorophore per well (here, 0.6–600 pmol siRNA-AF750 were used [4]) by using the same excitation/emission filter pair as used for FMT. FLI signal from the multi-well plate was quantified by using the 'well plate quantification tool' provided by the software. This database was then employed in FMT/CT reconstruction to calculate the picomoles of siRNA-AF750 present in respective areas defined by 3D-ROIs (regions of interest).

5 Data processing and visualisation

The 3D signal reconstruction and quantification is performed with the Living Image® Software (Version 4.5.2., PerkinElmer, Waltham, MA, USA) as per the software manual and relevant technical notes [3]. In brief, the post-acquisition image sequence (containing both FLI and CT data) is loaded and CT reconstruction is performed automatically (figure 3). Prior to further analysis, individual FLI images have to be checked to have fluorescence signal in the recommended range, i.e. 600–60 000 counts (figure 3(a)). Images above or below this range are noted and can be excluded from analysis later on. Units are then switched from counts to calibrated units (NTF efficiency). The 3D reconstruction is initiated after selecting the standard curve for siRNA-AF750 (see section 4.3) to estimate fluorophore-labelled siRNA

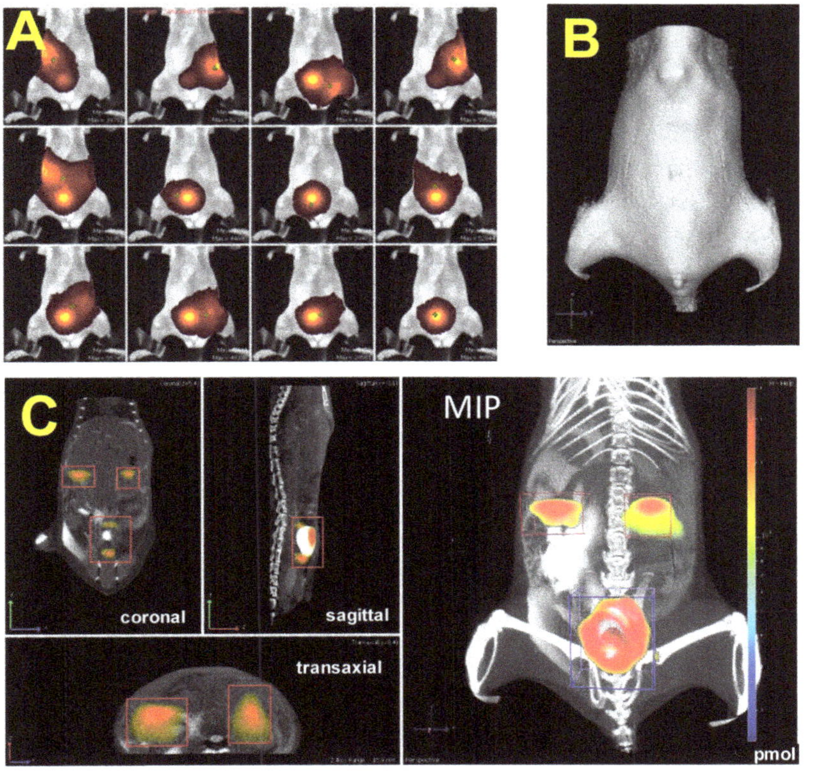

Figure 3. FMT/CT signal reconstruction and visualisation. A C57BL/6 albino mouse received siRNA-AF750 intravenously followed by Iopamidol intraperitoneally and was imaged by FMT/CT as described in the text. (a) 2D-fluorescence pictures with fluorescence recorded in transillumination (colour coded) overlaid by a reflected light picture. The green cross-hair depicts the position of the transilluminating light source. (b) Mouse surface topography reconstructed from CT data. (c) FMT/CT fusion showing indicated planes (left) and maximum intensity projection (MIP, right); fluorescence signal colour coded (in pmol units), CT in B/W, cuboid ROIs shown in all planes and MIP picture.

molecules per source giving both the localisation and quantification of the signal. After deselecting individual FLI images with a signal below 600 or above 60 000 units (see above), all remaining images are used for the reconstruction. This automated process can take up to several minutes depending on the data quantity and the computing power. From the CT data, a topographic map of the surface is generated (figure 3(b)) The readily reconstructed FMT data fused with CT is then presented in coronal, sagittal and transaxial view and as maximum intensity projection (MIP) with colour coded fluorescence signal in pmol units (figure 3(c)). To validate the accuracy of the reconstruction, measured and simulated light diffusion patterns (NTF efficiency maps) are compared and, when the difference is near zero for all data sets, the reconstruction results are accepted and saved. Further on, for visualisation purposes, threshold and voxel-size can be adjusted to discriminate between positive areas and background signal.

For quantification, 3D volumetric cuboid regions of interest (ROI) are placed covering the signal of each tissue or organ to be analysed (figure 3(c)). These 3D ROIs can be adjusted and positioned using ROI transform tools to visualise the signal from different perspectives and to adapt the 3D size and positioning of the ROI. This is repeatedly done by scrolling through different slices in all orientations (coronal, sagittal and transaxial) to get the whole signal area covered. Edited images can be exported as tiff files separately. This analysis is conducted for all time points and imaging positions. With the 3D ROIs correctly positioned and sized, the values are measured in 'total pmol' for each ROI and exported to an excel file for further analysis (figure 4(a)). The FMT/CT results can then be qualitatively validated by 2D epifluorescence FLI based organ imaging (figure 4(b)).

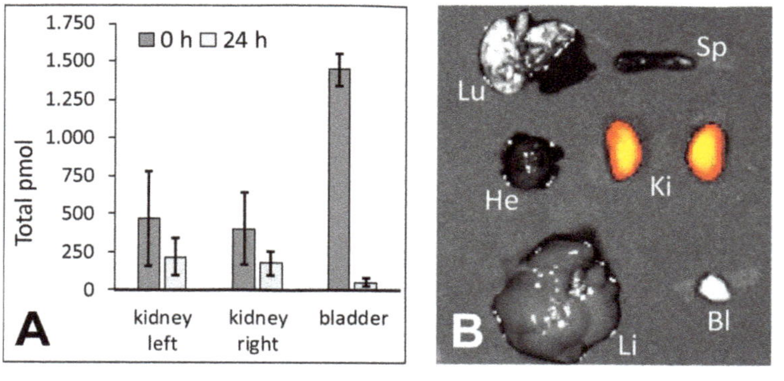

Figure 4. FMT signal quantification and *ex vivo* validation. C57BL/6 albino mice received siRNA-AF750 intravenously and were then imaged immediately after injection (0 h) and 24 h thereafter. Iopamidol was given intraperitoneally prior to every FMT/CT measurement. FMT/CT imaging was conducted 0 h and 24 h after siRNA-AF750 injection in supine position; (a) AF750 content in pmol measured in ROIs placed as in figure 3(c) in indicated organs immediately after injection (0 h) and 24 h thereafter (n = 3, +/– stddev). (b) 2D epifluorescence FLI organ imaging; Lu- lung, Sp- spleen, He- heart, Ki- kidney, Li- liver and Bl- bladder.

6 Conclusions

6.1 Strength and limitations

FMT/CT represents a convenient method for correlative imaging of fluorescence combined with morphological imaging. Major advantages are the ease of use and relatively low purchasing and running costs for the equipment. Also, safety requirements are low (no use of radionuclides, a fully contained X-ray source). The resolution (in the range of 1 mm for FMT) is sufficient for most applications and comparable to SPECT or PET. Although the CT is of medium resolution, soft tissue contrast can be significantly improved when using appropriate contrast agents. Sensitivity is dependent on the fluorophore used. With Alexa Fluor 750, for example, we obtained consistent results with amounts as low as 120 pmol dye/mouse [4], which is at least as sensitive as PET. Also, a rather broad range of fluorophores can be used with emissions ranging from far-red to near infrared. Notably, the absolute FMT signal quantification is a valuable method for pre-clinical pharmaceutical research. There are, however, some limitations. These include the occurrence of breathing artefacts, interference from food-related fluorescence in red/far-red wavelengths, and in the case of quantification the limitation to cuboid ROIs without the possibility for rendering. Thus, validation of FMT/CT results by organ imaging or other methods is recommended. Concerning hardware, laser excitation instead of using a tungsten lamp can improve the signal/noise ratio (e.g. FMT-imager series from PerkinElmer, see section 2.2)

6.2 Future developments

With appropriate software, organ segmentation is also feasible for FMT/CT [6]. Instead of CT, FMT can also be combined with MRI [7] or ultrasound [8]. In the example described here, FMT data are generated based on the surface topology assuming a homogeneous body, which limits the FMT resolution due to optical scattering in deep tissue depending on the tissue density and morphology. When harnessing the anatomical information obtained by CT or MRI, this can significantly improve accuracy and resolution by applying the so called kernel method [9].

References and further reading

[1] Stuker F, Ripoll J and Rudin M 2011 *Pharmaceutics* **3** 229–74

[2] Ale A, Ermolayev V, Herzog E, Cohrs C, de Angelis M H and Ntziachristos V 2012 *Nat. Methods* **9** 615–20

[3] Tseng J C, Vasquez K and Peterson J D 2015 Optical imaging on the IVIS SpectrumCT system: general and technical considerations for 2D and 3D imaging *Technical Note Pre-clinical In vivo Imaging* https://perkinelmer.com/lab-solutions/resources/docs/TCH_012007_01_IVIS-2D_3D_Imaging.pdf)

[4] Geyer A, Lorenzer C, Gehrig S, Simlinger M, Winkler J, Sami H and Ogris M 2017 *Int. J. Pharm.* **525** 359–66

[5] Geyer A, Taschauer A, Alioglu F, Anton M, Maier J, Drothler E, Simlinger M, Yavuz S, Sami H and Ogris M 2017 *Hum. Gene Ther.* **28** 1202–13

[6] Kunjachan S *et al* 2013 *ACS Nano* **7** 252–62

[7] Davis S C, Samkoe K S, Tichauer K M, Sexton K J, Gunn J R, Deharvengt S J, Hasan T and Pogue B W 2013 *Proc. Natl. Acad. Sci. USA* **110** 9025–30

[8] Su S P and Chiang H H K 2020 3D fluorescence tomography combined with ultrasound imaging system in small animal study *Future Trends in Biomedical and Health Informatics and Cybersecurity in Medical Devices. ICBHI 2019* ed K P Lin, R Magjarevic and P de Carvalho P (IFMBE Proceedings vol 74) (Cham: Springer) pp 166–73

[9] Baikejiang R, Zhao Y, Fite B Z, Ferrara K W and Li C 2017 *J. Biomed. Opt.* **22** 55001

IOP Publishing

Imaging Modalities for Biological and Preclinical Research: A Compendium, Volume 2
Parts II–IV: *In vivo* preclinical imaging, multimodality imaging and outlook
Andreas Walter, Julia G Mannheim and Carmel J Caruana

Chapter III.2.e

PET/CT/ultrasound

Bertrand Tavitian and Mailyn Pérez-Liva

We present a new hybrid *in vivo* imaging instrument combining positron emission tomography (PET) with computed X-ray tomography (CT) and ultrafast-ultrasound (US)-imaging (UUI). We describe the main characteristics of this instrument called PETRUS for PET-registered-ultrasound. It is presently the only instrument capable of acquiring fully simultaneous and co-registered image volumes of ultrasound and PET for preclinical research.

1 Introduction

Recent years have seen a blooming of preclinical and clinical *hybrid* instruments for *in vivo* imaging. The term *hybrid* designates the combination of two imaging modalities in the same machine, meaning that the image datasets (or image volumes) produced by each instrument are acquired in a synchronous manner, both in space (i.e. the two image volumes detected by the two fields of views can be superimposed) and, possibly, in time (i.e. acquisitions are simultaneous). Temporal simultaneity is optional, as image volumes may be acquired sequentially by moving the subject placed on an imaging bed from one gantry to the other and matching *a posteriori* the fields-of-view of both image volumes. However, with temporal simultaneity the fields-of-view of both instruments are spatially aligned during the acquisition process, ensuring that image volumes are acquired in strictly identical conditions, independently of the fluctuation of physiological parameters over time. Temporal simultaneity also greatly helps managing physiological movements, e.g. heart beats, respiration, etc. Naturally, simultaneous hybrid imaging comes with increased cost and complexity, complicates the scaling of the image volumes and often degrades image quality of one or both modalities. Technologically-wise, it is also more challenging, since the two types of energy sources used to create contrast in the body may interfere during the acquisition process and thus complicate the buildup of the

doi:10.1088/978-0-7503-3747-2ch22

Table 1. Hybrid imaging modalities.

Acronym	Clinical (c), preclinical (p)	Simultaneity (yes or no)	Electromagnetic wave	Pressure wave
PET-CT	c, p	No	Yes	No
PET-MRI	c, p	Yes, No	Yes	No
SPECT-CT	c, p	No	Yes	No
SPECT-MRI	c, p([a])	Yes, No	Yes	No
PET-US	p	Yes	Yes	Yes

[a] Under development

detecting systems. Essentially, the types of energy used to create a contrast in the body are either electromagnetic waves, that vary over a wide range of energies, or pressure waves (i.e. acoustic) waves. Table 1 shows the main preclinical and clinical hybrid imaging instruments: PET-CT [1], single photon computerised tomography with CT (SPECT-CT) [2] PET or SPECT with magnetic resonance imaging (MRI) [3] and PETRUS [4], which is the subject of this chapter.

Here we report PETRUS assembly, physical characteristics and a couple of *in vivo* applications in rodents [4–7].

2 Principles and setups of PETRUS

2.1 Physical principles

PETRUS is a hybrid imaging modality using two different types of energy waves in energy ranges for which there are barely any interactions at the macroscopic or mesoscopic scales. Moreover, PET is a molecular imaging technique (see chapter II.10 in this book), to which CT adds anatomical information, while ultrasound imaging can deliver anatomical, functional and molecular information in many different modes (see chapter II.6 in this book).

2.2 Typical setup

PETRUS (figure 1) is assembled from a compact PET-CT tomograph for small animals[1] and a clinical ultrasound scanner operated in research mode[2]. The ultrasound probe connected to the scanner is attached to the bed of the PET scanner by a robotic six-degree-of-freedom high-precision motorised micropositioner[3] (figure 1). Three critical points for acquiring ultrasound images inside a PET gantry will be described in the next sections:

1. replace hand manipulation by a robotised registered remote control displacement of the probe inside the PET gantry;

[1] nanoScan PET-CT, Mediso Ltd, Hungary.
[2] Aixplorer, Supersonic Imagine; Aix-en-Provence, France.
[3] the Hexapod H811 from Physik Instrumente, Germany.

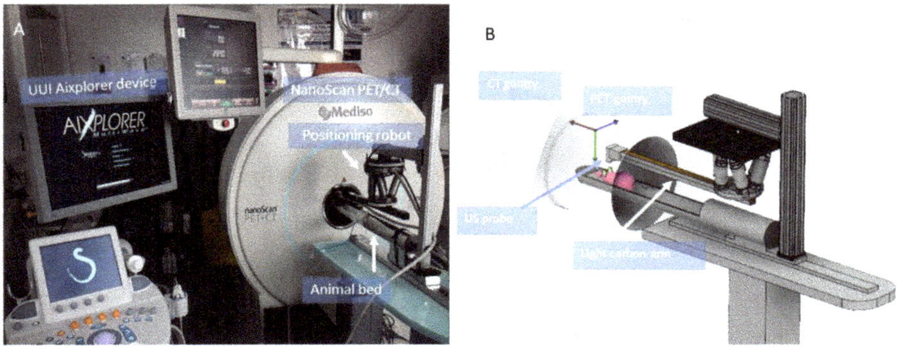

Figure 1. Description of the PETRUS imaging instrument. (a) Photograph of PETRUS setup. (b) Schematic representation showing the positions of the ultrasound probe and the animal inside the PET/CT gantry. Figure modified from [5], copyright IOP Publishing. Reproduced with permission. All rights reserved.

2. co-register precisely the ultrasound planes of acquisition within the PET acquisition volume and maintain co-registration accuracy during probe movements;
3. estimate the attenuation and/or distortion of the PET data induced by the presence of a probe in the PET-CT field-of-view.

Any type of ultrasound probe can be used provided that enough room is available inside the PET gantry to position the probe. Small miniaturised probes, whether custom-made or small-size commercial probes for rheumatology or pediatrics, are ideally suited. Manipulation of the ultrasound probe is realised by robotised registered remote control displacement inside the PET gantry. The ultrasound probe is attached to the motorised positioner by custom 3D-printed adapters, allowing the probe to be installed at an initial position located anywhere within the PET gantry. From its initial position, the motion range of the transducer inside the PET gantry covered approximately 36 mm × 40 mm × 40 mm, with minimum incremental motion of 0.2 μm.

3 Biomedical relevance

We used the *ultrafast ultrasound imaging* (UUI) mode, which is exquisitely well adapted to live imaging of fast moving rodent organs and performs excellently in terms of spatial resolution. The unique capability of UUI for acquiring images at thousands of frames per second in a large field of view allows one to study rapidly evolving phenomena or to compound a large amount of data to achieve very high contrast-to-noise ratios. Applications bridge both clinical and preclinical domains, and include the imaging of brain activation and the quantification of stiffness for cancer staging [8]. UUI has recently been generalised to three dimensions, allowing for the acquisition of thousands of imaging volumes per second. Ultrafast Doppler imaging provides images of the vasculature and blood dynamics with unprecedented resolution and sensitivity [8].

Coming to the rationale for *simultaneous* detection of several hallmarks by imaging, numerous fundamental studies have demonstrated the relationship between tissue vascularisation and metabolism. Exquisite cellular mechanisms maintain tissue homeostasis in an ever-changing environment in which local oxygen levels vary continuously. Vessels bring to tissues the substrates necessary to sustain metabolic needs, and tissue metabolism controls the expression of vascular youth factors.

3.1 Oncology

As cancer is a disease with many hallmarks [9], monitoring cancer during tumour growth and therapy implies monitoring multiple hallmarks. This trivial statement is in favor of imaging several cancer hallmarks simultaneously. Among the major cancer hallmarks [9], three are routinely explored by *in vivo* imaging: proliferation, dysregulation of glucose metabolism, and angiogenesis. Proliferation is reflected by tumour size measured by CT or MRI. Dysregulation of glucose metabolism, called aerobic glycolysis and described by Otto Warburg in the 1920s [10], is the rationale for tumour detection and staging based on PET imaging with FDG. The development of an abnormal vasculature by neo-angiogenesis/neo-vasculogenesis, demonstrated by Judah Folkman in the 1970s [11], may be explored by imaging tumoural blood supply *in vivo* using dynamic contrast-enhanced (DCE) MRI and CT.

In a series of preclinical studies we aimed to quantify in three dimensions the interplay between glucose metabolism and tumour vessels' architecture and function using PETRUS. We summarise here recent results obtained with this technology using genetically-modified tumours cell lines or tumour grafts expressing specific metabolic pathways implanted in mice, in longitudinal observational and treatment follow-up studies.

3.1.1 Bivariate analysis for tumour phenotyping

In a reverse-genetic approach, mutants of the CCL39 tumour cell line (noted wt) were genetically engineered to abolish either glycolysis (gly- cells), or mitochondrial respiration (res- cells) [12]. Mitochondrial respiration is a global name for the metabolic processes occurring in the mitochondria and requiring oxygen to convert the energy stored in nutrients into adenosine triphosphate (ATP). All else being equal in their genetic background, the three cell lines follow radically different energy metabolic pathways: exclusively anaerobic metabolism (res-); exclusively aerobic metabolism (gly-); and both aerobic/anaerobic metabolism (wt). The metabolic routes are maintained in tumours obtained by subcutaneous implantation of the cell lines.

PETRUS was used for non-invasive measurements of distinct tumour hallmarks in the same spatial coordinate system. Each individual voxel of the combined PET-CT-UUI dataset was defined by its spatial coordinates and by two quantitative parameters, one derived from PET and one from UUI. For instance, (i) standard uptake value (SUV) of FDG, representing the deregulation of energy metabolism and (ii) Doppler intensity, corresponding to functional perfused blood vessels. These values were then plotted to obtain bivariate histograms of the voxels' distributions along a metabolic axis (FDG: standard uptake values) and a vascular axis (Doppler signal intensity) over the 3D volume of each tumour (figure 2). PET-CT-UUI unambiguously separated tumour types according to their metabolic and vascular profiles: compared to the wt profile, purely glycolytic res- tumours exhibited a high FDG uptake with a low vessel density while, on the opposite, gly- tumours presented low FDG uptake and high vessel density (figure 2).

It is important at this stage to highlight the fact that the stratification into three classes was performed on a voxel-by-voxel basis. In other words, each point of the bivariate histograms correspond to the same $100 \times 100 \times 100$ m^3 voxel inside the tumour. Section 3.1.2 will show the advantage of a voxel-wise multiparametric exploration of tumours.

3.1.2 Multiparametric quantification and definition of early imaging biomarkers of resistance to sunitinib treatment

We followed mice bearing SDHB-/- tumours longitudinally using weekly PETRUS imaging [6]. In human patients, the SDHB mutation leads to loss of function of the enzyme succinate dehydrogenase (SDH) and is a bad prognostic factor for tumours of the parasympathetic and sympathetic systems, called paraganglioma [13]. The animals were separated in two groups, one receiving the anti-angiogenic drug sunitinib and the other receiving the vehicle. Over 20 quantitative parameters were derived from the serial longitudinal PET-CT-UUI images, including tumour volume derived from CT, metabolic parameters, e.g. mean and max standard uptake values, (SUV), mean rate of glucose utilisation (MRGlu), total lesion glycolysis (TLG), etc,

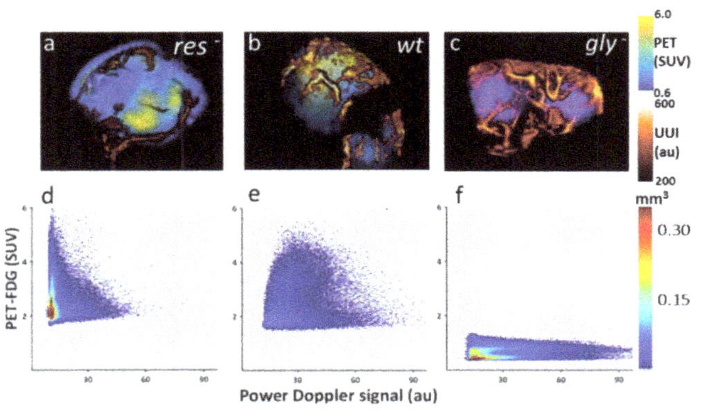

Figure 2. Tumour phenotyping based on bivariate FDG-PET/ultrafast-power-Doppler histograms in mice. (a)–(c) Fused images of FDG uptake and perfused vessels in tumours grown subcutaneously in nude mice with three variants of the Ras-transformed CCL39 cell line; left to right, respiration-deficient mutant (res-; a), wild type CCL39 (wt; b), glycolysis-deficient mutant (gly-; c). (d)–(f) Bivariate histograms corresponding to panels (a)–(c) showing the pixel-wise relationship between SUV and power Doppler signal (proportional to glucose metabolism and blood volume, respectively). The heat scale corresponds to the relative volume of tumour (equivalent to the number of voxels) characterised by a singular pair of PET-UUI values. Representative cases from triplicates for each mutant. Reproduced with permission of Springer from [4], copyright 2018.

derived from PET, and vascular parameters, e.g. vessel volume, total vessel length, number of nodes, tortuosity, etc derived from UUI. Searching for an early response biomarker capable of predicting resistance to sunitinib, we calculated the correlations between the changes of all imaged-derived parameters from pre-treatment to end of first week of treatment, and their values at end of treatment at week 6 (figure 3). A heatmap plot was constructed using the square of Pearson correlations (R2) between PET-CT-UUI parameters calculated by MatLab^{TM} software. Considering the small size of the sample (n = 12 mice), we fixed the significance level at $p = 0.01$ (R2 0.61).

Regarding the metabolic parameters, changes of mean SUV and maximum SUV after one week of treatment were poorly correlated to tumour size at week 6 ($R = 0.37$ and 0.34, respectively). In contrast, early changes at week 1 in total lesion glycolysis (TLG), a parameter that combines tumour volume with mean FDG uptake, were correlated with tumour size at week 6. Regarding the vascular parameters, the reduction of maximum vessel length at week 1 was correlated with the final tumour size at week 6 ($R = 0.78$), indicating that the larger the inhibitory effect of sunitinib on vessel length, the smaller the tumour volume at week 6.

In summary, the decrease in the maximum vessel length and total lesion glycolysis, measured after one week of sunitinib, were correlated with the endpoint tumour volume. The capacity of these parameters to predict tumour escape from

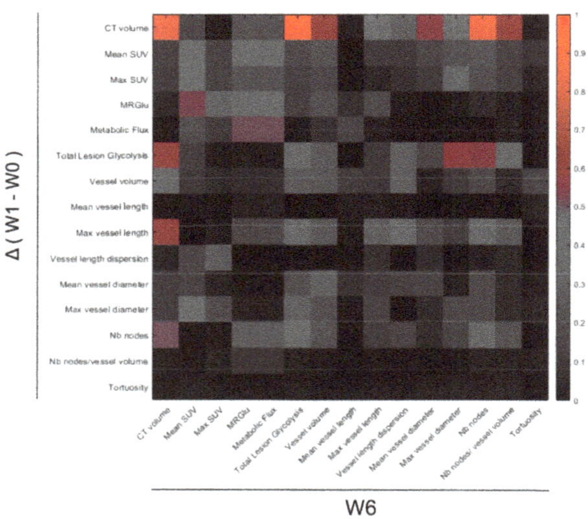

Figure 3. Early changes in PET-CT-UUI parameters may predict outcome of anti-angiogenic treatment. Heatmap plotting the square of the Pearson correlation coefficient values (R2). Parameters variations between W0 and W1 ((W 1 W 0)) plotted on the y-axis against the parameter values at W6 on the x-axis in sunitinib-treated mice: CT volume, mean and max SUV, MRGlu, metabolic flux, total lsion glycolysis, vessel volume, mean and max vessel length, vessel length dispersion, mean and max vessel diameter, number (Nb) of nodes. Given the small size of the sample (n = 12 mice), the significance level was fixed at $p = 0.01$ (R2 0.61, red coloured). Reproduced from [6] with permission from *Theranostics* (CC BY 4.0).

antiangiogenic treatment remains to be demonstrated in future studies in other tumour models, as well as in clinical settings.

3.2 Cardiology

Imaging a fast-moving organ like the heart calls for the best possible temporal and spatial resolutions. Assuming that the human heart at rest beats at 60 beats per minute (60 bpm) with a maximal left ventricular wall velocity of 10 cm s^{-1} [14], imaging the wall position with a spatial accuracy of 0.1 mm requires one acquisition per millisecond (1 kHz or 1000 fps). Similar calculations for preclinical studies of the mouse heart (600 bpm, 2–3 cm s^{-1}) [15] supposed to double the rate of acquisition to 2 kHz or 2000 fps. Conventional ultrasound is far from these values, but UUI can easily reach these rates of acquisition. In contrast, high temporal resolution PET imaging of the heart remains a technical challenge because of spatial and temporal resolution issues: over 1 mm and seconds to minutes for preclinical PET scanners, respectively [16]. Therefore, each cardiac PET image accumulates several heart cycles, leading to significant blurring, motion-induced artefacts, and imprecise radioactivity quantification. Gating raw PET data along the cardiac cycles synchronously with an electrocardiogram (ECG) compensates for cardiac motion (gated cardiac-PET [17]), but comes at the cost of lower counting statistics and contrast.

We assessed an ultrasound-based super-resolution (SR) technique in rodent hearts using PETRUS. We employed UUI-B-mode cine images to record with high-temporal framing (500 fps) the spatio-temporal heart deformation during a cardiac cycle (Mi(x)), where x is the position in the image domain and i defines the temporal framing. The SR method was applied to real datasets acquired with the PETRUS system from two successive imaging sessions of a 10-week-old female Wistar rat: (i) in normal conditions (baseline) and (ii) 4 h after surgically-induced myocardial infarction through a permanent ligation of the left anterior descending coronary artery (LADCA).

Figure 4 shows images of the end-diastole cardiac phase of a rat heart 30 min after FDG injection, obtained with gated-PET, motion corrected (MoCo) using the motion estimated from the UUI-B-mode, ultrasound-based SR and static-PET images, before and one week after ligature of the left descending coronary artery (LDCA). It shows clearly that the ultrasound-based SR images present less spillover out of the myocardium and identify metabolic remodelling of the ischemic myocardium corresponding to the vascular territory of the ligated LDCA. Interestingly, ultrasound-based SR shows a small but significantly higher than zero FDG uptake in the ischemic region, which is not seen using the other image reconstruction and correction algorithms. Interestingly, ultrasound-based SR corrections yielded an excellent sub-millimetric average spatial resolution for PET of 0.88 mm.

Overall, a proof-of-concept was obtained that ultrasound-based SR techniques based on co-registered, simultaneously acquired PET and ultrasonic data drastically enhance the quality of metabolic images of the rodent heart in terms of spatial resolution, SNR and contrast. In the infarcted rat heart, ultrasound-based SR

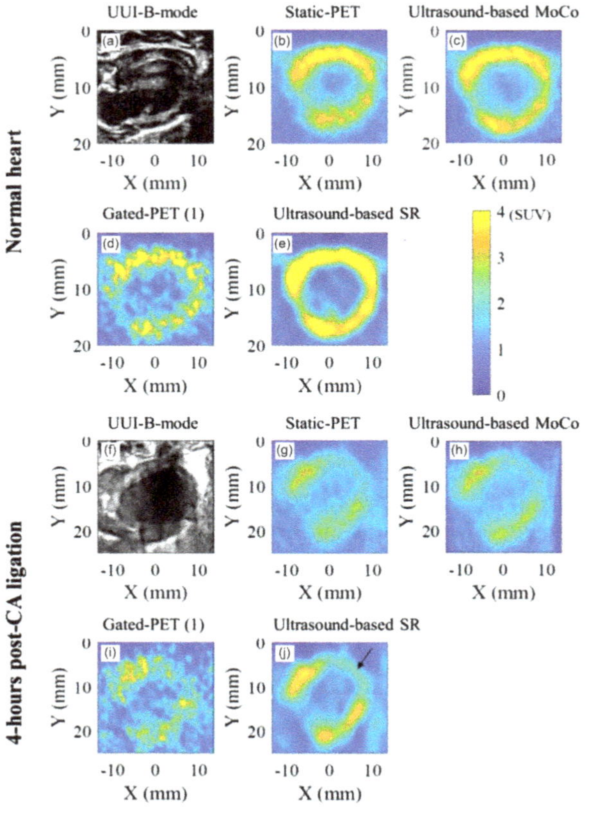

Figure 4. Multimodal PETRUS images for cardiac applications. Images represent a rat heart before (a)–(e) and 4 h after left-descending coronary artery ligation (f)–(j). (a) and (f), UUI-B-mode image in long-axis orientation and end-diastole phase. (b) and (g), static-PET images. (c) and (h), ultrasound-based MoCo images, (d) and (i), gated-PET end diastole frame, (e) and (j) ultrasound-based SR images. The black arrow in (f) points towards the ischemic area. Figure reproduced with permission of Springer from [7], reprinted under a Creative Commons License (CC BY 4.0).

improved the delineation of the metabolic defect due to the post-ischemic lesion. Therefore, reconstructing PET data with UUI-based SR brings a significant improvement in the quality of cardiac metabolic images in pathological situations characterised by cardiac glucose dysregulation.

3.3 Sample preparation

PETRUS enables *in vivo* measurements of live animals. The preparation of the animals is identical as for standalone US or PET systems.

4 Parameters of image quality

Ultrasound probes are made of dense materials that attenuate gamma-rays, which may cause artefacts on PET images. Using small probes limits attenuation. In a

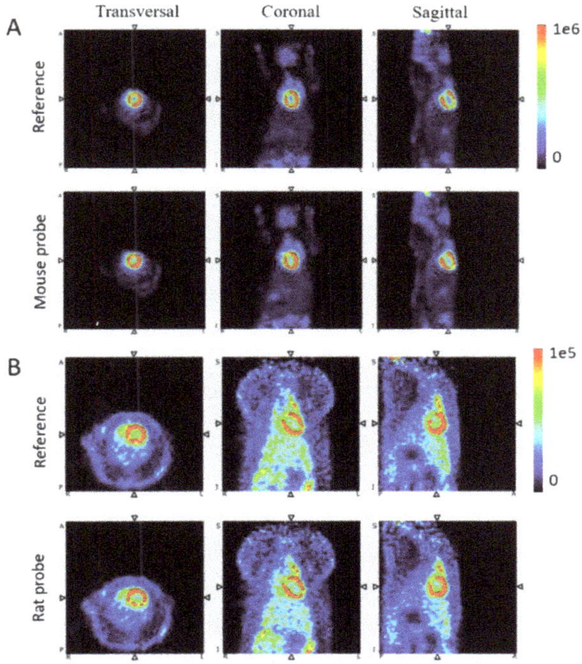

Figure 5. FDG uptake in rodents with PETRUS. Images of a mouse: (a) without probe (reference) and (b) with the probe over the heart of the animal. Images of a rat (c) without probe (reference configuration) and (d) with the rat probe over the heart of the animal. All images were reconstructed using 5 min acquisitions obtained after FDG injection at 83 and 90 min for the configurations with probe and without probe, respectively. Reproduced with permission from [5]. Copyright IOP Publishing. All rights reserved.

comparative study of the physical properties of the PET-CT with and without the ultrasound probe [5], we showed that overall, the presence of the probe inside the FOV PET/CT had a minor effect on the radioactivity concentration measurements in PET images, and did not degrade significantly the quantitative and qualitative PET data derived from the images. Typically, maximal discrepancies induced by the presence of the probe in the FOV were under 10% with a rat probe and under 7% with a mouse probe. *In vivo* dynamic studies in rodents shown in figure 5 demonstrate similar image quality with and without probe in the heart. Quantification of FDG uptake obtained by calculating time-activity curves (TAC) in volumes of interest covering the heart of the animals showed that, at any time point during cardiac FDG uptake, the TACs with and without probe were essentially superimposable in both species. The maximal discrepancies between the TAC with and without probe in mouse and rat were under 2% and 10%, respectively.

Detailed evaluation on the image quality can be found in the literature [5].

5 Data processing

The important point is to process PET, CT and ultrasound in the same spatial references, using affine transformations to link the coordinates systems of the three imaging modalities (PET, CT and ultrasound). Built-in PET-CT systems intrinsically co-register PET and CT volumes in a single coordinate system, with a precision better than 0.05 mm [4]. The hexapod robot controls the movement of the probe with sub-micron accuracy by a set of six variables referring to the positions of the six legs of the motorised micropositioner (figure 1). The driver of the controller converts the position of the six legs into translation (X,Y,Z) and rotational (V,U,W) motions with respect to the center of rotation representing the origin of the UUI image plane, O°, the center of the lens of the ultrasound probe. Once the imaging volumes are obtained by ultrasound and CT, rigid registrations and trigonometric calculations are used to determine the co-registration parameters. The precision of the co-registration process was estimated as 0.095 ± 0.028 mm [4, 5].

6 Conclusion

6.1 Strength and limitations

PETRUS can perform extensive analysis of quantitative parameters of tumours with different metabolic backgrounds and under various treatments. An application would be to predict the sensitivity of a given tumour type to a targeted treatment. Moreover, as it describes simultaneously the time course of several hallmarks of the response to treatment, it may help defining new treatment associations, such as addition of a second drug targeting specifically the mechanism of escape from the first drug.

PETRUS can benefit from the wide diversity of functional ultrasound, such as cardiac contraction parameters, changes in blood flow and in the elasticity of tissues. The analysis of those parameters co-registered in space and time with high-resolution metabolic maps may help to better understand the dynamics of evolution of ischemic processes, on from an early stage of the disease. PETRUS could also provide precise criteria to guide decision-taking processes of reperfusion therapies. In general, the multi-parametric and longitudinal vision of PETRUS may be an excellent tool for precision medicine, helping to elucidate and interpret different physiological and pathological processes with a much deeper and comprehensive knowledge of them.

The main limitation of PETRUS is related to the two-dimensional character of the ultrasonic images used, which restricts the complete exploration of phenomena taking place three-dimensionally. Other limitations include the difference in spatial and temporal resolutions between the modalities involved in PETRUS, and the relatively small field of view of ultrasounds compared to PET or CT.

To put it in a nutshell, ultrasound is progressing at a fast pace and invading evidence-based medicine. Combined with PET in PETRUS, it extends the molecular imaging field to new avenues of hybrid imaging.

6.2 Future developments

Clinical PETRUS may become a reality in the near future. Given the availability of the PETRUS components in the clinical field, its extension to human explorations seems natural. A future version of this device might also include collaborative robotisation and machine learning algorithms to perform automatic lesions detections and optimal regions exploration.

Acknowledgements

The authors express their deepest gratitude to those without whom this research would have not been possible. At the very forefront, Professor Mickael Tanter, a splendid scientist and a good friend, who shared his brilliant ideas with us and was a lighthouse for our studies. Franck Le Thimonnier who took it upon himself to fund a research line that was raising much skepticism. Anikitos Garofalakis, Jean Provost, Daniel Balvay and Thomas Viel who were instrumental in building the prototype. Our dear colleagues Judith and Charlotte Lepoutre who provided material and support; Joevin Sourdon, Béatrice Berthon, Caterina Facchin and Thulaciga Yoganathan, who dealt with the teething problems, and our graduate students Mariana Aramburo-Garcia, Alicia Arevalo-Garcia, Lingxi Bao, Geert Leenders, Jun Ma, Thi-Ngoc-Phuong ('Fong') Nguyen, François Peyre-Pradat, Konstantina Tzavella, Prashanth Vaddeboina, Xumeng Zhang. Very special thanks to Anaïs Certain for her enthusiasm, dedication and exceptional craftsmanship.

References and further reading

[1] Beyer T, Townsend D W, Brun T, Kinahan P E, Charron M, Roddy R, Jerin J, Young J, Byars L and Nutt R *et al* 2000 A combined pet/ct scanner for clinical oncology *J. Nucl. Med.* **41** 1369–79

[2] Hasegawa B H, Wong K H, Iwata K, Barber W C, Hwang A B, Sakdinawat A E, Ramaswamy M, Price D C and Hawkins R A 2002 Dual-modality imaging of cancer with spect/ct *Technol. Cancer Res. Treat.* **1** 449–58

[3] Schmand M, Burbar Z, Corbeil J, Zhang N, Michael C, Byars L, Eriksson L, Grazioso R, Martin M and Moor A *et al* 2007 Brainpet: first human tomograph for simultaneous (functional) PET and MR imaging *J. Nucl. Med.* **48** 45P

[4] Provost J, Garofalakis A, Sourdon J, Bouda D, Berthon B, Viel T, Perez-Liva M, Lussey-Lepoutre C, Favier J and Correia M *et al* 2018 Simultaneous positron emission tomography and ultrafast ultrasound for hybrid molecular, anatomical and functional imaging *Nat. Biomed. Eng.* **2** 85–94

[5] Perez-Liva M, Viel T, Yoganathan T, Garofalakis A, Sourdon J, Facchin C, Tanter M, Provost J and Tavitian B 2018 Performance evaluation of the pet component of a hybrid pet/ct-ultrafast ultrasound imaging instrument *Phys. Med. Biol.* **63** 19NT01

[6] Facchin C, Perez-Liva M, Garofalakis A, Viel T, Certain A, Balvay D, Yoganathan T, Woszczyk J, De Sousa K and Sourdon J *et al* 2020 Concurrent imaging of vascularization and metabolism in a mouse model of paraganglioma under anti-angiogenic treatment *Theranostics* **10** 3518

[7] Perez-Liva M, Yoganathan T, Herraiz J L, Porée J, Tanter M, Balvay T, Viel D, Garofalakis A, Provost J and Tavitian B 2020 Ultrafast ultrasound imaging for super-resolution preclinical cardiac pet *Mol. Imag. Biol.* **22** 1342–52

[8] Tanter M, Bercoff J, Athanasiou A, Deffieux T, Gennisson J-L, Montaldo G, Muller M, Tardivon A and Fink M 2008 Quantitative assessment of breast lesion viscoelasticity: initial clinical results using supersonic shear imaging *Ultrasound Med. Biol.* **34** 1373–86

[9] Hanahan D and Weinberg R A 2011 Hallmarks of cancer: the next generation *Cell* **144** 646–74

[10] Warburg O, Wind F and Negelein E 1927 The metabolism of tumors in the body *J. Gen. Physiol.* **8** 519

[11] Folkman J, Merler E, Abernathy C and Williams G 1971 Isolation of a tumor factor responsible for angiogenesis *J. Exp. Med.* **133** 275–88

[12] Pouysségur J, Franchi A and Pagès G 2001 pH_i aerobic glycolysis and vacular endothelial growth factor in tumour growth *Novartis Found. Symp.* **240** 180–96

[13] Favier J, Amar L and Gimenez-Roqueplo A-P 2015 Paraganglioma and phaeochromocytoma: from genetics to personalized medicine *Nat. Rev. Endocrinol.* **11** 101

[14] Ashrafpoor G, Bollache E, Redheuil A, De Cesare A, Giron A, Defrance C, Azarine A, Perdrix L, Ladouceur M and Diebold B *et al* 2015 Age-specific changes in left ventricular diastolic function: a velocity-encoded magnetic resonance imaging study *Eur. Radiol.* **25** 1077–86

[15] Luo J and Konofagou E E 2011 Imaging of wall motion coupled with blood flow velocity in the heart and vessels *in vivo*: a feasibility study *Ultrasound Med. Biol.* **37** 980–95

[16] Moses W W 2011 Fundamental limits of spatial resolution in pet *Nucl. Instrum. Methods Phys. Res., Sect.* A **648** S236–40

[17] Lang N, Dawood M, Büther F, Schober O, Schäfers M and Schäfers K 2006 Organ movement reduction in PET/CT using dual-gated listmode acquisition *Z. Med. Phys.* **16** 93–100

Section 3

Correlative multimodality imaging across scales

Chapter III.3

Correlative multimodality imaging across scales

Carles Bosch, Matthia A Karreman, Alexandra Pacureanu, Andreas T Schaefer and Andreas Walter

Biomedical research aims at gaining a mechanistic understanding of diseases and entire organisms. This aim is significantly facilitated by current trends in bioimaging, which combine complementary modalities to gain holistic structural, functional and chemical information of the entire sample and place molecular details into their overall spatio-temporal multiscale context. Correlative multimodality imaging (CMI) creates a composite and zoomable view of exactly the same specimen and region of interest (ROI) by (sequentially) correlating two or more modalities. There are two prominent examples: correlated light and electron microscopy (CLEM) in biological research, and hardware-fused hybrid imaging setups (such as PET/CT) in (pre)clinical research. While they are more and more applied to tackle biomedical research, the development of advanced implementations is still in its infancy. Specifically, CMI approaches that bridge *in vivo* and *ex vivo* imaging across scales beyond CLEM and hardware-fused hybrid imaging are lacking—despite their great potential for research and diagnostics. In this chapter, we address the state-of-the-art and challenges of CMI and present two advanced showcases of holistic multiscale imaging.

1 Introduction

CMI is an emergent imaging approach that allows gaining both holistic and multiscale information about a single specimen. CMI provides structural, functional and chemical information (figure 1) and visualises the sample across all relevant scales (figure 2). It aims at combining high lateral resolution with high penetration depths and scanning a wide field of view in a reasonable time to allow the correlation of complementary parameters acquired across the entire sample. No single imaging modality alone can decipher the inner working of a cell, let alone entire organisms.

doi:10.1088/978-0-7503-3747-2ch23

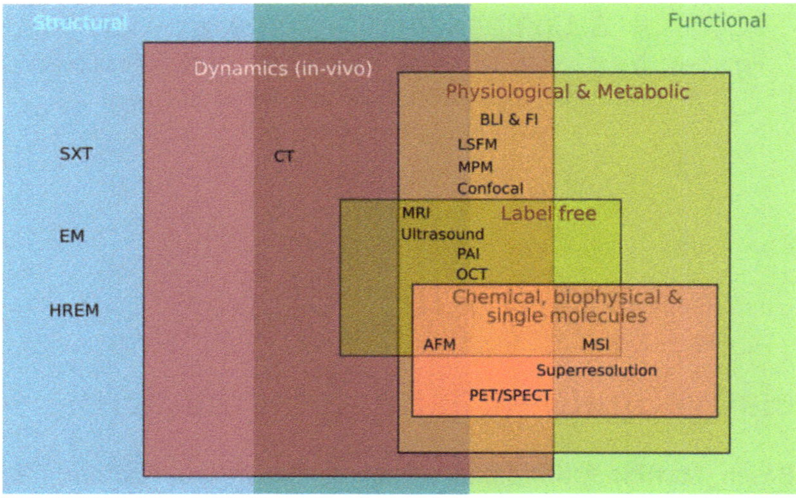

Figure 1. Holistic imaging accessing all relevant information spaces from structural to functional imaging. Representation for selected modalities: SXT (soft X-ray tomography), HREM (high resolution episcopic microscopy), EM (electron microscopy), CT (computed tomography), BLI and FI (bioluminescence and fluorescence imaging), LSFM (light sheet fluorescence microscopy), MPM (multiphoton microscopy), confocal (microscopy), MRI (magnetic resonance imaging), ultrasound, PAI (photoacoustic imaging), OCT (optical coherence tomography), AFM (atomic force microscopy), MSI (mass spec imaging), superresolution (microscopy such as STORM, PALM, STED), PET/SPECT (positron emission tomography/single photon emission computed tomography). Adapted from [1].

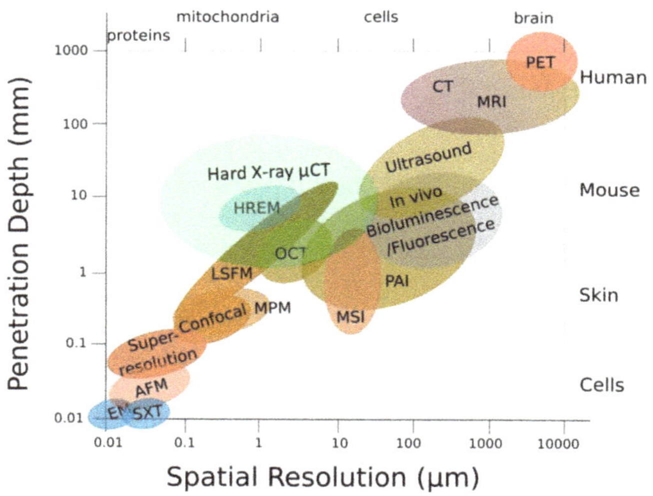

Figure 2. Multiscale imaging. Penetration depth usually comes at the expense of lateral resolution. Same abbreviations as in figure 1. We specifically include hard X-ray microCT ranging from table-top to synchrotron-based implementations since it is a crucial technology showcased below. Adapted from [1].

To gain such insights, imaging technologies need to be combined and, importantly, correlated. Correlative imaging refers to visualising exactly the same specimen with two or more complementary modalities that—combined—create an informative, composite view of the sample. CMI approaches in biological and preclinical research mostly focus on the combination of two modalities with two prominent examples for each field: (1) CLEM in biological research [2] (compare chapter III.1.a), and (2) hybrid hardware-fused imaging platforms in (pre)clinical research and diagnostics [3], such as SPECT/PET/CT (compare chapters III.2.a–c). Novel CMI pipelines with more than two modalities beyond hybrid (pre)clinical setups or CLEM are still in their infancy since they require substantial optimisation and method development to ensure compatible sample preparation protocols and precise correlation of the diverse imaging data sets. Nevertheless, CMI is highly dynamic and pushing the frontiers of imaging towards more and more complex and advanced imaging combinations. These correlative workflows and solutions range from advanced super-resolution cryo-CLEM [4] to novel CMI pipelines to bridge preclinical (*in vivo*) imaging and biological (*ex vivo*) microscopy (e.g. [5, 6, 7]) and zoom in from organ levels in a living sample to individual cellular structures, or to add localised spectroscopic biophysical or chemical information to the acquired structural and functional parameters [8].

Due to the abundance of potentially beneficial imaging combinations, CMI has been used to tackle a variety of research questions and will continue to broaden the accessible biomedical information significantly. In this chapter, we specifically showcase two holistic and multiscale CMI studies that involve more than two imaging modalities and succeed in integrating *in vivo* dynamics with *ex vivo* high resolution.

2 Showcases and setups

2.1 Multimodality microCT-guided correlative microscopy

Critical biological events, such as pathological or developmental processes, can be visualised by *in vivo* microscopy (IVM) or by fluorescence microscopy (FM) of tissue sections. However, FM has a limited resolution and fails to reveal the ultrastructural context of the event of interest. Therefore, many biomedical studies require correlating FM to electron microscopy (EM) or volume EM (vEM). Here, the main challenge is to retrieve the volume of interest in the EM-processed sample. Therefore, we developed a multimodal microCT-guided correlative microscopy approach (figure 3), which can provide the position of the ROI, identified with FM, in the EM-sample with about 5 μm accuracy, so that it can effectively be approached and studied at high resolution [5, 9].

2.1.1 Setup

Upon identification of the event of interest using IVM or FM imaging of tissue sections, the ROI is marked on the surface of the sample via near-infrared branding (NIRB) (figure 3) [10]. The NIRB markings are visible on the fixed sample with a stereomicroscope, which allows dissecting a small biopsy containing the volume of interest, which is subsequently processed for EM. The EM processing protocol

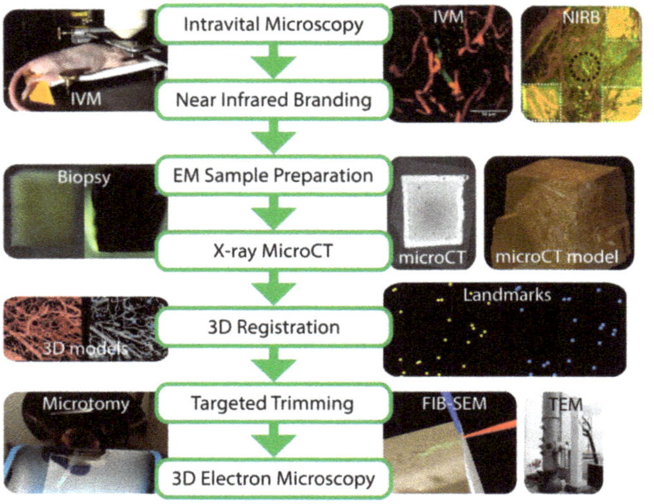

Figure 3. Multimodal microCT-guided correlative microscopy workflow. The diagram shows the critical steps of intravital multimodal microCT-guided correlative microscopy illustrated by targeting a metastatic tumour cell (green) in the mouse brain. Abbreviations: *in vivo* imaging (IVM); near infrared branding (NIRB); electron microscopy (EM); focused ion beam scanning electron microscopy (FIB-SEM); transmission EM (TEM).

includes infiltration with heavy metals, which stabilise the sample and generate strong image contrast in EM and microCT (figure 3). Following subsequent infiltration and polymerisation of the resin, the sample-block is trimmed around the biopsy to generate a small sample. Next, the sample is imaged with a benchtop microCT, which uniquely allows revealing the biopsy inside the resin block. Importantly, microCT imaging of the EM-processed sample can reveal structural details, such as cell nuclei and the vasculature (figure 3). Since some of these features, such as blood vessels, are also visible with FM, these can then be used as endogenous landmarks to perform the correlation and 3D registration (figure 3). Importantly, different cells and tissue types reveal different features in microCT, and in some cases the ROI itself can already be distinguished.

Using 3D image processing software, such as Amira, 3D models of the imaged volumes are generated by semi-automatic segmentation of the data sets. The position of the ROI, provided in the IVM volume, can then be mapped with respect to the outlines of the EM-processed resin block, which can be retrieved from the microCT volume. Using digital measuring tools, the distance in X, Y and Z from the surfaces of the sample block to the ROI is determined. The EM-processed sample is then trimmed in an ultramicrotome to accurately approach the ROI. Upon reaching the position of the event of interest, the sample is ready to be prepared for (v)EM [5, 9].

2.1.2 Biomedical relevance and application range

The multimodal microCT-guided correlative microscopy approach described here was originally developed to enable correlation of *in vivo* imaging of metastatic events to high-resolution 3DEM. Here, invasion of metastatic tumour cells in mouse skin and the extravasation of tumour cells in mouse brain were investigated [5, 9]. Moreover, microCT-guided correlative microscopy was used on vibratome sections to identify specific cells, based on electron-dense staining of the ROI or introduction of artificial multimodal landmarks. Since electron-dense staining—mandatory for commonly used 3DEM-processing protocols—generates microCT imaging contrast in many different tissue types, this technique can be employed for different applications in biomedical research. The procedure described here relies on correlation of endogenous landmarks in the tissue that are visible with both microCT and FM. Therefore, no electron-dense labelling or introduction of artificial landmarks are required, which makes this a versatile method that finds applications in the targeting and imaging of rare and/or transient pathological and/or developmental events.

2.1.3 Sample preparation

Multimodal microCT-guided correlative microscopy can be performed following *ex vivo* microscopy on living tissues or after FM of thick tissue slices, such as ~50–200 μm vibratome sections. For larger animals, such as mice, perfusion fixation is recommended. Following *in vivo* imaging, mice can be perfused with formaldehyde (FA) and glutaraldehyde (GA). Importantly, when FM needs to be performed on fixed tissue sections, it is recommended to first perfuse with FA and only perform FA/GA fixation of the biopsy after FM imaging, to prevent quenching of the fluorescent signal or autofluorescence by GA-fixation. For EM processing, it is helpful to select a 1–2 mm^3 biopsy containing the ROI, which is asymmetric so that it is easier to position it in the correct orientation within the resin block. Sample preparation for vEM and microCT involves infiltration with 1% osmium tetroxide (OsO_4) and 1.5% potassium ferrocyanide, followed by another fixation by 1% OsO_4, and staining with 1% aqueous uranyl acetate. However, if desired, other EM-processing protocols can also be employed, as long as sufficient image contrast is generated in the microCT setup. Next, the sample is stepwise dehydrated and infiltrated with epoxy resin, followed by polymerisation into a resin block. The infiltration and speed of the protocol can be significantly improved through microwave-assisted processing [5, 9].

Following microCT imaging and targeted microtomy, the sample is ready for (v)EM, which can be performed through imaging (serial) sections with TEM or TEM tomography, serial-block face imaging and/or focused-ion beam scanning electron microscopy (figure 3).

2.2 X-ray microscopy in CMI approaches for neuroscience

Systems and circuit neurosciences aim at providing a mechanistic understanding for how neural circuits (the assembly of, and connections between neurons) generate neural activity and ultimately behavior. Typical mammalian circuits span millimeters but the resolution required to reliably identify synapses, the contact points between neurons, is only few to tens of nanometers. This discrepancy in length scales poses significant experimental challenges on how to combine the measurements of neural activity *in vivo* (e.g. using two-photon fluorescence imaging methods) with techniques capable of resolving densely packed nanometer structures, such as EM. Recent advances in vEM methods such as FIB-SEM, single-section TEM camera arrays, serial block-face SEM or multibeam SEM have made it possible for small to medium circuits (embedded in volumes of up to a few ~(100 μm)^3 to be imaged at a resolution that can resolve single synapses. However, in very few cases this structural insight has been combined with prior *in vivo* functional imaging of the neural circuit's activity patterns [11] (figure 4(b–d)). Even for circuits significantly smaller than brain regions in mammals (in the millimeter range), combining two such distinct imaging modalities—one often *ex vivo* (figures 4(c–d1)), one in small fixed samples (figures 4(c–d4))—requires sophisticated strategies for correlation, involving manual identification of distinct landmarks or—in the absence of such landmarks—creation of artificial landmarks, using e.g. NIRB [10]. A key challenge is that *in vivo* imaging and EM require different sample preparations and operate at different length scales (live animal versus EM-ready trimmed sample of below 800 μm). Besides, the tissue properties that provide contrast are highly distinct (fluorescence versus heavy-metal staining).

2.2.1 Setup

To help bridge this gap, we and others [12, 14] have used X-ray CT: it operates on length scales compatible with *in vivo* imaging (sample sizes of up to several cm), and it can be employed in tissue after heavy-metal staining for EM, with the heavy metals providing similar contrast for X-ray imaging and transmission or scanning EM. This facilitates the identification of shared landmarks between *in* and *ex vivo* imaging, and allows efficient EM data acquisition with high accuracy and subsequent registration of the *in vivo* imaging and EM datasets to a common 3D framework. We term this use of X-ray imaging a 'bridge use'.

Benchtop microCTs can achieve a resolution of 5–10 μm in samples spanning several mm in size, thus facilitating efficient targeting. However, with such resolution one can only delineate limited features of the morphology of neural circuits, such as cell layers, large cell bodies or neuropil structures [15, 16]. Most cell bodies and all neurites (the neural processes extending from neurons to make or receive connections: axons and dendrites) remain essentially undetectable. Thus, a large fraction of the neural structures in the imaged volume are not detected in these correlative approaches involving *in vivo* imaging, CT and EM, and their identification is limited to volumes that can be readily probed with EM. As these EM volumes rarely extend over more than a few 100 μm, the structures retrieved of most neurons in those studies remain incomplete. Synchrotron X-ray imaging approaches have recently emerged as a possible way to provide sub-cellular context to EM data over extended volumes [12, 13]. High-brilliance and high-coherence X-ray beams created at synchrotrons allow for a variety of absorption and phase contrast

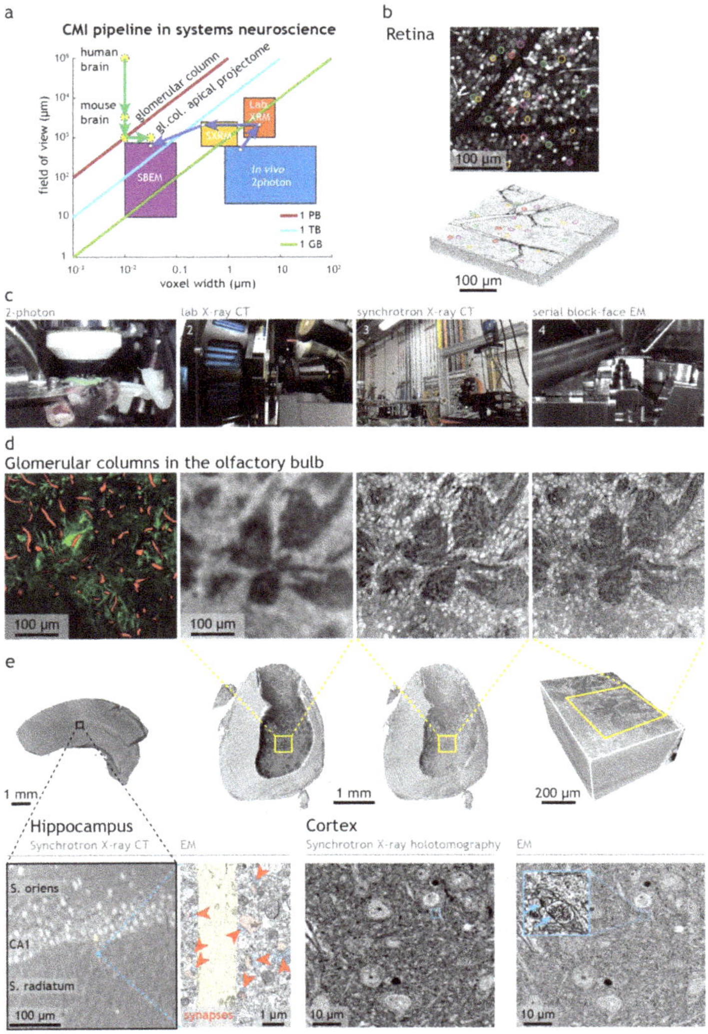

Figure 4. Dual correlative microscopy pipeline used in systems neuroscience. (a) CMI pipelines (blue arrows, data sets in white nodes) help to obtain a comprehensive understanding of tissue structure and function by combining information gathered from complementary techniques. Systems neuroscience research can benefit from CMI approaches since computational principles of the brain can be extracted (green arrows) from studying the structure and function of complete neural circuits. (b) Some direction-selective mouse retinal ganglion cells are colour-coded by their preferred moving stimulus direction based on *in vivo* Ca^{2+} imaging data acquired using 2-photon microscopy. Below, the same neurons were acquired using serial block-face EM, which revealed their connectome. (c) Equipment employed in a CMI pipeline: (c1) 2-photon microscope, (c2) lab X-ray microscope, (c3) synchrotron X-ray tomography (SXRM) beamline I13–2 at Diamond, (c4) serial block-face electron microscope. (d) Details of data sets in the mouse olfactory bulb, obtained using these imaging techniques for the same structure. Below, rendering of the whole data set. (e) SXRM can provide subcellular context and is compatible with subsequent EM. Panels from literature: (b) adapted from [11], (a, d, e) adapted from [12], cortex adapted from [13]. Reprinted with permission from Springer.

techniques that enable the acquisition of volume information in a non-destructive way. In its simplest implementation, parallel-beam free-space propagation X-ray tomography can be used to obtain volume reconstructions of EM-stained tissue over length scales of several millimeters with a resolution below 1 μm. Not only can cell bodies be reliably visualised (figure 4(d)) but also many neurites such as apical dendrites in the mouse olfactory bulb or in the hippocampus (figure 4(e)). Moreover, as synchrotron X-ray imaging is largely non-destructive, the tissue can subsequently be used for EM analysis of targeted areas (figure 4(e)). Here, X-ray imaging not only facilitates the alignment and registration of *in vivo* imaging to EM data but also provides quantifiable subcellular context over a wide range of several millimeters. We refer to this approach as 'context use' of X-ray imaging. Because it requires high spatial resolution, for now, this use is restricted to synchrotron sources and advanced lab sources of X-rays.

2.2.2 Biomedical relevance and application range

Recent advances in X-ray nano-focusing optics, nano-positioning systems, and cryogenic imaging, have fostered the development of higher spatial resolution X-ray phase contrast techniques, such as holography [13] and ptychography [17]. Combined with tomography, these techniques can generate 3D images of brain tissue with isotropic spatial resolution enabling resolution of most neruites, and they might eventually reach sufficiently high resolving power to identify synapses (figure 4(e)). These approaches can be combined with *in vivo* imaging as well as subsequent EM analysis.

While at this point data collection time limits the imaged volume at 30 nm voxels to ~1.2 million cubic microns for holography (seven cylindrical data sets, 60 μm diameter and height) and to ~0.02 million cubic microns for ptychography (three data sets, 20 μm diameter and height) per day, the transition to fourth generation synchrotrons, improvements in instrumentation and development of optimal staining methods has the potential to transform synchrotron X-ray imaging into a tool for investigation of neural circuit structure that can be directly combined with functional imaging of extended circuits.

2.2.3 Sample preparation

Contrast in various imaging modalities is usually generated by distinct sample properties. Making imaging modalities compatible within CMI approaches requires tradeoffs between contrast generators and identifying a conserved set of landmarks.

Fluorescence multiphoton microscopy of brain tissue *ex vivo* relies on fluorescent contrast in live specimens, which introduces multiple challenges. Two main types of conserved landmarks have proven useful in this CMI pipeline: blood vessels describe structurally unique patterns that are conserved across imaging modalities and that can be resolved *ex vivo* by infusing fluorophores into the bloodstream. Simultaneously, conserved histological landmarks can be recorded from tissue autofluorescence.

Contrast in X-ray imaging is generated either by the attenuation or the phase shift induced by the sample in the transmitted beam. Phase contrast offers higher sensitivity, particularly at the high X-ray energies that are required to image large samples. The typical staining strategies used for EM rely on heavy metals that adhere to lipids and proteins. These heavy metals generate both absorption and phase contrast in X-ray imaging, and it has been shown that EM-stained samples are compatible with X-ray microscopy. However, optimal staining of soft tissue samples to enhance X-ray phase contrast specifically has not been thoroughly explored yet. Further research in this area will help to resolve small sub-μm structures across samples spanning several millimeters reliably and efficiently with X-rays—making this type of microscopy, and its integration into CMI pipelines, a very promising tool to answer many of the questions posed by systems neuroscience.

3 Parameters of image quality

Due to the large variety of imaging technologies that can be involved in any CMI approach, universal parameters that affect image quality are difficult to define. The final image is here referred to as the combined and correlated information of the diverse imaging technologies involved in the CMI pipeline, and hence crucially depends on their correlation accuracy. There are two main aspects that will affect correlation (apart from the quality of the individual images acquired by each separate modality): (1) sample handling and preparation, and (2) hard- and software solutions for image correlation including re-identifying the ROI after platform relocation and visualisation of correlated image data.

(1) Each (*ex vivo*) imaging technology has its unique sample preparation protocols that might even differ substantially within a modality dependent on the desired contrast and resolution (such as cryo-EM and in-resin EM at room temperature or AFM under vacuum, atmosphere or liquid conditions). For correlation, the specific protocols need to be carefully adapted not to compromise image quality. A well-known example is the tradeoff in CLEM at room temperature between preserving fluorescence or generating EM contrast as fluorescent proteins for FM are quenched by standard EM sample protocols using epoxy resins and osmium tetroxide. For *in vivo* imaging, the specimen needs to be fixed or transferred to the next imaging platform as quickly as possible after image acquisition to rule out significant structural changes at the micro- and macroscale in between the imaging steps to ensure capturing and correlating the same event of interest across all modalities.

(2) Final data processing, i.e. the image correlation of involved diverse data sets, is facilitated by resolution matching of the technologies and by using fiducial markers or endogenous landmarks within the sample, such as blood vessels. Markers that can be visualised in different imaging technologies will also help to re-identify the ROI after relocation between imaging platforms.

Matching the FM resolution to that of the microscopy technique with the highest resolution (EM, SXT, AFM) can be for example achieved by integrating super-resolution FM. For CMI pipelines that bridge *in vivo* and *ex vivo* imaging, intermediate (mesoscopic) resolution steps are commonly inserted. An emerging example in this context is microCT, which is also used as an intermediate modality between in- and *ex vivo* imaging to create a 3D template of the sample, scan distortions and morphological changes after fixation, and help identifying the ROI.

While there are many fiducial markers that can be visualised both in FM and EM including quantum dots or dye-labelled nanoparticles, correlative markers that are robust and withstand electron bombardment and high X-ray doses and that can be tracked both in preclinical modalities and microscopy at high resolution are lacking. An alternative to fiducial markers, NIRB is used to expedite correlation when using CT as an intermediate modality specifically between volume FM and EM—as illustrated in section 2.

4 Data processing

In CMI, image correlation (i.e. registration and fusion of images) is the biggest bottleneck in data processing. Image registration is the process of computing the transformation that links two images or volumes to overlay corresponding structures. This involves first the manual or automatic identification of features of interest (segmentation) and secondly computing the matching transformation. These transformations might need to take local deformations into account and can be classified into rigid and non-rigid.

Image fusion is the visualisation of multimodal imaging data and requires the knowledge about the image transformation to match features of interest. Fusion can be achieved using (1) image and volume rendering (with ray-casting algorithms for example), (2) region-of-interest rendering (such as iso-surface rendering), or (3) slicing of a 3D volume to reduce it to a two-dimensional fusion problem. Usually, a combination of these methods is applied for visualisation.

Image correlation represents an often unmet challenge due to the potential variety of imaging modalities involved that can differ in field of view, contrast and resolution, and, consequently, features of interest might vary in their appearance substantially. For visualisation, this might require interpolation, the addition of new pixels to the lowest resolution modality to superimpose them with the highest resolution modality. Additionally, correlation might be challenged by images that do not have the same dimensionality (temporal, spatial or spectral data). Setting up solutions to assessing the accuracy of correlation quantitatively will be of crucial importance for CMI approaches to be routinely used in biomedical research. For further information, please compare the dedicated chapters (chapters III.4.a–g, *Part III section 4 Multimodal image processing*).

5 Conclusions

5.1 Strength and limitations

CMI will be the method of choice to explore cells, organisms and diseases mechanistically and holistically by deciphering their molecular mechanisms within their native context across all relevant spatial and temporal scales. It spans the relevant resolution range from milli- and micrometers down to nanometers (figure 2), and provides complementary information about dynamics, function, morphology and localisation within a close-to-native 3D ultrastructural context (figure 1). CMI and image processing routines will allow us to zoom in from a scene in a living organism to individual cellular structures and molecular information within a specific microenvironment. An important prerequisite for the successful implementation of beneficial CMI pipelines is bringing together the necessary, highly interdisciplinary expertise across diverse imaging modalities from sample preparation to image processing.

For the routine implementation of CMI, several bottlenecks need to be addressed, including: (1) sample handling and preparation procedures that can be compatible across modalities without compromising data quality; (2) soft- and hardware solutions to relocate the same ROI after transfer between imaging platforms including correlative probes and fiducial markers; and (3) automation of software solutions to correlate and handle complex, multiscale, multimodal and volumetric image data including reconstruction, segmentation and visualisation. Due to these challenges and lack of gold standards, despite its enormous potential, CMI is hardly accessible to biomedical researchers. The technological challenges are additionally aggravated by the limited availability of in- and *ex vivo* imaging technologies to a single researcher and the lack of commercial solutions to facilitate CMI workflows. Since each CMI setup needs to be individually optimised and might involve cutting-edge modalities, CMI can be very time-consuming and expensive—and throughput is currently limited.

5.2 Future developments

CMI will play a vital role in the future of bioimaging and in life sciences. Its increasing importance is already reflected by major European initiatives, such as the European Society for Hybrid Imaging, Austrian BioImaging and its Correlated Multimodal Imaging Node, or COMULIS, an EU-funded COST network that aims promoting CMI and advancing its technological implementation as a versatile tool in biomedical research [18]. CMI will be the method of choice to explore multiple spatial and temporal scales for a mechanistic understanding of living tissues and organisms. Due to the variety of beneficial combinations of imaging technologies, CMI is a highly dynamic field that will continue to combine more complex and advanced microscopy and imaging technologies, including advanced non-commercial setups, such as soft X-ray tomography in biological imaging. In future, CMI will continue to go beyond established hybrid setups and CLEM and include more than two complementary modalities. Main aims of CMI will be to: (1) integrate (pre) clinical *in vivo* imaging and biological (*ex vivo*) microscopy to zoom in from a living sample to individual close-to-native ultrastructure; and (2) adding localised spectroscopic or chemical information to the acquired structural and functional parameters.

References and further reading

[1] Walter A *et al* 2020 *Front. Phys.* **8** 47
[2] de Boer P, Hoogenboom J P and Giepmans B N 2015 *Nat. Methods* **12** 503–13
[3] Beyer T, Freudenberg L S, Townsend D W and Czernin J 2011 *Insights Imaging* **2** 161–9

[4] Tuijtel M W, Koster A J, Jakobs S, Faas F G A and Sharp T H 2019 *Sci. Rep.* **9** 1369
[5] Karreman M A *et al* 2016 *J. Cell Sci.* **129** 444–56
[6] Boido D *et al* 2019 *Nat. Commun.* **10** 1110
[7] Keuenhof K *et al* 2020 *Methods in Cell Biology: Correlative Light and Electron Microscopy IV* 16 (Amsterdam: Elsevier) pp 389–415
[8] Svirkova A, Turyanskaya A, Perneczky L, Streli C and Marchetti-Deschmann M 2018 *Analyst* **143** 2587–95
[9] Karreman M A *et al* 2017 *Methods Cell. Biol.* **140** 277–301
[10] Bishop D *et al* 2011 *Nat. Method* **8** 568–70
[11] Briggman K L, Helmstaedter M and Denk W 2011 *Nature* **471** 183–8
[12] Bosch C *et al* 2021 Functional and multiscale 3D structural investigation of brain tissue through correlative *in vivo* physiology, synchrotron micro-tomography and volume electron microscopy *bioRxiv* https://doi.org/10.1101/2021.01.13.426503
[13] Kuan A T *et al* 2020 Dense neuronal reconstruction through x-ray holographic nano-tomography *Nat. Neurosci.* **23** 1637–43
[14] Starborg T *et al* 2019 *Ultramicroscopy* **201** 58–67
[15] Dyer E L *et al* 2017 *eNeuro* **4**
[16] Bushong E A *et al* 2015 *Microsc. Microanal.* **21** 231–8
[17] Shahmoradian S H *et al* 2017 *Sci. Rep.* **7** 6291
[18] Walter A, Kleywegt G J and Verkade P 2021 *Methods in Cell Biology: Correlative Light and Electron Microscopy IV* 17 (Amsterdam: Elsevier) pp 417–30

Section 4

Multimodal image processing

Section Editor: Nataša Sladoje

IOP Publishing

Imaging Modalities for Biological and Preclinical Research: A Compendium, Volume 2

Parts II–IV: *In vivo* preclinical imaging, multimodality imaging and outlook

Andreas Walter, Julia G Mannheim and Carmel J Caruana

Chapter III.4.a

Introduction—computer-assisted analysis of multimodal image data

Nataša Sladoje and Joakim Lindblad

One consequence of the availability of an ever increasing variety of powerful imaging techniques (a number of them being described in this book, and also recently comprehensively presented in [1]) is that the amount of data to be visually processed becomes unmanageably large, while at the same time complex and difficult to interpret. Up until now, the heterogeneity of data and the difficulty of analysis have led to that far too little of the collected data is actually analysed [2]—their manual, visual analysis is typically deemed too time-consuming and tedious. However, with the very rapid development of the field of computerised image analysis, supercharged by the ongoing deep learning revolution, there is great potential to develop methods to successfully analyse such large and heterogeneous multimodal data automatically. We are witnessing that modern image analysis methods, heavily relying on the latest machine and deep learning techniques, are becoming increasingly useful in the fields of medicine and biomedicine, following closely their extraordinary success in other domains of application, such as computer vision and remote sensing.

Image processing and image analysis are wide and very active fields of research, with major, and rapidly growing, conferences such as CVPR or ECCV, attracting annually thousands of participants from both academia and industry, and highly prestigious journals, such as IEEE TPAMI and TIP, publishing novel results on both theoretical development and practical usage of generally applicable image processing methods. The scientific relevance of the fields of image processing and analysis within life sciences is reflected both by the existence of specified conferences (such as ISBI, MICCAI, or NEUBIAS) and specialised journals (examples are IEEE TMI or Elsevier Medical Image Analysis) publishing new methods aimed particularly for processing and analysis of medical and biomedical images, but also

doi:10.1088/978-0-7503-3747-2ch24

in the large number of image analysis related articles in broader life science journals (e.g. Nature Methods). For several decades, there has been a general trend within imaging, for life sciences as well as other fields, to acknowledge the growing importance of computerised processing of image data. As a consequence, the boundary between imaging and image processing and analysis is progressively getting more and more diffuse. For many modern imaging modalities, significant image processing is an integral and essential part of the modality. This has led to the creation of the field of *computational imaging*, in which computing is tightly integrated with the sensing system in order to, through advanced algorithms, form the images of interest. Originating in the seventies, with computed tomography (CT) and magnetic resonance imaging (MRI), the field continues to reach Nobel prize levels, e.g., through the computational microscopy techniques of super resolution localisation microscopy [3]. Such a tight level of integration of software and imaging has only partly reached the topics of multimodal and correlated imaging and analysis. To combine a range of different imaging techniques and realise their full potential, still largely requires the skilful combined usage of a number of software tools and approaches.

Aiming at quantification, interpretation, and higher level understanding of the heterogeneous information originating from a range of modalities, automated methods typically address different sub-problems towards these high-level goals. Common operations include: image restoration and image enhancement (removing noise and artefacts, and enhancing the quality and resolution of the acquired images), image registration (aligning different images of the same or similar objects, acquired at different times, and/or by different imaging devices), image segmentation (identifying and spatially localising objects or regions in images, of interest for further analysis), fusion and visualisation (generating a comprehensive, potentially interactive, representation of the acquired imaging data). Some of them will be described in more detail in this section, with no ambition for a complete coverage of the increasingly rich and very dynamic field.

In the following chapters of this section, several key aspects of computer assisted (multimodal) image analysis are presented in more detail. Chapter III.4.b is devoted to *image registration*, describing main approaches to the task, as well as challenges experienced when these are adjusted to a multimodal setting. Measures of similarity between images coming from very different modalities, constraints that can improve quality of the alignment, as well as evaluation of the correspondence/alignment established by registration, are discussed. Examples of registration of multimodal multidimensional image data are presented. Chapter III.4.c gives a brief overview of the basic concepts and terminology of *machine learning*, focussing on the main learning-based approaches used in multimodal image segmentation, registration, and fusion. Deep convolutional neural networks, largely responsible for the outstanding results in (multimodal) image processing and analysis nowadays, are introduced, and the main building blocks in deep network architectures are briefly elaborated on. The introduced concepts are referred to in some of the subsequent sections. The section includes illustrative examples of applications in multimodal biomedical imaging and analysis. Chapter III.4.d discusses *image segmentation* approaches, first defining the segmentation task and related terminology, and then presenting specificities and needs imposed by multimodality of data. Several recent segmentation algorithms and software packages are presented, including those based on convolutional neural networks and deep learning (and co-learning), with a particular focus on user-friendly open-source solutions. Evaluation strategies for the generated segmentation results are discussed. Chapter III.4.e presents goals, challenges, and techniques of multimodal *data visualisation*. Facilitating interpretability of the data by a human user, visualisation often serves as a dimensionality reduction technique, simplifying the multimodal image content, while maximising its informative power. The multiplicity of modalities increases the complexity of the problem. Several approaches to interactive (correlative) image data visualisation are presented, illustrated by examples and practical recommendations. Chapter III.4.f puts into focus the highly relevant issue of multimodal *data compression*. Considering the huge, and continuously increasing, amounts of acquired data, but also the requirements imposed by sophisticated applications in which the level of relevant informative detail to be preserved is application dependent while at the same time beyond human visual perception, the selection of suitable data compression and information management approaches is becoming a challenge. Some guiding principles and suitable techniques to enable efficient data storage and transmission, but at the same time ensure sufficient quality of analysis, are elaborated on. Finally, chapter III.4.g discusses some of the main challenges of the field, popular approaches in addressing these challenges, and ongoing community efforts to enhance communication, collaboration, and knowledge exchange in this highly multidisciplinary, and rapidly developing field. It also concludes the section of the book.

Acknowledgements

Perrine Paul-Gilloteaux (Université de Nantes, CHU Nantes, Inserm, CNRS, SFR Santé, Inserm UMS 016, CNRS UMS 3556, F-44000 Nantes, France), **Raphaël Marée** (Institut Montefiore, Université de Liège, 4000 liège, Belgique), and **Arrate Muñoz-Barrutia** (Biomedical Imaging and Instrumentation Group, Universidad Carlos III de Madrid) have reviewed some of the sections and provided a number of valuable comments and suggestions. Their contribution to the overall quality of this section of the book is highly appreciated.

References and further reading

[1] Walter A *et al* 2020 *Front. Phys.* **8** 47

[2] CA15124 2015 A new Network of European BioImage Analysts to advance life science imaging (NEUBIAS) (MoU) https://cost.eu/actions/CA15124/

[3] Betzig E, Patterson G H, Sougrat R, Lindwasser O W, Olenych S, Bonifacino J S, Davidson M W, Lippincott-Schwartz J and Hess H F 2006 *Science* **313** 1642–5

IOP Publishing

Imaging Modalities for Biological and Preclinical Research: A Compendium, Volume 2
Parts II–IV: *In vivo* preclinical imaging, multimodality imaging and outlook
Andreas Walter, Julia G Mannheim and Carmel J Caruana

Chapter III.4.b

Multimodal image registration

Alain Pitiot

Image registration consists in estimating spatial correspondences between two or more images. In multimodal registration those images come from different modalities, e.g. PET and MRI, or immunohistochemistry and histology. Registering multimodal images is especially beneficial when considering images from the same organism, for which we then obtain a more complete characterisation. In the case of human patients, that, in turns, opens up avenues for diagnosis aid and treatment planning. Other applications include longitudinal monitoring of structure growth, atlas building, and automated segmentation.

However, because multimodal images present different appearances (e.g. in terms of contrast, sharpness, colour or dimension), they pose a number of specific challenges, which affect every component of the registration process: the similarity criterion, the deformation model and the regularisation scheme, and the optimisation algorithm.

In this chapter, we introduce registration, highlight the specificity of the multimodal case, discuss the associated issues, and comment upon potential solutions before presenting a couple of applications.

1 Introduction

Image registration can be generally defined as the process of aligning two or more images. It makes it possible to combine information across modalities, patients, and/or time, and is therefore a pivotal step in a large variety of image analysis pipelines.

More formally, given two input images, registering the source image to the target image entails finding the transformation that maximises the similarity between the transformed source image and the target one (or, equivalently, between the inversely transformed target image and the source one). As such, it can be decomposed into three elements: a similarity criterion, a deformation model and an optimisation scheme.

doi:10.1088/978-0-7503-3747-2ch25

The similarity criterion estimates the adequacy of a match between two images based on the comparison of selected image features: e.g. intensities for iconic (intensity based) registration approaches, or anatomically meaningful invariants such as structure boundaries for geometric approaches. The deformation model describes the set of admissible transformations, which is usually dependent upon the application. Finally, the optimisation algorithm traverses the transformation space parameterised by the deformation model in search of the transformation that optimises the similarity criterion.

The registration process consumes the input images and typically produces three different types of outputs:

- a set of deformation models (e.g. affine transformation matrices, dense deformation fields) establishing spatial correspondence between corresponding (homologous) locations across the input images (this is sometimes referred to as image matching);
- a set of warped images obtained by applying the estimated matrices or deformation fields to the input images (image warping);
- a single image formed by registering the input images to each other and merging the warped images together to produce an average image (image fusion).

Applications are plentiful. For instance, fusing together images of the same organ from the same patient but acquired with different imaging devices, each giving a specific insight into the disease, can help with diagnosis and treatment planning. Image matching enables the tracking of anatomical structures over time (longitudinal analysis): e.g. mapping the growth of brain structures during childhood, or monitoring the evolution of an invasive cancer tumour over the course of therapy. It also enables the estimation of morphological differences across groups (cross-sectional analysis): e.g. between genders, between disease and control groups, or between knockout and non-knockout animal models (see III.4.b section 4.1). Registration is the core technology for the building of statistical anatomical atlases, which are instrumental in the study of normal/abnormal anatomical variability and in the modelling of populations. It also helps plan surgical interventions and evaluate their results, by matching pre-, per-, and/or post- surgery data: e.g. preoperative MRI and/or CT with per-operative US. Finally, among many other applications, it can be used to segment anatomical structures, typically by registering the newly acquired images to an atlas where the structures of interest have already been outlined.

Image registration is a very active field of research, with dozens upon dozens of new techniques being introduced each year. We refer the reader to the references in the Further Reading section below for systematic reviews.

2 Challenges in a multimodal setting

Multimodal image registration is a special case of image registration where the input images come from different modalities. As mentioned above, registering multimodal images of the same organism (intra-subject) provides a more holistic view of the situation, i.e. a fuller characterisation of the structures of interest, which is of considerable help when modelling normal anatomy and in diagnosis. As an illustration, consider a patient suffering from cancer and the benefits of fusing a PET scan providing functional information about the malignancy of the tumour with an MRI scan outlining the structure of the tumour environment.

However, dealing with multimodal images almost invariably means dealing with images with different appearances. The input images might differ in terms of:

- contrast: e.g. MR images typically highlight soft tissues, whereas CT scans, in the absence of a contrast agent, mostly show harder, denser ones, such as bones;
- sharpness and resolution: e.g. CT scans generally offer well-defined edges across tissues, whereas PET images tend to be much more diffuse; histological images present much higher resolution and functional specificity than neurological images of the same structures;
- colour: e.g. a BerEP4 immunohistochemistry (IHC) image will highlight the epithelium in brown, whereas it will be a mix of red and magenta and blue in an H&E histology image;
- dimension: e.g. IHC images are 2-D, whereas most MR scans are 3-D, etc.

Such diversity not only makes it difficult to devise a generally applicable similarity criterion but it also negatively impacts the optimisation process, even in those specific cases where the modalities of the input images are fixed and known. Indeed, the more different the input images, the more complex the similarity criterion, and, in turn, the more difficult it is to navigate the transformation space towards an optimum solution. Therefore, evaluating registration results is also especially challenging in a multimodal setting.

3 Principles and main approaches

From an analytical standpoint, registering a source image, I_S, to a target image, I_T, both of them defined over an image domain Ω, translates into estimating the transformation, T, that maximises an objective function (energy) of the form:

$$E(I_T, I_S, T) = S(I_T, I_S \circ T) + \lambda \times R(T)$$

S is the similarity measure that quantifies the level of alignment between the images.

$T: \Omega \rightarrow \Omega$, $T(x) = x + d(x)$ maps points in the target image to their homologous points in the source image. d is the deformation field.

R regularises the deformation field and, together with the deformation model, embodies *a priori* information about the registration problem.

λ is a scalar value that balances the respective influence of S and R.

3.1 Similarity criterion

Similarity criteria come in two main flavours: geometric and iconic (intensity-based).

Geometric approaches rely on the matching of homologous landmarks, manually, semi- or fully-automatically extracted from the input images. The landmarks typically correspond to salient points of the input images that are anatomically meaningful: e.g. at the corners or along the edges of clearly visible structures. Once the landmarks have been extracted they are fully characterised, usually by summarising information about the landmark's neighbourhood in a feature vector.

In iconic approaches the similarity criterion takes the whole image domain into account. While this provides a richer, more complete characterisation of the match, it also substantially increases the complexity of the registration process.

In a mono-modal setting, the similarity criterion already has to contend with difficulties such as: (a) the tessellated nature of medical images whereby very similar looking structures, e.g. cells or glands, are present all over the image, which makes it difficult to adequately match them across the images; and (b) potential non-bijective correspondences whereby areas in one image may not have homologous areas in the other: e.g. when a region in one image has been cropped out in the other, or when registering two consecutive IHC slices where cells present in the first image are absent in the second, or when tearing occurred during the histological sample preparation.

In a multimodal setting, the difficulty of crafting an adequate similarity criterion is compounded by the fact that the images have different appearances. For geometric approaches, the accuracy of the landmark locations, which is so crucial to the quality of the registration, is diminished, especially when the input images have vastly different sharpness or contrast (e.g. PET and CT). For iconic approaches, the criterion must account for the consequences in terms of image intensity of the differences in terms of acquisition: since different physical principles were involved during acquisition (e.g. the magnetic properties of protons for an MR scan, the light absorption properties of tissue for histology), the intensity relationship between the resulting images is much more difficult to model.

3.2 Deformation model, regularisation, and optimisation

On a continuum from merely translating the input images (parametric model with one parameter) to moving all voxels arbitrarily and independently from each other (non-linear, non-parametric model), the deformation model captures the family of transformations that best fit the application at hand. It reflects *a priori* information about the correspondence between the anatomical structures underlying the images and a compromise between descriptive completeness and computation efficiency.

Owing to the somewhat local nature of the process estimating point matches, an appropriate regularisation scheme is also required to discipline an otherwise unruly displacement field (i.e. to make it less noisy, more smooth) and to enforce application-dependent constraints so as to guide the optimisation process towards an application-acceptable solution. This is crucial when the deformation model is non-parametric and, consequently, does not constrain the transformation space, which, in that case, is solely defined by the regularisation scheme.

Accommodating multi-modal images will almost invariably make for a non-linear, non-convex objective function, with differences in resolution and sharpness alone yielding countless equivalent matches. As a consequence, no closed-form solutions exist and the optimisation algorithm will likely reach only a local maximum of the parameter space (it is also often the case in mono-modal settings, but the issue is all the more severe that the images are more different). This reflects the ill-posedness nature of the registration problem in general: a solution may not exist, or it may not be unique, or it is substantially affected by the initial conditions.

Furthermore, since, with very few exceptions, registration algorithms are asymmetric and, therefore, swapping the source and target image does not yield the inverse transformation (mostly because of the optimisation scheme getting stuck in a different local maximum), it becomes important to select which is which carefully. It is especially important to take into account the application when considering multimodal images with different resolutions since the warped images will potentially be sub- or super-sampled, e.g. when registering a high-resolution IHC image to a much lower resolution MR scan, or vice-versa.

3.3 Dealing with multimodal images

A variety of methods have been introduced to deal with the specific issues posed by multimodal registration. They generally rely upon (a) using information theoretic measures as similarity criterion, or (b) reducing the multi-modal problem down to a mono-modal one by either simulating one modality from the other or mapping both modalities to a common domain.

Information theoretic approaches, chief among them normalised mutual information, NMI, have proved particularly apt at registering 3-D neurological modalities, e.g. PET, CT, or MR. The initial shortcomings of the original MI approaches, such as their lack of overlap invariance or of spatial awareness (i.e. they assumed a stationary intensity relationship between the input images), were later identified and properly addressed. The success of NMI paved the way for all manners of statistical criteria (see [1] for a review of f-measures).

The alternative approach is to reduce the multi-modal registration problem to a mono-modal one in the hope of alleviating the severity of the issues identified above. Indeed, the more similar the modalities, the simpler the similarity criterion and, consequently, the smoother and more convex the transformation space. This reduction is ordinarily performed in one of two ways:

- When sufficient information about the physical acquisition process of the input images is available for that process to be modelled, it becomes possible to simulate what one of the input images would look like had it been acquired in the same modality as the other. We then obtain two images of the same modality and, consequently, a simpler registration problem. In more recent years, learning approaches to simulation have been introduced and are leveraging available public databases of pre-registered multi-modal images to produce promising results.
- Rather than transforming one image into the other, a popular technique is to convert both of them into a third modality, thereby the registration becomes

mono-modal. For instance, for histology and IHC images, colour deconvolution is a convenient way to extract from the input images similar looking channels.

Neither of those approaches are however helpful when dealing with differences in dimensionality, e.g. when registering 2-D histology to 3-D MR. Despite recent progress, no convincing, general method is available to date. Most approaches tend to be ad hoc and rely on intermediate modalities (e.g. 2-D histology images are first registered to 2-D blockface images, which are themselves registered to 3-D MR scans), on 3-D reconstructions (e.g. the 2-D images are first reconstructed into a 3-D volume, which is then registered to the 3-D volume), or, when all else fail, on manually placed landmarks.

4 Examples of application

We illustrate the above considerations on a couple of multimodal registration applications.

4.1 Correlating genotype and phenotype in rat models of hypertension (2-D histological images to 3-D MR scans)

Most common human diseases, such as hypertension, have complex phenotypes: their pathogenesis is multifactorial and involves a variety of genes as well as environmental contributions. The genetic dissection of phenotypes is typically conducted by characterising the statistical relationships between phenotypes and genotypes.

In this project, we were interested in anatomically phenotyping two closely genetically related inbred strains of rats: Brown Norway (BN) and spontaneously hypertensive rats (SHR), an animal model for hypertension [2].

To do so, a number of BN and SHR rats were scanned *post mortem* in a high-field, 7T MR scanner. The brains were then fixed, cut axially, and histologically processed using a Nissl stain to highlight neuron bodies. As mentioned in section 3.3, we then reconstructed a 3-D volume from the consecutive 2-D stained sections (figure 1(c)) as a means to facilitate the registration of histology and MR data. On this occasion, we used our own topology-preserving, rigidity-adaptable registration approach [3] in order to handle the nonlinear shrinkages and distortions that occurred during the histological process. The 3-D histological volumes were then affinely and non-linearly registered to the corresponding 3-D MR volumes.

By registering *post mortem* histological data with MR, we could establish a one-to-one spatial correspondence between the modalities, with the MR volume acting as an anatomical framework. This enabled the extraction of 3-D, geometrically correct phenotypes from histology. It also made it possible to analyse structures extending across more than one 2-D slice, and to compute the volume of regions of interest that could only be defined on the histological slices. Figure 1 illustrates one of the findings: a genetically linked morphological difference in the folia of the cerebellum of SHR rats.

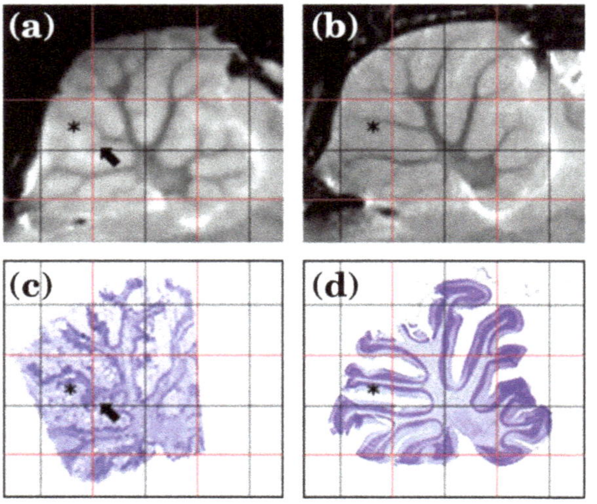

Figure 1. Comparison of the cerebellar folia of a representative BN (left) and SHR (right) rat: (a, b) 2-D sagittal section virtually cut through 3-D post-mortem MR scans; (c) 2-D sagittal section virtually cut through a 3-D reconstructed histological volume registered to the 3-D MR scan; (d) 2-D section cut sagittally through the SHR rat brain and registered to the 3-D MR scan.

4.2 Characterising the immune response to colorectal cancer (IHC to IHC)

Colorectal cancer (CRC) is a type of cancer that emerges from the glandular, epithelial cells in the large intestine (colon), the final part of the digestive tract. It is the fourth most diagnosed cancer in the world, with over two million new cases and one million deaths worldwide every year. CRC starts with the formation of polyps, in the wall of the colon or rectum. While most are benign, some of them later develop into cancerous tumours.

CRCs comprise a heterogeneous group of diseases driven by a vast array of mutations and mutagens. Tumours are either microsatellite instable (MSI) or microsatellite stable (MSS). In MSI tumours, the genes responsible for DNA repair, the mismatch repair genes, MMR, are not functioning correctly. This makes for a highly unstable tumour micro-environment. MSS tumours (80% of cancer cases) are highly mutated and not MMR deficient.

MSI tumours usually attract a high immune response, whereas MSS tumours do not. The aim of this project was to characterise that immune response [4].

To do so, consecutive sections cut out of resected MSI and MSS tumours were immunohistochemically stained for CD3 (a marker of immune T-cells), CD20 (a marker of immune B-cell) and BerEP4 (a marker of the epithelium). For each tumour, the IHC sections were non-linearly registered to each other using an in-house approach. The latter dealt with the large differences in terms of contrast and colour (figure 2, first row) by colour deconvolution of the IHC images and considering the haematoxylin channel, thereby reducing the multi-modal registration problem down to a mono-modal one. The haematoxylin maps were registered to each other using successive runs of our topology-preserving, rigidity-adaptable registration

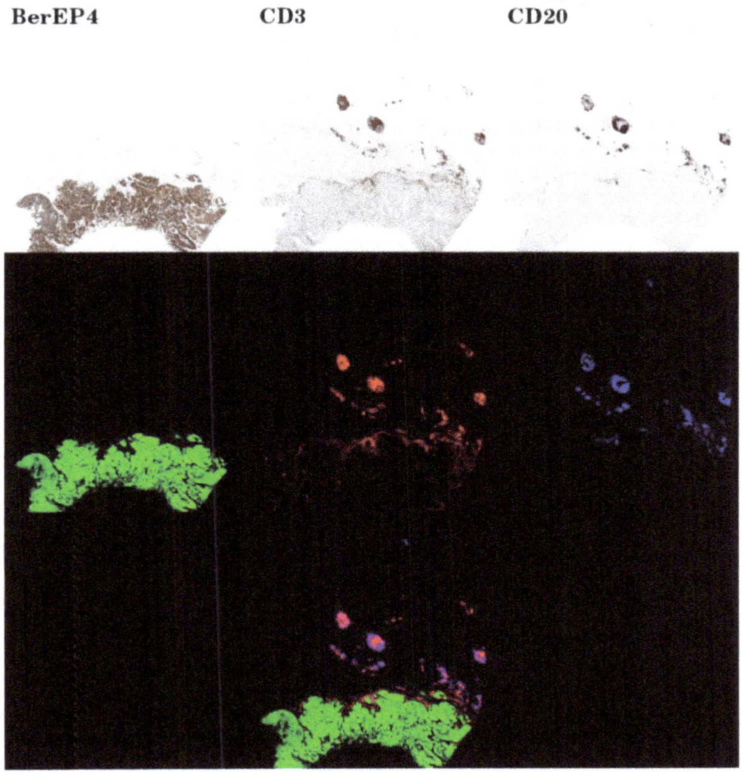

Figure 2. MSI tumour: (first row) IHC images registered to each other, (second row) thresholded, colour-coded images, (third row) proteomic map.

approach [3]. By decreasing the radius with each run, we could first tackle global displacements, then regional ones, and, finally, local ones. That made it possible to precisely register IHC sections despite potential large differences between them, such as missing tissue, tears, holes and floating parts. The registered images were then thresholded (second row) and merged into a colour-coded map highlighting the region positively expressing various combination biomarkers (third row). We could then spatially quantify various immune features, such as the density of T-cells or B-cell in the tissue or in the epithelium, and statistically compare them across MSI and MSS.

5 Conclusions

Despite considerable progress over the past few decades, multi-modal registration is still very much an open problem, and particularly so for non-parametric transformation models and inter-subject (cross-sectional) applications.

As ever, validating registration results is not trivial and a gold standard for non-parametric results is still missing despite more and more annotated datasets coming online [5].

Developing more adequate similarity function seems to be the approach most likely to bear fruit: learning approaches already showed promising results a decade ago, and have been revitalised by the advent of deep learning.

References and further reading

[1] Pluim J P, Maintz J A and Viergever M A 2003 *IEEE Trans. Med. Imaging* **22** 986–1004

[2] Pitiot A, Pausova Z, Prior M, Perrin J, Loyse N and Paus T 2007 *Hum. Brain Mapp.* **28** 555–66

[3] Pitiot A and Guimond A 2008 *Med. Image Anal.* **12** 16–25

[4] Susanti S, Pitiot A and Ilyas M 2019 *J. Pathol.* **249** 42

[5] Rohlfing T 2011 *IEEE Trans. Med. Imaging* **31** 153–63

[6] Maintz J A and Viergever M A 1998 *Med. Image Anal.* **2** 1–36

[7] Sotiras A, Davatzikos C and Paragios N 2013 *IEEE Trans. Med. Imaging* **32** 1153–90

IOP Publishing

Imaging Modalities for Biological and Preclinical Research: A Compendium, Volume 2
Parts II–IV: *In vivo* preclinical imaging, multimodality imaging and outlook
Andreas Walter, Julia G Mannheim and Carmel J Caruana

Chapter III.4.c

Learning-based approaches for multimodal imaging

Stephan Winkler, Andreas Haghofer and Hannah Janout

In the context of image analysis, machine learning techniques are nowadays often used to automatically identify objects in images. These algorithms have also become very important in the context of biomedical imaging. Using machine learning, we are able to find models that, e.g., detect cells or distinguish between types of cells or tissues. In the context of multimodality imaging, these models can for example be used to simulate fluorescence staining for nuclei extraction, which would otherwise require the use of additional imaging modalities. Especially artificial neural networks (ANNs) and deep learning are commonly used methods in biomedical image processing. In this chapter, we discuss the basics of machine learning and recent approaches that use deep neural network structures in the context of multimodality imaging.

Furthermore, multimodality imaging often requires image registration to find the perfect alignment of two (or more) images, and image fusion. In this chapter, we describe machine learning-based approaches for both fields.

1 Introduction: machine learning in image analysis

1.1 Machine learning basics

The field of machine learning has become one of the fastest-growing areas in the computer science community, supported by rapid developments in specialised hardware and software resources as well as the availability of increased amounts of data. Independent of the actual application context, all machine learning algorithms have the common goal to identify the relationships of variables within the presented database or to learn a specific behavior.

doi:10.1088/978-0-7503-3747-2ch26

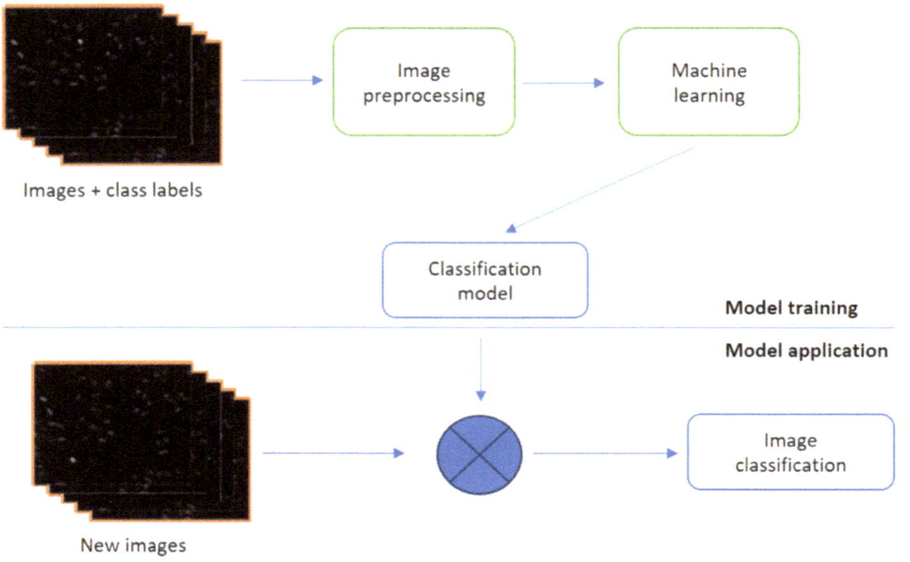

Figure 1. Image analysis and machine learning. In the training phase, models are identified; in the test phase, new images are presented, and their labels/classes are predicted.

As shown in figure 1, the most common modelling processes start with a dataset that is representative for the real-world problem which has to be solved. In the model training phase, models are identified based on labelled images. The training algorithm learns a function that predicts the label (in most cases the class) of an image on the basis of the image pixels. In the test phase, new images are presented, and their labels/classes are predicted.

All **supervised modelling** techniques [1] use data that provide information both about the features and the target values (e.g., the class) for each sample. The main idea of this approach is to generate a model $y[i] = f(x_1[i] \ldots x_n[i])$, where $y[i]$ represents the target value of the sample i, which is to be predicted using the (independent) features x_1 to x_n.

For the analysis of images of biological samples, supervised learning is often applied for learning classifiers that distinguish between classes or states. In the following, we specifically discuss supervised machine learning techniques. In supervised learning, the modelling algorithm is given a set of samples with input features as well as target/output values—i.e., the data are labelled. The goal then is to find a model (or a set of models) that is (are) able to map the input values to the designated target variable(s). That is, for each target variable y, the goal is to find a model that allows us to define y as a mathematical function of all available/allowed variables x: $y = f(x) + e$ (where e denotes the error term).

During this supervised modelling process, we distinguish the following phases: Training, validation, and testing.

- In the training phase, models are learnt on the basis of available training data.

- The validation data is often perceived as a data set that is available during the training phase and used for evaluating the learnt models and for detecting overfitting, which means that the models are more complex than necessary. Especially in the presence of noise this is a major problem, as overfitting can occur by choosing a too complex model structure and fitting that model too tightly to the given (noisy) data, so that the model will perform very well on training data, but rather poorly on test data. The opposite of overfitting is underfitting, which means that the model is not able to learn the complexity of the system's behavior and performs poorly on the given training data.
- After completing the whole training phase and selecting the models on the basis of the training or validation data, the quality of the models and thus also of the applied algorithms is assessed on test data which are not used during the whole training and validation process. This approach provides an unbiased impression of the model's real-world performance.

In contrast, **unsupervised learning** [2] does not use pre-labelled data for learning models. Instead, these algorithms find patterns, for example clusters (groups) of data entries, without human supervision (input labels). Each cluster should mainly contain entries with a certain feature similarity, which represents a measurable difference compared to entries of another cluster.

1.2 Artificial neural networks and deep learning

Artificial neural networks (ANNs) are networks of simple processing elements, which are called neurons and are in most cases structured in layers. As shown in figure 2, these fully connected neurons are typically represented as nodes within a graph. All neurons of one layer are connected to all neurons of their neighboring layers. The first layer consists of input neurons, which receive the input variables x, and the last layer of output neurons, which define the network's output y. For each neuron the behavior is clearly defined, in most cases as a transfer function Φ of the

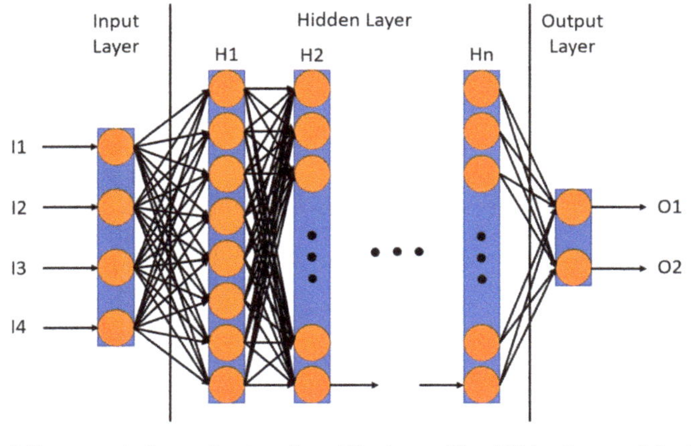

Figure 2. A fully connected neural network architecture with n hidden layers. Adapted from [3].

weighted sum of the connected neurons' values. That is, for neuron n the output y is defined as $y = \Phi(\sum w_i \times x_i)$, where x are the output values of the nodes in the previous layer that are connected to n. Thus, these are the inputs for n, and w are the weights for these inputs. The transfer function Φ can, for example, be implemented as a sigmoid function or another nonlinear activation function. During the training phase, the weights are trained (optimised), in most cases using backpropagation algorithms.

Over recent years, deep neural networks have been used very successfully in the context of numerous research projects and applications. In general, a deep neural network is an artificial neural network that consists of multiple (hidden) layers between the input and output layers [4]. In particular, the development of novel types of activation functions for the neurons and improved backpropagation algorithms has enabled the development of deep neural networks, which can consist of dozens of layers. As network models grow bigger and bigger, the number of parameters increases, so that more and more training data as well as computing resources are necessary for training and using them.

The convolutional neural network (CNN) is a certain type of neural network that is widely used in computer vision. As shown in figure 3, convolutional neural networks consist of various layers of different types to prepare the input data for the classification layers. The core elements, which differentiate these networks from other neural network architectures, are the convolution layers. These convolution layers consist of kernels, which are translated over the given image or the previous layer following the idea of sliding windows, which means that they are applied to all parts (sub-areas) of the given data. The kernels are used during the training phase for extracting key elements (features) which are recognised in the image. These key elements might be certain objects that are to be recognised, or parts of these objects.

Every time the kernel is applied to a part of the image, the dot product of the convolutional kernel and the windowed pixels is passed on to the activation function [4]. The extracted information of the convolution layer is then passed to the so-called pooling layers that act as selection filters as they select the maximum value within a predefined window or calculate the average within the window values.

This procedure is repeated until the final pooling layer is reached, where only the most relevant features are left. This layer usually acts as the first layer of a fully connected neural network which performs the actual classification. These last layers

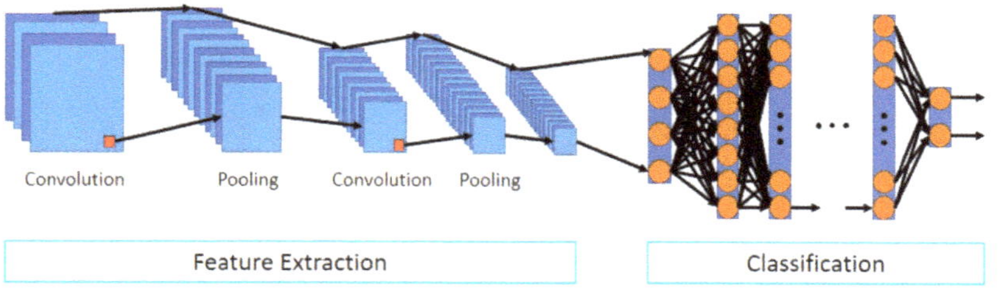

Figure 3. Visualisation of a convolutional neural network architecture with feature extraction and classification. Adapted from [5] with permission from MDPI.

of the network are responsible for weighting the learnt features depending on the class the input belongs to.

A concept that is very important and of growing scientific interest in this context is **transfer learning**, which refers to the approach that models (or parts of models), that were learnt solving a given problem, are applied to solve another (different, but somehow related) problem. Using CNNs this can be achieved by re-using the convolution layers, which were learnt to detect certain objects or parts of objects, and then adjusting the fully connected classification network using the new problem data.

2 Use case for state-of-the-art learning models for the analysis of biological images: cell nuclei segmentation using U-Nets and multimodality imaging

In this section, we describe a use case for deep learning in multimodality biomedical image analysis, namely the identification of models for detecting cell nuclei using deep learning and multimodality imaging. Cell segmentation is a crucial part within biomedical image analysis research. A very important neural network model in this context is the so-called U-Net [6], which has become the state of the art in cell segmentation. As an evolution of the traditional convolutional neural network, this network is able to localise and distinguish cells within an image by classifying each individual pixel as part of a cell or background.

As shown in figure 4, this neural network uses a so-called encoding and decoding architecture: First, the image is encoded by various levels of convolution layers followed by max pooling, which allows the network to learn detailed structures within the image. After that, the following layers decode the information including the output of the previous encoding layers. Instead of the previously mentioned max pooling during the encoding, these decoding layers use specialised upscaling convolution layers. The idea of this architecture is to use the resulting feature maps (convolution results after applying the transfer function) of the encoding for the image reconstruction during the decoding. This rather complex architecture published in 2015 represents the state-of-the-art method for biological cell localisation and segmentation tasks.

Figure 5 shows an example of the application of a U-Net structure for learning cell detection models. The left image represents the original microscopy image. The center image is an example of a mask which defines the information where the cells are located. This mask is used as input to train the U-Net and has to be provided by the user; in most cases this is initially done by manually segmenting the cells. The right image represents the actual output of the U-Net. Each pixel is represented by a value between 0 and 1 which corresponds to how certain the model is regarding the classification of the pixel. This output should now contain segmented cell areas where the values are near 1.0 and background areas with values near 0.0.

This segmentation capability of the U-Net architecture can be especially useful in the analysis of data from multimodal imaging. For example, it is often required to use DAPI staining for the fluorescence microscopy imaging of the cell nuclei in

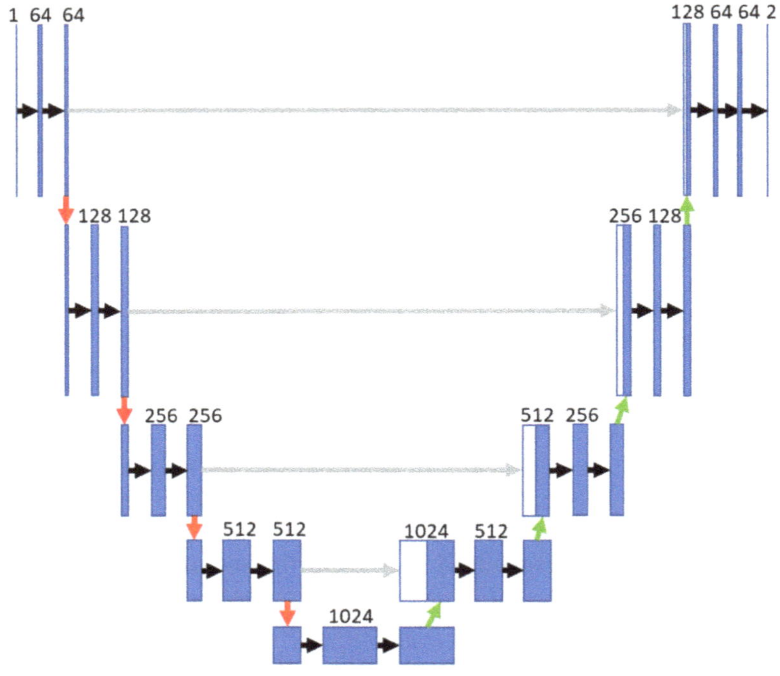

Figure 4. U-Net architecture, adapted from [7]. The given input images are first encoded using multi-channel convolution layers; the number of channels increases as indicated by the numbers above the layers. In a second step, this complex feature information is reduced by the decoding layers with decreasing numbers of features.

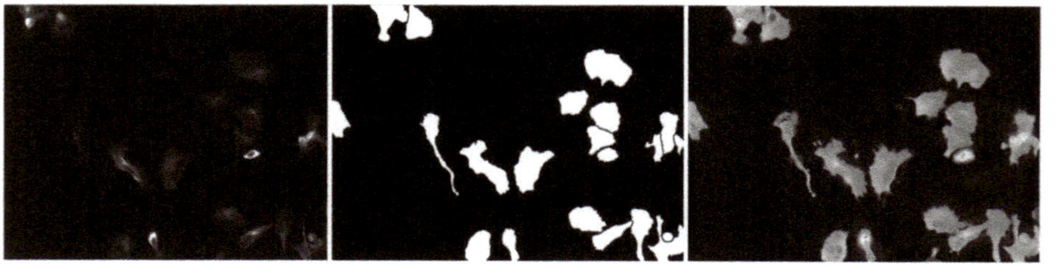

Figure 5. Analysis of fluorescence microscopy images of cultured neuroblastoma cells [8]. Left: original microscopy image; centre: labelled pixels for training nets for detecting cells; right: output of a U-Net trained for detecting cells in microscopy images.

addition to imaging techniques such as bright-field or phase contrast imaging. This staining can be estimated (simulated) by a U-Net which was previously trained on the segmentation of the cell nuclei. Using the DAPI stained cell nuclei images as labels and the bright-field cell images as input, this neural net architecture can provide a binary representation of the cell nuclei in the bright field microscopy image.

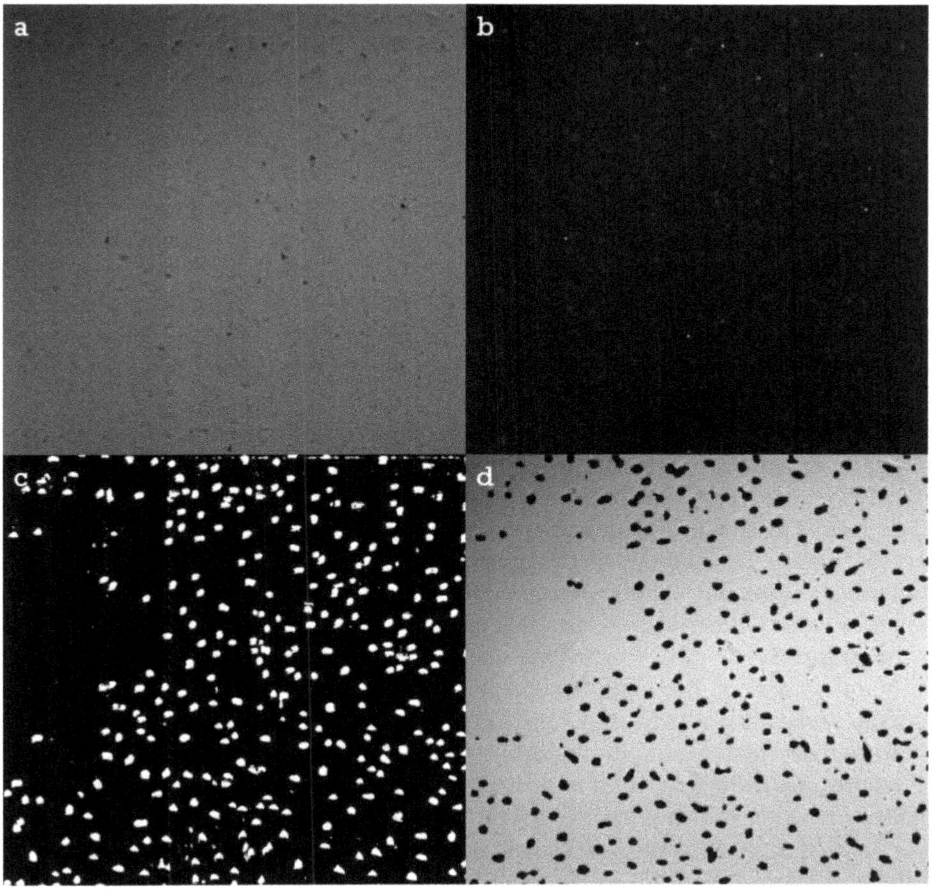

Figure 6. Cell nuclei detection using a U-Net architecture. (a) Raw bright-field image of cells, (b) DAPI stained cell nuclei, (c) binary labels generated out of the DAPI image, (d) segmentation of the cell nuclei in the brightfield image using the previously trained U-Net. Courtesy of Medical University of Vienna (Austria), Department of Vascular Biology and Thrombosis Research Center of Physiology and Pharmacology. Adapted from [7], copyright (2015) with permission of Springer.

An example for the application of this approach is illustrated in figure 6. We see that this neural network architecture is in fact able to correctly estimate the staining of the cell nuclei. Additionally, it provides a clear separation of the cell nuclei from the background by using a binary format. This binary representation of the cell nuclei can then also be used for the localisation of the nuclei within the initial bright-field image.

3 Machine learning in image registration and image fusion

Especially in the context of multimodality imaging, diverse combinations of images are used. The analysis of these combinations is often challenging, as the images might not be aligned or may show different parts of the analysed tissue or objects.

Therefore, image registration and image fusion are often key factors to success in biomedical research. In this section, we describe the use of machine learning, especially deep learning, in image registration and image fusion.

3.1 Image registration

As described in detail in the previous chapter on multimodality image registration, image registration is the process of geometrically aligning two images. The aligned images show the same objects, or at least parts of it, which differ in the sense of their modality, time, or viewpoint. This alignment creates a more complex representation and more detailed understanding of the imaged object, as each added imaging technique increases the amount of information to generate a more accurate view of a scene [9, 10].

Due to the wide variety of images and the influence of noise or image distortions, finding a universal registration method is widely considered impossible [11]. Furthermore, the type of information used for alignment differs for each application and dataset. This is why a wide variety of registration algorithms has been developed over the years. In general, these methods are either feature-based, for images containing prominent and distinctive objects or area-based for images without distinct objects, where the algorithms have to rely on the differentiation of intensity or colouration [10, 11]. Regardless of the registration type and the extracted information used for the alignment, registration consists of four steps: detection of an image's relevant features, matching of those found features within multiple images, transformation model estimation, and image resampling and transformation [9, 10].

Recently, algorithms for automatic image registration using deep learning have been developed in order to increase the performance of registration algorithms and decrease the costs (with respect to runtime as well as human expert resources). In general, there are three different types of deep learning based approaches for registration:

On one hand, the **deep similarity metrics approach** uses deep learning to determine similarity metrics, focussing on the maximisation of similarity between the input images. The determined similarity metrics are then given to a classical intensity-based registration algorithm [12]. An example of a pipeline used for the determination of deep similarity metrics is shown in figure 7. On the other hand, **supervised transformation estimation** focusses on the development of models for the determination of transformation methods that align the images in an optimal way (regarding the similarity metric). These networks require detailed annotated images, i.e. the ground truth, which are used by the network's loss function during training [12].

Finally, **unsupervised transformation estimation** avoids the need for manual annotations. For example, spatial transformer networks (STN) are used to perform the deformations used in registration [13]. Unsupervised transformation estimation is further divided into two classes, namely similarity metric based versus feature metric based: similarity metric based approaches use similarity metrics with regularisation to define their loss function, whereas feature metric based approaches use a feature extractor that maps the fixed and moving image to a feature space for the prediction of transformation parameters [12].

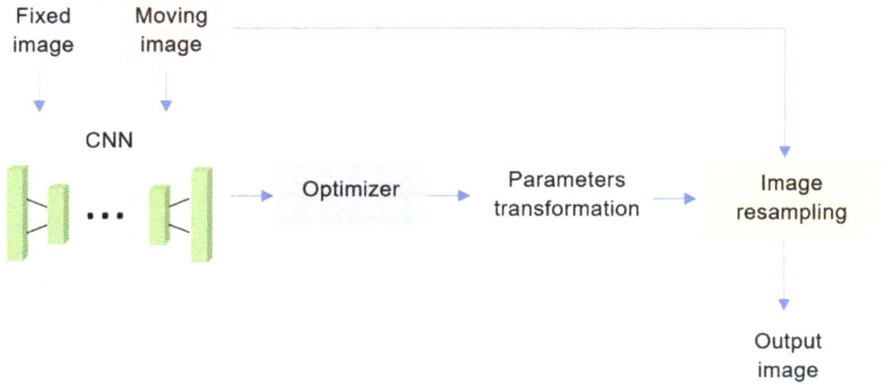

Figure 7. Registration pipeline including a deep learning network, which determines the input images' similarity in an intensity-based registration framework. Adapted from [12], copyright (2020) with permission of Springer.

3.2 Image fusion

Similar to image registration, image fusion describes the combination of data from different sources. The aim is to obtain additional information of greater quality by combining different images and merging their structures [9, 14, 15]. It is a very useful approach to extract significant features from data, remove unnecessary artefacts or noise, and decrease the overall data volume. The input images for image fusion can be divided into two groups, namely single and multi-sensor fusion systems, depending on the number of sensors the images stem from [16].

A basic image fusion algorithm consists of the following steps: In the first step, image registration (as shown in figure 7) is applied to align the given images. The next step consists of feature extraction, which for example can be learnt using CNNs, creating a feature map for each image. Then the image pixels are labelled through a decision operator, creating a decision map. Should the feature and decision maps not refer to the same objects, semantic alignment is applied. This semantic alignment refers to the conversion of the two maps to common objects and is needed as only inputs referring to the same objects can be fused [17]. The aligned images and feature maps are transformed to a standard scale, which represents the input for the last step, namely image fusion, combining the previously acquired information into one image [14].

In general, there are two classes of image fusion algorithms:

- **Spatial domain fusion** merges and manipulates local spatial features to acquire a fused image [16].
- **Temporal domain fusion** describes the blending of images acquired from several different frames taken over a specified time interval. It renders events of varying duration and preserves moving content in a single output image [16].

In the medical domain, image fusion is an everyday practice used to enhance the information acquired from different modalities. Examples of these modalities are

MRI and PET, MRI and SPECT, and MRI and CT, all of which can be aligned and further analysed applying image fusion [14].

4 Conclusion

After decades of research, machine learning methods are nowadays able to produce models that can detect and classify a huge variety of objects in images. In biomedical research, machine learning has enabled us to solve data-based problems that were yet unsolved. Due to the rapid development of hardware infrastructures as well as easy-to-use software frameworks, machine learning algorithms are nowadays no longer only applicable in computer science research but have become an essential part in various technical and medical problem-solving processes. Implementations are also available online, for example Keras (the Python deep learning API), TensorFlow, and HeuristicLab. Nevertheless, as model structures become more and more complex, there is an increasing need for huge amounts of correctly labelled images that serve as training input data.

References and further reading

[1] Cunningham P, Cord M and Delany S J 2008 Supervised learning *Machine Learning Techniques for Multimedia* (Cognitive Technologies) ed M Cord and P Cunningham (Berlin, Heidelberg: Springer) pp 21–49

[2] Duda R, Hart P and Stork D 2001 Unsupervised learning and clustering *Pattern Classification* 2nd edn (New York: Wiley)

[3] Bre F, Gimenez J M and Fachinotti V D 2018 *Energy Buildings* **158** 1429–41

[4] Heaton J 2012 *Introduction to the Math of Neural Networks* (Heaton Research, Inc.)

[5] Alom M Z, Taha T M, Yakopcic C and Westberg S 2019 A state-of-the-art survey on deep learning theory and architectures *Electronics* **8** 292

[6] Falk T *et al* 2019 U-Net: deep learning for cell counting, detection, and morphometry *Nat. Methods* **16** 67–70

[7] Ronneberger O, Fischer P and Brox T 2015 U-Net: convolutional networks for biomedical image segmentation *Medical Image Computing and Computer-Assisted Intervention – MICCAI 2015. MICCAI 2015* ed N Navab, J Hornegger, W Wells and A Frangi (Lecture Notes in Computer Science vol 9351) (Cham: Springer) pp 234–41

[8] Yu W, Lee H K, Hariharan S, Bu W Y and Ahmed S 2012 Mus musculus, neuroblastoma. Cell Images Library, CIL:40217, [online, http://cellimagelibrary.org/images/40217]

[9] Ranganathan S, Gribskov M, Nakai K and Schönbach C 2018 *Encyclopedia of Bioinformatics and Computational Biology* (Amsterdam: Elsevier)

[10] Woo J, Stone M and Prince J L 2014 Multimodal registration via mutual information incorporating geometric and spatial context *IEEE Trans. Image Process.* **24** 757–69

[11] Zitová B and Flusser J 2003 Image registration methods: a survey *Image and Vision Computing* (Amsterdam: Elsevier) pp 977–1000

[12] Haskins G, Kruger U and Yan P 2020 Deep learning in medical image registration: a survey *Mach. Vision Applic* **31** 8

[13] Jaderberg M, Simonyan K, Zisserman A and Kavukcuoglu K 2015 Spatial transformer networks *Advances in Neural Information Processing Systems* (Curran Associates, Inc.) pp 2017–25

[14] EL-Ghani A A A, Fatma, Elmogy M and Atwan 2015 Current trends in medical image registration and fusion *Egypt. Inform. J.* **17** 99–124
[15] El-Hoseny H M, Abd Elrahman W, El.Rabaie E-S M, Faragallah O S and Abd El-Sami F E Medical image fusion: a literature review present solutions and future directions *Menoufia J. Electron. Eng. Res.* **26** 321–50
[16] Masood S, Sharif M, Yasmin M, Shahid M A and Rehman A 2017 Image fusion methods: A survey *J. Eng. Sci. Technol. Rev.* **10** 186–95
[17] Mitchell H 2012 Radiometric normalization *Data Fusion: Concepts and Ideas* (Berlin: Springer) pp 31–50

Chapter III.4.d

Multimodal image segmentation

Jaromir Gumulec and Tomas Vicar

Image segmentation is an integral part of image analysis workflow enabling researchers to extract information from images and thus to evaluate image data more objectively. Multimodal imaging often introduces new combinations of image data and therefore, no exclusively designated segmentation workflow exists. In this chapter, the segmentation pipeline applicable to multimodal datasets is described, the advantages of segmentation over manual annotations are discussed, and limitations of segmentation in the multimodal setting are mentioned. Particular focus is given to the open-source tools with graphical user interface requiring no coding skills. Moreover, deep-learning approaches are discussed side-by-side with model-based segmentation methods, and their suitability is mentioned for various scenarios.

1 Introduction

Image segmentation is a process of dividing the image into multiple simplified meaningful segments whose pixels share some characteristics. Traditionally, there are three tasks in the segmentation process (figure 1); first, object(s) are distinguished from the image background in a process of semantic segmentation. Secondly, individual objects in the field of view are detected. During the third step—the instance segmentation—objects are segmented from each other.

The aim of this chapter is to describe the segmentation approaches applicable to multimodal datasets, to show the advantages and disadvantages of automated segmentation over manual segmentation. Segmentation allows a large degree of robustness of the experiment; well-designed segmentation pipeline for particular dataset enables the extraction of more information from images, providing more objectivity for an experiment.

doi:10.1088/978-0-7503-3747-2ch27

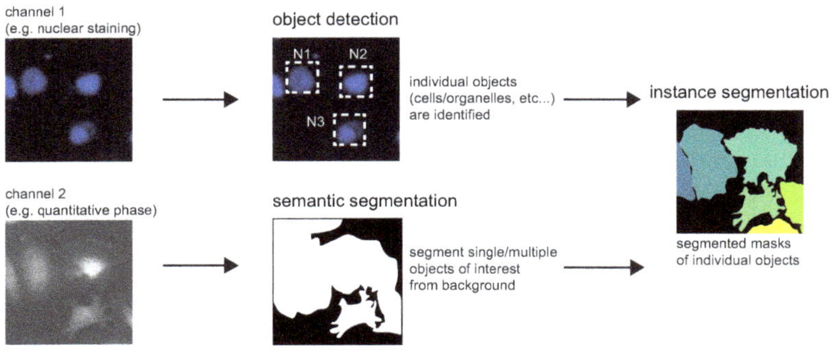

Figure 1. Combination of three image segmentation approaches for segmentation of cells in microscopy images. Example on multimodal fluorescence/quantitative phase imaging of PNT1A cells.

In this text, both model- as well as learning-based approaches are discussed, and for each segmentation step, it is mentioned to what extent it is necessary to master programming skills.

2 Challenges in a multimodal setting

Specifically, in a multimodal setup, there is a question whether all individual imaging modes will be used for segmentation; for example, for CLEM just the fluorescent layer might be used to create a segmentation mask, whereas in MRI both T1- and T2-weighted images might be used.

Traditional graphical user interface-based (GUI-based) approaches are usually designed to segment one single modality. Moreover, correlative microscopy means numerous combinations of modalities, and the solutions optimised for a particular combination do not hold for most others. Nevertheless, as fluorescence microscopy is multichannel by nature, segmentation approaches originally intended for this technique might be employed; such are mentioned in section 3.8.

Although the segmentation based on multiple modalities is rather advantageous (segmentation based on two modalities will often be more accurate than segmentation based on one modality), precisely spatially aligned (registered) images are however crucial and accordingly, the image registration is a distinctive task in multimodal imaging itself (see chapter III.4.b, *Multimodal image registration*). When done accurately, the usage of multiple channels may provide a refinement of segmentation performance (CT/T1/T2 MRI, fluorescence/brightfield microscopy might be combined with this regard).

3 Principles and main approaches

There are multiple tasks in image segmentation applicable for various scenarios (figure 1). Individual objects can be identified (individual cells/organelles, etc) in an 'object detection', resulting in bounding boxes or object centroids (seed points). Next, instance segmentation creates segmented masks of these individual objects by a combination of semantic segmentation and object detection. Individual masks then conveniently provide information about object detection together with background separation. This two-step approach can be beneficial in specific scenarios, e.g., segmentation of multichannel microscopy images of cells (cytoplasm and nucleus, etc), where semantic segmentation and object detection can be performed on different image channels.

Over the last decade, a wide spectrum of approaches has been developed. In recent years, several approaches based on (deep) learning algorithms were published for all above-mentioned steps of the segmentation process. Apart from them, there is still a spectrum of model-based approaches, which may appear to be inferior regarding the achieved accuracy. Conversely, the advantage of model-based approaches is in their interpretability—they are often intuitive. These approaches may also be more appropriate in scenarios with a limited amount of data.

3.1 Neural networks

Deep learning approaches became state-of-the-art in the contemporary image analysis tasks across scales. To date, these methods find the application in all the steps of the image segmentation process, as well as in other image analysis tasks.

To date, a major success in medical image analysis was recorded for convolutional neural networks (CNNs) and recurrent neural networks (RNN). For medical or microscopic images, these techniques are typically employed in a supervised way, where the user defines the ground truth (marks/annotates cell/background, tumour/healthy tissue, etc.). However, manual annotation is often necessary to be performed on large amounts of data, which is expensive and introduces some level of subjectivity. On the other hand, there is a spectrum of unsupervised networks for which the user annotation is unnecessary. Novel data generation approaches like Variational autoencoder (VAE) or Generative adversarial networks (GANs) began to be widely used. From a segmentation point of view, these approaches are rather marginal, the importance of GANs lies in data augmentation and training data generation in specific scenarios (e.g. unsupervised generation of the binary mask from one of the image channels).

U-Net [1] is the most commonly used CNN architecture for all stages of image segmentation. Compared to classification architectures, both input and the output of this network is an image. The output is, for example, a distance map, which can be used for both detection (achieved with local maxima detection) and semantic segmentation (achieved with thresholding), and these can be combined.

In order to increase the adoption of DL in the research community of life scientists, the current trend is to enable robust solutions in an accessible way, for instance, as a GUI for Fiji or as a web-based GUI solution, like a most commonly used convolutional network for segmentation—U-Net for Fiji plugin or ImJoy [2].

Compared to traditional methods, DL networks are 'black-boxes' by principle, and therefore the network's decision process remains obscure. Despite significant interest in the reverse-engineering of the networks, their highly complex reasoning is currently still a subject of research.

Another issue of DL is a risk of overfitting. Because the number of training samples is in most cases limited compared to the number of learnable parameters, there is a risk of the situation, where the network model does not learn the general parameters applicable for real data but instead learns something specifically for the training dataset (e.g. noise). This phenomenon—overfitting—can be recognised by a difference in accuracy between training and validation dataset. To reduce this, several algorithms were employed, including early stopping, L2 regularisation, dropout, etc.

3.2 Data pre-processing

Pre-processing of the data may improve the segmentation performance of model-based approaches. For the purposes of image segmentation, techniques depending on the physics behind the modality (knowing the noise model, or point spread function of the imaging system) are used to enhance specific features of interest, transforming the image into another (e.g. enhancing spots or filaments with specific filters).

For instance, brightfield microscopy techniques are characterised by specific point spread function (phase contrast images contain halo and shade-off, differential image contrast microscopy introduces non-uniform shadow-cast artefacts). The consequent segmentation performance will increase if these artefacts are removed so that the cells look like 'blob-like' elements. Such image enhancement techniques are further discussed in Vicar *et al* 2019 [3].

In the case of digital pathology hematoxylin-eosin stained slides, colour normalisation (colour deconvolution) is used, correcting for the variable amount of hematoxylin and eosin (or different staining) between the images. Such an approach reduces the dataset variability and leads to an easier segmentation. In the case of fluorescence confocal microscopy, deblurring (deconvolution) is widely used to improve image quality.

The pre-processing step is not of such importance in DL segmentation workflows; nevertheless, if the artefacts (noise, uneven illumination, etc.) are present in all the datasets to which the models will be applied, they need to be present in the training dataset, for the network to learn to omit these artefacts.

3.3 Ground truth: manual annotation for learning strategies

The crucial first step for the machine-learning-based techniques lies in the labelling of a part of a dataset. Learning efficiency is highly dependent on this single step; however, manual labelling of large datasets might be time consuming, labelling of 3D datasets slice-by-slice in particular. Effective learning on limited data is still a challenge in the machine learning field. To deal with this problem, several strategies exist, including: (a) increase of labelling accuracy by multiple experts; (b) crowd-sourcing; (c) unsupervised learning (e.g. from another image channel from multi-modal data); (d) transfer learning (usage of pre-trained network); and (e) data augmentation. Multiple-expert annotations may provide improvements. However, this approach does not guarantee an improvement in learning accuracy in situations when there is no consensus between experts. As shown by numerous studies, 'labelling noise' may occur, and then an approach to deal with such uncertainty must be put in place, further complicating the network design. An extreme case of such a solution is the use of crowdsourcing. At the same time, however, there is a need for efficient filtering of incorrect annotations further increasing complexity.

An effective approach nowadays is the reuse of pre-trained models (transfer learning). The pre-trained network is used as a feature extractor and is further fine-tuned on a small amount of relevant data. As the retraining is in most cases possible (only) via scripts or command lines, such an approach is not always accessible to a general and broad research community. Nevertheless, as shown in ImJoy, the retraining might be employed in the GUI directly, thus completely omitting these limitations.

Data augmentation is an approach to artificially increase data set size. Basic approaches include applying different geometrical transformations (translation, rotation, deformation), addition of artificial noise or illumination etc., to the existing images. More advanced augmentation techniques include GANs (mainly conditional GANs). GANs are composed of two networks (generator and discriminator), where the generator is trying to predict real-looking images, while the discriminator is a classifier trained to recognise between real and generated images. Both networks are trained together in a competitive manner, where the generator tries to 'fool' a discriminator and the discriminator serves as the loss function for the generator. As a result, a well-trained model will generate realistic looking artificial images.

However, multimodal data might provide a workaround to avoid lengthy manual annotation: one of the channels of multimodal data might be converted to a binary mask and thus used as a ground truth. On the tissue level, information extracted from PET images might be used for segmentation of CT images; on a microscopic scale, fluorescence channel might be helpful for segmentation of underlying electron or light microscopy images.

3.4 Semantic segmentation

In the traditional 'bottom-up' segmentation workflow, the semantic segmentation is a first step of the complex process. In this step, the image is divided into semantically meaningful classes (two or more, for instance cell nucleus, cytoplasm, and background). In this case, individual cells are not separated from each other, as this is a task of a consequent instance segmentation process.

There is a wide spectrum of methods for semantic segmentation, which can be divided into three main categories: (a) model-based (non-learning based), (b) pixel classification methods, (c) deep learning-based methods. Most widely used model-based methods include simple thresholding techniques (applied to original or feature image) and more advanced techniques like level set-based approaches or Graph Cut-based methods. Very popular learning-based methods are based on classification of every pixel, where the most well-known is a so-called 'Random Forest classifier' (an ensemble of decision trees). For every pixel, several local features are extracted (most often with some filter) and these features are used for its classification.

These methods do not (in general) perform as well as deep-learning approaches; however, they require much less training data and are faster to train, which allows the existence of interactive segmentation tools like Ilastik (ilastik.org), Trainable WEKA segmentation (imagej.net/Trainable_Weka_Segmentation) or Labkit (imagej.net/Labkit). These tools allow the user to label some pixels which are used for interactive classifier training. Today, the most powerful techniques are based on deep learning. There are many different architectures which are suitable for semantic segmentation (SegNet, DeepLab, DeconvNet), where output is an image, but the most popular architecture is U-Net [1].

3.5 Object detection

The goal of object detection is to localise different object instances, where most often output of detection algorithm is either bounding box around object or point at the object centroid.

Range of model-based approaches is based on detection of 'blob-like' objects (individual cells, yeasts in microscopy, circular tumour in CT). Laplacian of Gaussians (LoG) is one of the widely used methods for blob-detection, which works on a matched filter principle. LoG is blob-shaped and convolution (correlation) with this filter therefore enhances the blob-shaped objects. Additionally, LoG filters of different sizes can be combined to fit differently sized objects. Other blob-detecting methods include radial symmetry transform and radial voting, where gradient vectors with a magnitude above defined threshold (magnitude of the gradient is typically high on cell borders) are used to construct a parametric image characteristic of highest values in the object centers, further detectable by maxima detector. Specifically, each gradient vector increases the value of the parametric image at the distance of the expected object radius in the gradient's direction—the expected object center.

For more complex-shaped objects, template matching might be suitable. It is based on the same matched filter approach as LoG, but instead of a blob-like function, a cropped image template of an object of interest is used. In the case of maximally stable extremal region detector, the image is thresholded with all possible thresholds. From those, the threshold is used where a change of the threshold value will affect the area of the individual detected objects minimally. All previously mentioned approaches are available as Fiji plugins (LoG implmented in TrackMate).

DL-based approaches also exist for this task; there are several bounding-box regression methods including Faster R-CNN, Yolo and Single Shot detectors (SSD). The priority of those approaches is real-time detection speed, being highest for SSD at the expense of detection accuracy. These techniques are also advantageous for the detection of heterogeneous objects. When the aim is to detect a single class of objects (e.g. tumour, cells, vesicles), seed point prediction with U-Net is more suitable. For example, an image with small Gaussian functions at a position of seed points is used as a desired output of the U-Net network.

3.6 Instance segmentation

Instance segmentation (such as, for example, individual lesion segmentation in MRI/CT/PET datasets) is an approach where each individual instance of a class of the foreground objects (i.e. individual lesions) is separated from all other instances. Therefore, the performance of this segmentation step depends on the performance of the semantic segmentation and object localisation. In some scenarios (e.g. in cytology images), semantic segmentation and object detection can be combined to perform instance segmentation with simple seeded watershed (marker-controlled watershed) transform. Watershed is a technique where each regional minimum in the image is 'flooded' until two 'water sources' are met. At a 'meeting' position, a 'barrier' is created. The resulting set of barriers constitutes a watershed by flooding that corresponds to a final segmentation of the image. Seeded watershed is a modification, where instead of each local minima, only seed positions are used for flooding.

Although the segmentation pipeline is doable in these three separate steps in GUI software like Fiji, the final synthesis is rather complex. Direct application of U-Net network, which simply predicts foreground with separation line between the objects leads to insufficient results as small errors in the separation line lead to a false connection of the objects. The basic method to overcome this problem is to add more weight to the cell boundary in the loss function or to predict object centers and borders as two different classes (centers lead to the identification of individual objects and borders refine the segmentation). Such an approach is employed in U-Net-based StarDist (imagej.net/StarDist) [4], where the separation of individual objects is achieved by prediction on a normalised distance transform image instead of binary foreground. Additionally, for every pixel, distance to the border in several directions is predicted. After prediction, an individual cell is identified, for details see [4]. StarDist has also a Fiji plugin for prediction of pre-trained models.

A U-Net architecture with a different desired output for instance segmentation was employed in Cellpose (cellpose.org), where spatial gradient vectors of 'simulated diffusion from cell center' are used as a desired output together with binary masks. The binary masks are then used for semantic segmentation, and predicted vectors are used for single-cell separation. This approach provides a higher level of accuracy compared to similar methods like StarDist (figure 2).

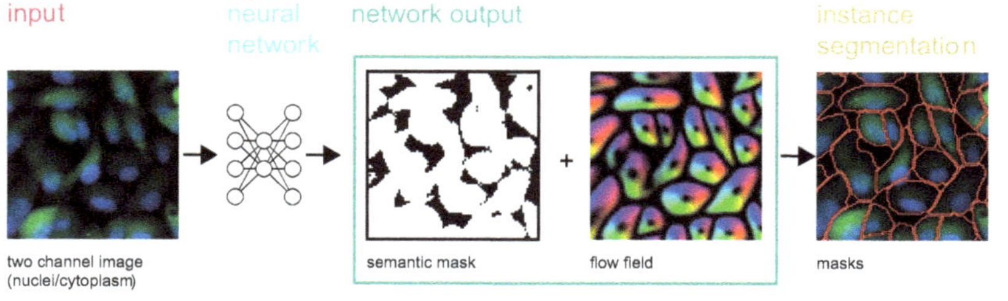

Figure 2. Cellpose segmentation process. Neural network is trained to predict flows in cells which are combined to a flow field. This is subsequently used to create the masks. Example of epifluorescence microscopy of PC-3 cells stained with Hoechst 33342 and Calcein AM, 4x.

Another popular approach for deep-learning instance segmentation is Mask R-CNN employing a powerful bounding box R-CNN regression, where segmentation of each object is predicted inside the bounding box.

3.7 Segmentation quality measures

Image segmentation may consist of a number of steps. For every such step, there is a specific way to measure segmentation quality. As multiple measures exist with various limitations or with mathematical relations between them, here we suggest general and most widely used measures: precision (sensitivity), recall (positive predictive value) and intersection over union (IoU, Jaccard Index). Due to the high rate of background pixels in typical cases, true negative pixels are not considered in these metrics. If we define a set of pixels of segmentation result X and ground-truth (GT) segmentation mask Y, then:

$$\text{precision} = \frac{X \cap Y}{Y} \tag{1}$$

$$\text{recall} = \frac{X \cap Y}{X} \tag{2}$$

$$\text{IoU} = \frac{X \cap Y}{X \cup Y} \tag{3}$$

This can be easily calculated for sematic segmentation problems, where in the case of more than two classes, we can simply calculate the average of these values over classes. In the case of instance segmentation, we need to find correspondence between segmented objects and GT objects. In such a case, we suggest using the same approach as in [5]—the SEG score. For every GT object, a segmented object with the largest IoU is found. Next, average IoU of all GT objects is calculated. If IoU for any GT object is smaller than 0.5, then IoU for this object is set to 0. This ensures that each GT object can be paired with only one segmented object.

3.8 GUI-based segmentation tools

To date, there is a spectrum of powerful segmentation tools which might be a good starting point. In this text, particular emphasis is given to open-source software solutions suitable for multimodal data; there are three major platforms—Icy (icy.bioimageanalysis.org), Fiji (imagej.net/Fiji), and CellProfiler (cellprofiler.org), all enabling a high level of functionality far beyond segmentation tasks. While Icy and CellProfiler are focused more (but not strictly) on microscopic data, Fiji enables specific tools across scales. These platforms enable the users to use both simple and fast traditional image segmentation workflows, but also to integrate deep-learning-based approaches. For this use-case, U-Net is implemented/available in CellProfiler, Fiji and ImJoy.

Although deep-learning-based environments listed below are equipped with a GUI, it is not possible to perform further training through this interface. In most cases, a modification of pre-defined code is possible in a convenient way in a Jupyter notebook interface. Although the user input is done by a direct coding, no *de novo* code preparation is needed, just parameter modification according to included documentations.

Icy is a complex open community platform for bioimage analysis which among other functions enables image segmentation. The aim of this platform is to provide easy-to-use workflows for users through dedicated plugins. For the segmentation tasks, several plugins are available: Spot detector is suitable for detection of bright spots in fluorescence images based on a wavelet transform. Plugin Active Contours automatically segments the boundary of a nucleus or cell starting from an approximate ROI, and further refining the segmentation. Furthermore, this platform also enables HK-Means and Potts segmentation algorithms with learning-based approaches still in development.

CellProfiler is an open-source cell image analysis software. Although originally intended for a single-cell segmentation, it nowadays also enables segmentation and tracking of bigger scale objects. In a recent version, U-Net segmentation is also implemented. Its 'IdentifyPrimaryObject' module combines thresholding and watershed. Additionally, the module uses a degree of smoothing and enables seed point removal below a defined distance.

Deep learning approaches have been employed in several environments. Dominating in this regard is the U-Net network, which is in various modifications accessible through a graphical user interface in multiple solutions. U-Net Fiji plugin is accessible for researchers without any experience in deep learning. As for most DL approaches, it uses GPU. It runs on a cloud service or on a single GPU-equipped computer/server. It enables detection and semantic segmentation for 2D as well as for 3D datasets with one or multiple channels and comes with a step-by-step protocol which explains how to annotate data.

To overcome the need of a large amount of annotated data, which is typical for most learning approaches, the U-Net Fiji plugin comes with a pre-trained model (on microscopy data) which makes it sufficiently effective after retraining/fine-tuning on two annotated images in most scenarios, or up to ten images for difficult tasks. The deep-learning nature of this approach provides a high level of flexibility on a heterogeneous spectrum of data across scales. When the pre-trained model does not provide sufficient segmentation effectivity for particular data, there is still an option to train the network from scratch using Jupyter notebook.

U-Net-based segmentation is also a part of ImJoy [2] (imjoy.io), an open-source cloud browser-based deep-learning platform. It aims to make the deep-learning approaches open to a broad community of life scientists. GUI is employed in all the steps of the process, including retraining. The functionality is far beyond segmentation due to the existence of multiple plugins, which also enable porting of software libraries such as Python, NumPy, scikit-learn or DL frameworks. For the segmentation task, the process starts with an ImageAnnotator plugin, which also enables annotation of multichannel data (e.g. nucleus and cell in fluorescent images). The subsequent segmentation task might then be done with the U-Net DL network within the ImJoy environment (U-Net Segmentation plugin).

The Cellpose (cellpose.org) is a cloud framework capable of predicting a segmentation with a network pre-trained on a heterogeneous set of microscopic data with further training possibilities in a Jupyter notebook.

The principle of cloud computing is also employed by the DeepCell (deepcell.org) via Google Cloud. It is a web-interface ecosystem for single-cell segmentation and tracking allowing the researchers to easily deploy and scale a deep learning platform. Users can set up their own 'Kiosk', which then enables them to drag-and-drop and process images. Alternatively, segmentation and tracking can be performed on a running version directly on the website. It works with 2D and 3D data; however, as of the date of writing this chapter, it is not compatible with multichannel images.

A similar principle is also employed by CDeep3M (github.com/CRBS/cdeep3m2), designed to work with 3D datasets of light, X-ray, and electron microscopy. Compared to the previous one, CDeep3M employs Amazon cloud machine.

Users might also employ commercial software solutions for their task. The commercial developers put significant effort into enabling comfortable and reliable work with big data. Regarding the segmentation capabilities, there is an evident effort to implement deep-learning approaches (DRV technologies Aivia, Olympus ScanR). Nevertheless, most of the commercial packages are based on relatively simple techniques like thresholding to be as robust as possible. Imaris (Oxford Instruments) provides solutions primarily aiming on 2D/3D fluorescent, electron, microscopy, X-ray, and CT imaging. It enables spot detection and tracking and simple workflows for small-scale manually aided segmentation. Arivis (Arivis AG) is suited for a similarly wide spectrum of data. It employs threshold-based, membrane-based, and blob-finder-based tools for segmentation. With the use of Arivis Vision4D, machine-learning on multichannel data is also possible. This software is intended in particular for light sheet, two-photon, or similar imaging producing big data.

For segmentation of crowded cells in microscopy, a spectrum of tools employing different segmentation techniques exists (Fogbank, FastER, CellX or Farsight toolkit). Of those, Fogbank is the most universal for the task (for detailed comparison see the Vicar *et al* review [3]), but these tools are suitable for single-channel microscopic data only and thus still of limited use in a multimodal setting. A high level of flexibility is enabled by using learning-based techniques like Ilastik [6] or Trainable WEKA segmentation. These approaches are primarily used on 2D or 3D data and, due to the learning-based nature, and effective GUI, can easily be deployed for a semantic segmentation across scales (instance segmentation needs to be performed consequently with another tool).

4 Examples of application

4.1 T1 and T2-based MRI

The advantage of multimodality for segmentation was described in the multimodal brain tumour segmentation challenge where a combination of native (T1) and T2 Fluid Attenuated Inversion Recovery (T2-FLAIR) volumes of human brain was used for classification of three tumour zones—enhancing tumour core, cystic/necrotic tumour core, and tumour edema. The segmentation of these tumour zones was possible by using a combination of these imaging modalities with an application of U-Net. For example, the best-performing 2018 approach [7] utilised multimodal segmentation, where four-channel 3D MRI images were concatenated to the network input. The authors used a near-standard U-Net network and tested the effects of pre-processing, post-processing as well as a modification of network width and depth. Interestingly, most of these affected the results significantly, except for an additional variational auto-encoder branch serving as a regularisation. Small improvements were also achieved by a test time augmentation (mirror flipping of an input image) and averaged result of an ensemble of 10 models.

4.2 Digital pathology slides

In digital pathology, the slides are characterised by their very large size (100 000 × 100 000 pixels). A large annotated dataset of 400-whole hematoxylin-eosin stained slides provided within the CAMELYON16 challenge enabled usage of very deep networks.

As for MRI segmentation (described above), the aim of this challenge was to reduce the workload and subjectivity of a physician—a pathologist in this case. Instead of single-cell pixel-wise segmentation, identification of the tumour zones is needed. Classification of small patches is performed with standard classification architectures (e.g. GoogLeNet, ResNet of VGG-Net). As a classification network is used, standard pre-training on ImageNet can be applied. However, CAMELYON16 results showed that pre-training on such a different dataset may be rather harmful, and if sufficient data is available it is better to omit it and train from scratch. Compared to model-based methods, image colour normalisation did not improve the segmentation performance.

4.3 Multichannel microscopy

Such extensive amounts of primary data, however, cannot always be created in multimodal setup, which might limit the employment of deep-learning methods. To provide the users with a starting point for segmentation of their multimodal datasets, CellProfiler ‘IdentifyPrimaryObjects’ and ‘IdentifySecondaryObjects’ might be used to detect objects from cell nuclei (primary objects) and cell segmentation masks from secondary objects (actin staining).

4.4 Neural networks-based generally applicable segmentation tools

Ilastik is a good example of a generally applicable semantic segmentation tool. Although not primarily intended for multimodal data, its functionality has been shown on drosophila brain electron microscopy slices and many other datasets.

Finally, a good starting point for a deep-learning-based image segmentation with a possibility to train on new data is ImJoy browser-based platform.

5 Conclusions

Image segmentation enables one to locate/delineate relevant parts (objects) in the image. The application of the segmentation approach to multimodal data can be a challenge as a clearly defined tool for particular image modalities does not exist. However, the expansion of machine-learning approaches in recent years introduced a significant level of flexibility, and subsequent creation of GUIs for these approaches enable dissemination of segmentation to a broad research community. In this regard, tools like CellProfiler, ImJoy, Ilastik, or Icy are good starting points. Nevertheless, no 'one-size-fits-all' solution exists for given data, so it is often necessary to try a certain spectrum of segmentation tools. Still, in certain situations (complicated dataset, big data), further adjustments and implementations (requiring coding skills) are necessary. This chapter provides just a basic introduction to the field. Readers are recommended to continue with the references listed below.

References and further reading

[1] Falk T *et al* 2019 *Nat. Methods* **16** 67–70

[2] Ouyang W, Mueller F, Hjelmare M, Lundberg E and Zimmer C 2019 *Nat. Methods* **16** 1199–200

[3] Vicar T, Balvan J, Jaros J, Jug F, Kolar R, Masarik M and Gumulec J 2019 *BMC Bioinf.* **20** 360

[4] Schmidt U, Weigert M, Broaddus C and Myers G 2018 *Medical Image Computing and Computer Assisted Intervention MICCAI* 265–73

[5] Ulman V *et al* 2017 *Nat. Methods* **14** 1141–52

[6] Berg S *et al* 2019 *Nat. Methods* **16** 1226–32

[7] Myronenko A 2018 arXiv 1810.11654

[8] Litjens G, Kooi T, Bejnordi B E, Setio A A A, Ciompi F, Ghafoorian M, van der Laak J A W M, van Ginneken B and Sánchez C I 2017 *Med. Image Anal.* **42** 60–88

[9] Moen E, Bannon D, Kudo T, Graf W, Covert M and Van Valen D 2019 *Nat. Methods* **16** 1233–46

[10] Wang S, Yang D M, Rong R, Zhan X and Xiao G 2019 *Am. J. Pathol.* **189** 1686–98

[11] Zhou T, Ruan S and Canu S 2019 *Array* **3-4** 100004

IOP Publishing

Imaging Modalities for Biological and Preclinical Research:
A Compendium, Volume 2

Parts II–IV: *In vivo* preclinical imaging, multimodality imaging and outlook

Andreas Walter, Julia G Mannheim and Carmel J Caruana

Chapter III.4.e

Visualisation for correlative multimodal imaging

Noeska Smit, Katja Bühler, Anna Vilanova and Martin Falk

The field of visualisation deals with finding appropriate visual representations of data so people can effectively carry out tasks related to data exploration, analysis, or presentation using the power of the human visual perceptual system. In the context of biomedical imaging data, interactive visualisation techniques can be employed, for example, to visually explore data, as image processing quality assurance, or in publications to communicate findings. When dealing with correlative imaging, challenges arise in how to effectively convey the information from multiple sources. In particular, the information density leads to the need for a critical reflection on the visual design with respect to which parts of the data are important to show and at what level of importance they should be visualised. In this chapter, we describe several approaches to interactive imaging data visualisation in general, highlight several strategies for visualising correlative multimodal imaging data, and provide examples and practical recommendations.

1 Introduction

Interactive visualisation systems in general employ visual representations of data to help people perform their tasks more effectively [1]. The main aims of visualisation include exploring data, analysing data, or communicating findings. Combining data from multiple sources, allows for integration of complementary information from the different modalities, for example, placing functional activity in an anatomical context [2]. In correlative imaging visualisation plays a key role in supporting tasks related to both image processing and analysis. Visualisation is used to look at acquired data after initial acquisition as a form of quality assurance, to visually check whether an image registration has been performed successfully, and to present key findings in publications via figures. Input to the visualisation pipeline can be the imaging data itself, derived data, e.g., segmentation information, and non-spatial

doi:10.1088/978-0-7503-3747-2ch28

related data, e.g., genetic information. The output of the visualisation pipeline consists of either interactive, animated, or static visualisations (images). Such visualisations can be either 2D or 3D and presented via a variety of media, all the way from images in a paper to a fully immersive virtual reality setting. In addition, special systems for visual data analysis may be developed, often referred to as visual analytics, that can be used in combination with other analysis methods, such as statistical analysis. A main intended output of any developed visualisation is increased insight and understanding for the person looking at the visualisation.

2 Challenges in a multimodal setting

The aim of correlative imaging in the life science is the acquisition of complementary information on, e.g., complex biological events, structures and/or the multi-omics environment of a sample, using a combination of imaging techniques. Combined visualisation of spatially aligned multimodal and multiscale data is an important means to provide first insight into the data in its spatial context and often the basis for the identification of regions of interest and further quantitative analysis. However, straightforward approaches like simply overlaying aligned imaging data might not always result in meaningful visualisations. In advanced correlative microscopy the data to be combined might have highly differing characteristics and information content resulting in several challenges when designing an informative visualisation of the combined data with high visual quality.

2.1 Different sampling grids

The data to be combined might be acquired at different spatio-temporal scales, varying in-slice resolution, or with varying levels of anisotropy, i.e., while a pixel within a slice is most of the time a perfect square (i.e. isotropic), the distance between acquired slices often differs from the pixel extension. In the case of 3D imaging, acquisition of different images might be done following different main axes.

In particular for 3D data, many visualisation systems cannot directly combine differently sampled image data, but require the resampling of all data to a joint reference space. This is often done already during the image alignment process. However, each resampling process delivers an interpolated/approximated version of the original data, i.e., introduces errors. Thus, the choice of resampling methods and the joint reference space must be carefully done to ensure a high quality of the depicted information.

2.2 Data size

In particular in the case of microscopy data, the size of each dataset might already be large. Data size becomes even more of an issue when combining several datasets. These might exceed easily the memory of available computers, requiring either downsampling or reduction of the data or specialised visualisation tools.

2.3 Information fusion

Fusing information from several imaging data sets into single visualisation is non-trivial. Occlusion of relevant information might occur, if the visualised content of several images overlaps at the same spatial location which might be the case even if in reality this information would be non-overlapping, e.g., due to limited spatial resolution of imaging techniques. The combination of too much information in a single image might lead to cluttering, distracting from relevant content. Both, occlusion and clutter, can be avoided by reducing shown image content to relevant information and the choice of an appropriate information fusion method, as described below.

2.4 Visualisation design

The visual appearance of the final visualisation has to be carefully designed to optimise the contained information and its perception by the reader to the context in which it is used, i.e., purpose (illustration for a publication or visualisation in an interactive application for a research task?), users (experts or the general public?) and tasks (getting an overview or analysing in depth?). For correlative imaging, this further extends to the decisions on what information is treated as context, i.e., background, what is the core information; what data reduction and abstraction methods will be used; which scale(s) are relevant and how information from different scales can be combined and finally also what is achievable with the available applications.

3 Principles and main approaches

Visualising imaging data in a correlative setting requires knowledge of advanced visualisation techniques due to the challenges that arise outlined above. Often, only the combination of different methods and visualisation paradigms lead to satisfactory results. We present here an overview on the most important approaches and refer the reader to the references listed below for further reading.

3.1 2D slice-based visualisation

A straightforward way of visualising imaging data is to present the data as a 2D image. This is a very natural way to represent imaging data, especially when the imaging data is acquired in this form as well. When 3D data is acquired, a 2D slice-based representation may still be a meaningful way to visualise the data. In this case, a suitable plane needs to be chosen that allows for a good view on the region of interest. Typically, such a plane is chosen along the axis that has the highest in-slice resolution. One way to select an appropriate plane is through user interaction. Here, the user can freely scroll through the slices to find interesting images. By scrolling through the 2D slices, users can also develop the skill to mentally construct a model of what the 3D data looks like; a skill that, for example, radiologists are highly proficient in. In medical imaging, standard planes that are visualised are the axial, coronal, and sagittal plane [6]. In addition, arbitrary planes through the data volume can be chosen. The process of visualising slices that are not aligned with the original acquisition is called multiplanar reformation or reconstruction. More advanced 2D visualisations can be obtained through a process called curved planar reformation, where a plane is reconstructed along a traced structure, for example a vessel. For multimodal imaging data, multiple imaging data slices can be visualised in a single image by picking a different colour mapping for each modality. An example of this is visualising CT data in a greyscale colourmap, with PET data visualised as a colour overlay [2].

3.2 2D slice-based visualisation

An extension to the idea of employing a 2D slice-based visualisation, is to combine the 2D image with a 3D reconstruction of a surface of interest. An example of this is the Online Anatomical Human (http://anatomy.tudelft.nl/). Here, the imaging context of the reconstructed structures is visualised as context. One challenge with such an approach is that occlusion may arise, either through the 3D surface occluding part of the 2D slice information, and/or through the 2D slice blocking the view on the sliced 3D surface. In a multimodal context, the slice-based visualisation could show one modality, while the surface is reconstructed based on another modality.

3.3 3D volumetric visualisation

When dealing with stacks of correlative images or volumetric data, volume visualisation can provide a better spatial understanding. While the interactive analysis and exploration of 2D slices can be realised by panning and zooming, the visualisation of image stacks entails additional challenges besides navigating in three dimensions. First, the extreme data sizes; image stack dimensions can easily exceed 100 000 by 100 000 pixels by 100 slices. Secondly, image data is dense and requires special consideration to introduce transparency and hide certain features while preserving native colours due to their distinct meaning, e.g., the stain in histopathology. And lastly, the exploration process needs to be facilitated as the image data itself provides little to no spatial clues for orientation especially in close-up views.

Large-scale volume rendering typically involves subdividing the image stack into blocks using a level-of-detail scheme. Depending on the view parameters and visible area on the screen, matching blocks are loaded from disk and displayed. Using such an approach, a 3D stack of histopathology data can be interactively explored in native resolution including frequent changes between low and high magnification [3] (see figure 1).

Colour mapping can directly be applied when the image data contains only one channel. By mapping values to opacity, regions of little interest can be rendered transparent. However, when dealing with full-colour data, i.e., images containing multiple colour channels, opacity mapping is not that straightforward. A direct mapping to opacity is rendered impractical by the large number of colours and small colour variations as well as noise in the data. A colour and opacity mapping based on colour similarity instead allows for a robust interface with minimal user

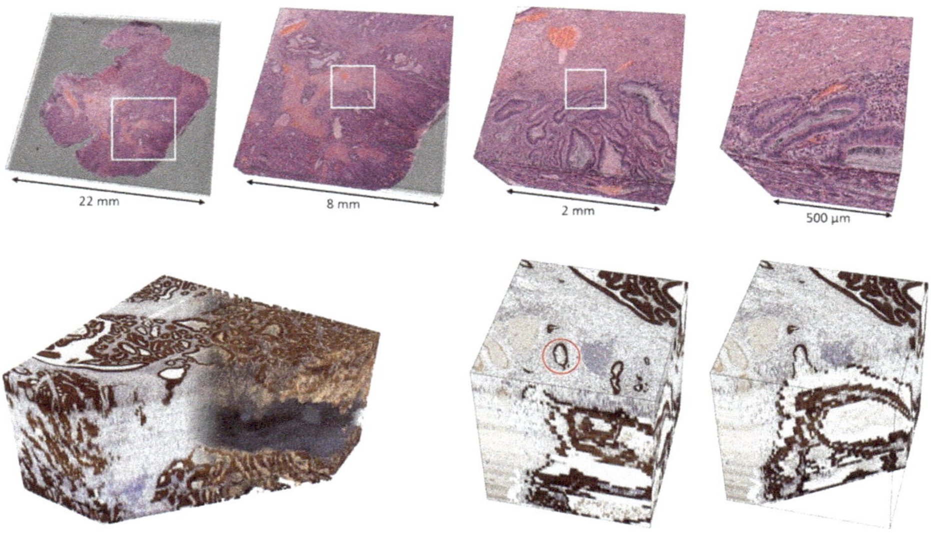

Figure 1. 3D image stack of a colorectal adenocarcinoma. At a resolution of 60 000 by 60 000 pixels and 100 slide images the dataset captures both macroscopic and microscopic details (top). Reducing the opacity of certain types of tissue can expose the intricate 3D structures hidden inside (bottom left). Clipping of the dense image stack close to a suspected tumour region (red circle) reveals a benign glandular structure instead (bottom right).

interaction requiring only a base colour and a similarity threshold. Thus, it is possible to remove the background of slide images by suppressing colours similar to white and gain insight into the hidden 3D structures (see figure 1, top and bottom left). Volume rendering can be combined with clipping planes for in-depth exploration (see figure 1, bottom right).

3.4 Spatio-temporal visualisation

Some of the imaging modalities can also be measured over time in order to analyse processes, for example, 2D images over time, or even imaging volumes that dynamically change over time. There are several ways of dealing with such time-lapse data. One approach is to visualise a single time step of the time series at once. Here, any of the visualisation techniques mentioned above may be employed, either to show the most important event in the data, or to display the data as an animation showing the time-steps sequentially. Alternative visualisation options are to attempt to visually summarise the data over time by aggregating information from the temporal component into a static summary visualisation. Bringing in multiple correlated spatio-temporal datasets further complicates efficient visualisation of spatio-temporal data, but successful visualisations have been made showcasing one modality dynamically as an animation, while another modality is shown as a static reference. An example of this is showing functional MRI activity over time on top of static structural MRI data for the brain [2].

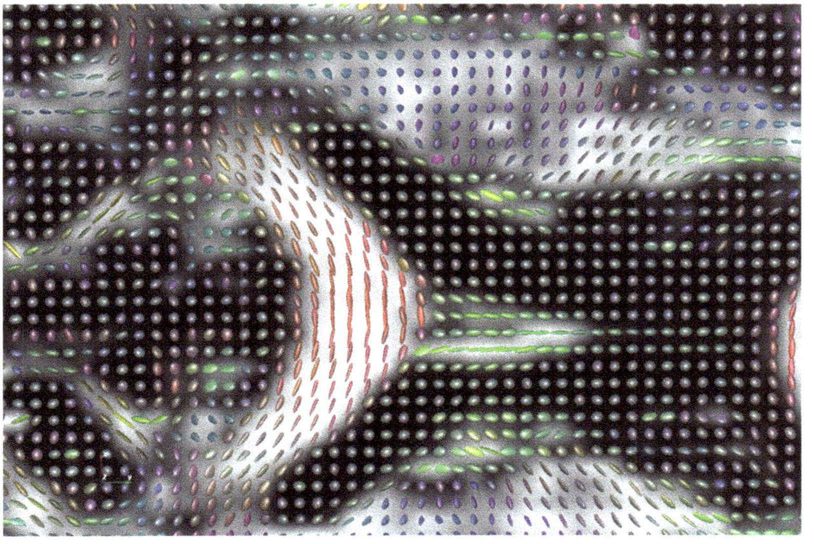

Figure 2. Tensor ellipsoid glyphs visualisation of a 2D transversal slice of a diffusion tensor magnetic resonance imaging data set of a human brain. Hue coding indicates orientation, saturation indicates anisotropy of the diffusion. Image generated with viste (https://sourceforge.net/projects/viste/).

3.5 Glyph visualisation

A glyph refers to a geometrical or graphical representation whose properties (e.g., colour, shape, scale) encode different data attributes. Glyphs are a common and effective means for multidimensional data visualisation given their versatility [7]. Furthermore, glyphs can effectively be compared and can show patterns, especially when pre-attentive characteristics of the glyphs are used to encode the attributes. In multimodal imaging data, a glyph represents local information, i.e., the information at a single position in the image. Glyphs are often designed to smoothly aggregate and represent areas to adapt to the visualisation scale. A simple example of glyph visualisation is using an arrow as glyph to represent a 2D/3D fluid velocity vector. Glyphs are common to show complex imaging information like tensors from diffusion weighted imaging (see figure 2) or stress and strain on the heart muscle. Glyph-based visualisations can suffer from clutter and occlusion issues, especially on 3D visualisations. These problems are often solved through aggregation or selection. Glyphs are commonly placed in regular grids whose regular patterns may distract from underlying data patterns, which can be mitigated by random or data-driven glyph placement.

3.6 Visual analytics

Non-spatial data is data that does not intrinsically have a 2D/3D spatial interpretation. In multimodal imaging, we can analyse the high-dimensional space defined by the range of the images independently of the spatial location of the domain. Visual analytics solutions for high-dimensional spaces is an active field of research [9].

Visualisations for high-dimensional data include scatterplot matrices (SPLOMS) and parallel coordinates, which are among the most scalable visualisations. However, these methods are limited in effectiveness when very high amounts of samples and dimensions are present in the data. In these cases, the visual analytics paradigm is an adequate choice: machine learning, visualisation and interaction are combined to generate effective insight. Multiple views showing different aspects of the data are combined interactively. Dimensionality reduction (DR) views are one of the most commonly used components. DR methods map the high-dimensional data to a low-dimensional space, e.g., 2D, preserving characteristics of the high-dimensional space. The choice of DR method depends often on the application and the task. Nonlinear methods that preserve cluster structure are used for exploration, e.g., t-Stochastic Neighbour Embedding. These methods facilitate the exploration and discovery of unknown groups of elements and have been applied successfully in many biomedical applications, such as for example Cytosplore (https://www.cytosplore.org/).

If the data is spatially resolved, there are two main ways to include the spatial information in the visual analysis [10]. The first way consists of having the visual exploration of the non-spatial and spatial domain in separate views. These views are linked through interactive selection and filtering approaches. The high-dimensional domain view and interaction with this view provide initial means to classify the space. This classification is mapped into the spatial domain and it is then refined by identifying correlations in the spatial domain and vice versa. This method is effective when there is a reference image that the user can understand, or a well-defined classification of the high-dimensional space. However, this is challenging when the information in the spatial domain characterisation is defined by difficult to classify high-dimensional space.

An alternative way to include spatial information is by increasing the dimensions of the high-dimensional space, for example by adding the spatial coordinates as weighted dimensions. The new space can make use of any existing visual analysis mechanisms for high-dimensional spaces. However, the semantics are lost and there is no intrinsic trivial way to combine the spatial and the feature domain without introducing parameters whose value is difficult to define.

3.7 Pixel/voxel level fusion

Pixel/voxel level fusion combines two or more images to a new image by directly computing the new pixel/voxel value out of spatially corresponding pixel or voxel values without considering the surrounding area. The combination can be done, e.g., by adding, by averaging, by selecting minimum or maximum values, or even by a logical operator. In figure 3, a visualisation employing pixel level fusion is showcased.

3.8 Feature level fusion

Feature level fusion combines imaging data employing extracted local image features like edge strength, structure tensors or measures evaluating the presence of vessel like structures at a location.

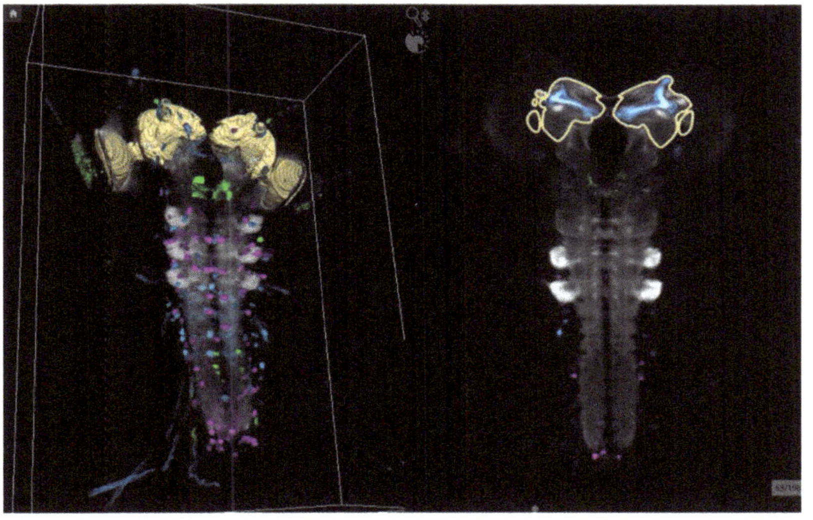

Figure 3. Mixed pixel level and decision level fusion (screenshot from www.larvalbrain.org).

3.9 Decision level fusion

Decision level fusion builds on higher level information already extracted from the images and found relevant for the visualisation, like segmentations and semantic annotations of structures of interest and related quantitative values. Decision level fusion does not use, or only partially uses, raw original data to compile the final visualisation, but uses the derived data instead. This data can be still spatial and visualised in the original space defined by the imaging data, like a segmented structure in its anatomical context or glyphs representing quantitative values at a certain location. An example of this is visible in figure 4, where segmented areas from Gal4 images are combined with glyphs [4]. If pushed to the extreme, decision level fusion might also result in a more abstract visualisation of the combined data leaving the original image space to focus on extracted core information.

4 Examples of application

Many visualisation research software applications have been developed for specific correlative data, users, and research tasks. Examples of this are ImaCytE [5] and Larvalbrain. ImaCytE is a tool for analysing spatially resolved high-dimensional data from imaging mass cytometry. It provides a visual analysis workflow and enables the user to discriminate cell types and analyse cellular microenvironments. Larvalbrain (www.larvalbrain.org) provides an example for the correlative visualisation of a collection of Gal4 images of neurons in context of a standard brain derived from a set of nc82 stainings (see figure 3).

In addition, visualisation functionality has been integrated into many of the software tools that are already in use for imaging research. We here briefly outline

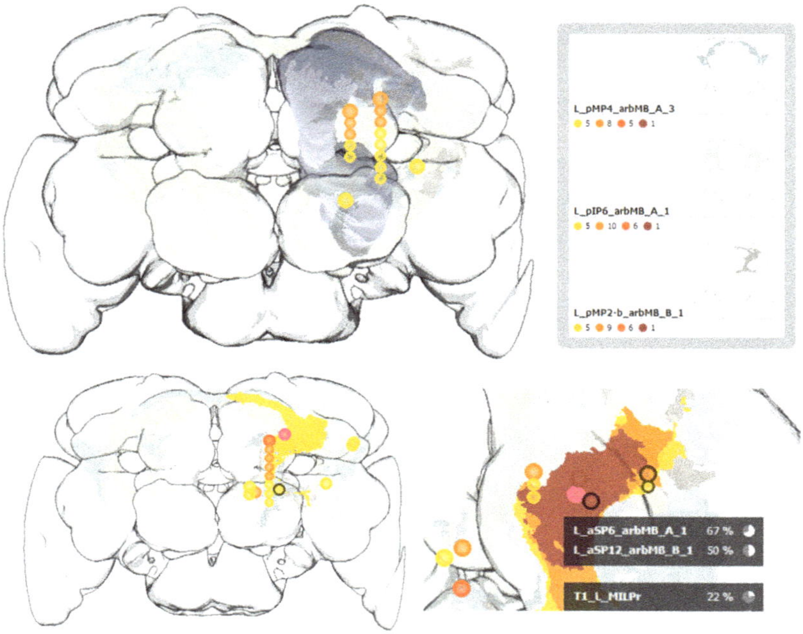

Figure 4. Example for decision level fusion: arborisation areas of segmented neurons extracted from Gal4 images of the adult drosophila melanogaster, colours depict overlapping areas and glyphs represent quantitative values [4], copyright John Wiley & Sons.

some of the freely available software tools that are ready to be used, though this is not an exhaustive list:

- Tomviz (https://tomviz.org/) is a software tool specifically targeting visualisation of transmission and scanning transmission electron microscope data, which supports image slicing, surface rendering, volume rendering, as well as simple animations.
- FIJI/ImageJ (https://imagej.net/) is a general software tool for image processing that also includes visualisation options, similar to what Tomviz provides. In addition, ImageJ enables volume-rendering of multi-channel data, including time-lapse data.
- Icy (http://icy.bioimageanalysis.org/) is a community platform for bioimage informatics. Visualisation options include volume rendering via VTK. The platform integrates ImageJ and its functionality.
- 3D Slicer (https://www.slicer.org/) is a platform for medical image processing and visualisation, which includes functionality for visualisation of time-varying volumes.
- ParaView (https://www.paraview.org/) is a general purpose data analysis and visualisation platform with powerful visualisation capabilities for a variety of domains.

5 Conclusions

Visualisation can be of great value in exploration, analysis, and communication of correlated multimodal imaging data. Many of the visualisation techniques that were originally developed with single modalities in mind, can also be employed in the visualisation of multiple correlated imaging modalities. However, this gives rise to additional challenges in developing efficient visualisation techniques for the data, tasks, and users at hand. More visualisation research is needed on visualising multimodal spatio-temporal data, as well as how to effectively visualise combinations of more than two imaging modalities at once. In addition, many of the visualisation research software prototypes are not ready for general use by users from other domains, while many of the established visualisation platforms do not focus on visualising correlated multimodal imaging data explicitly. Further visualisation research and software development is needed and ongoing in this area with the aim to establish high quality software for efficient visualisation of correlative imaging data that can be used by non-visualisation researchers and practitioners alike.

References and further reading

[1] Munzner T 2014 *Visualization Analysis and Design* (Boca Raton, FL: CRC Press)

[2] Lawonn K, Smit N N, Bühler K and Preim B 2018 *Comput. Graph. Forum* **37** 413–38

[3] Falk M, Ynnerman A, Treanor D and Lundström C 2019 *IEEE Trans. Vis. Comput. Graph.* **25** 1008–17

[4] Swoboda N, Moosburner J, Bruckner S, Yu J Y, Dickson B J and Bühler K 2016 *Comput. Graph. Forum* **36** 160–71

[5] Somarakis A, Van Unen V, Koning F, Lelieveldt B P and Höllt T 2019 *IEEE Trans. Vis. Comput. Graph.* **27** 733–43

[6] Smit N and Bruckner S 2019 Towards advanced interactive visualization for virtual atlases *Biomedical Visualisation* ed P Rea (Advances in Experimental Medicine and Biology vol 1156) (Cham: Springer) pp 85–96

[7] Ropinski T, Oeltze S and Preim P 2011 *Comput. Graphics* **35** 392–401

[8] Beyer J, Hadwiger M and Pfister H 2015 *Comput. Graph. Forum* **34** 13–37

[9] Kehrer J and Hauser H 2012 *IEEE Trans. Vis. Comput. Graph.* **19** 495–513

[10] Sorger J, Ortner T, Piringer H, Hesina G and Gröller M E 2015 A taxonomy of integration techniques for spatial and non-spatial visualizations *Vision, Modeling and Visualization (VMV)* (Graz: The Eurographics Association) pp 57–64

IOP Publishing

Imaging Modalities for Biological and Preclinical Research: A Compendium, Volume 2

Parts II–IV: *In vivo* preclinical imaging, multimodality imaging and outlook

Andreas Walter, Julia G Mannheim and Carmel J Caruana

Chapter III.4.f

Data compression algorithms for biomedical images

Francisco Cunha, Sérgio Faria, João Parracho, Lucas Thomaz and Pedro Assunção

Computational image analysis and automation assume an increasing importance in today's biomedical workflows. The massive adoption of imaging protocols, combined with the advent of cutting-edge modalities, has been pushing data throughput at bioimaging facilities to unprecedented levels. Based on this, a natural requirement for exploiting the full potential of multimodal frameworks consists of simple and fast access to the wide range of acquired pictorial data. For this purpose, the design of more efficient data structures, as well as the adoption of good information management practices applied to bioimage data, constitute very relevant initiatives. One such intervention is data compression and this text provides an overview and description of the main image compression methods, as well as an exposition of the potential of certain image formats according to specific application requirements, while ensuring the preservation of the data integrity.

1 Introduction

Image compression refers to the set of image processing techniques devised to convert the acquired image data into more efficient data representations. By contributing for storage savings, data compression enhances the value of the hardware allocated to image archiving, as well as simplifying bandwidth requirements for transmission purposes. The question of the need for image compression in clinical and preclinical settings becomes especially relevant in a time that many bioimage protocols easily overload the installed storage infrastructure of imaging facilities, and the alternatives provided by cloud-based services typically involve considerable long-term costs for institutions. Given such limitations, image compression may constitute a very relevant tool in the bioimaging ecosystem.

doi:10.1088/978-0-7503-3747-2ch29

In the following subsections, the most common encoders and respective fundamentals will be discussed, as many of them are currently implemented in a wide range of imaging systems that simply encapsulate their logic through proprietary file formats.

2 Challenges in a multimodal setting

In the scope of the review of image compression algorithms that will be discussed, it is important to note that those are designed to operate on a single-modality basis, i.e. they are not prepared, by definition, to jointly encode multimodal sequences.

Nonetheless, at this time, image domain transfer and image translation constitute hot research topics that, in the short term, can deliver important foundations for the development of coding tools specifically designed for the joint compression of paired multimodal sequences. By allowing one to virtually synthesise multimodal data [1–5], it is expected that these data-driven approaches contribute to generalise the predictive coding that most compression algorithms implement, so that a supplementary dimension, which refers to the multimodality, can be handled and thus reduced.

3 Principles and main approaches

In short, compression systems involve a paired group of independent algorithms aiming to reduce the amount of some input data (coding) and to reconstruct the compressed information (decoding). Generally, the compressed information is converted into a bitstream (sequence of bits) or a file, according to a specified semantic, prior to being stored or transmitted. There are two distinct compression modalities—lossless and lossy—which differ in terms of their reversibility, i.e. the ability to fully recover the original data after compression. If not well-constrained, the information loss resulting from the (lossy) compression process may lead to the corruption of analysis outcomes (automated or not), thus making lossless compression the preferred option for bioimaging applications. It should be mentioned, however, that some applications may allow relaxing some distortion-related requirements, thus enabling the adoption of visually lossless compression schemes, which provide increased storage savings, by introducing little, non-perceptible loss of visual information.

Image compression techniques exploit the redundancy existing between adjacent pixels in the spatial, temporal, or frequency domain. These techniques may be applied to individual pixels or groups of pixels, thus allowing one to discard all the information allocated to represent redundant data. The effectiveness of image compression algorithms depends not only on the used techniques, but also on the amount of data redundancy, and therefore on the image content. However, given that distinct compression algorithms differently model such redundancy, for a given image, variable compression is achieved by distinct compression methods. Furthermore, it must be noted that the compression efficiency offered by lossless methods is far below that offered by lossy algorithms given the requirement for reversibility of the former. An intuitive metric for measuring the compression efficiency of a given compression algorithm is the ratio of the original image size to the compressed image size—compression ratio (CR). In CR units, lossless coding performance typically ranges between 1.5 and 3, whereas lossy compression algorithms may reach values starting from 10 to 100's, depending on the tolerable distortion of the reconstructions, with the visually lossless methods comprehended in the lower tens.

Two important techniques implemented in lossy schemes that enable such high CRs are transform-based coding and quantisation. While the latter refers to the process of mapping a larger domain of image intensities into a narrower one, transform coding briefly constitutes an initial processing step aimed to convert images into more compressible representations, by projecting the image signal into different spaces that typically verify some sparseness and compactness requirements (e.g. frequency-domain). In many image encoders, the energy compactness of the transform is exploited by the quantisation process to eliminate the irrelevant information, usually not perceived by the human visual system. Examples of some transform-based encoders are the JPEG and JPEG 2000 algorithms that exploit frequency-based representations obtained through the discrete cosine transform (DCT) and the discrete wavelet transform (DWT), respectively.

In lossless compression frameworks, in contrast, quantisation is not allowed since it originates information loss. Additionally, transform-based coding is constrained to reversible transforms and its overall performance is limited as quantisation cannot be applied to the transformed coefficients. Instead, lossless algorithms exploit the assumptions of similarity and causality in image data, in order to model image intensity statistics through predictive coding, which ultimately acts as a decorrelation technique. Predictive coding constitutes a set of techniques by which the values of specifically sampled reference pixels serve to estimate (predict) the intensity of other pixels in the neighbourhood. Since the differences between the predicted and the original values usually correspond to more sparse and compact numerical distributions, they can be represented using a fewer number of bits, thus providing storage savings. Although no universal solution is guaranteed for image compression, optimal compression for a given input with a predictive method is provided by information theory principles, thus making image compression the problem of designing data models that optimally reduce the amount of required data to encode given information.

4 Examples of application

Several standardisation efforts in the field of digital imaging have been made to provide interoperability between different picture archiving and communication systems. Particularly, standardisation bodies such as the International Telecommunications Union (ITU-T) and the International Standards Organisation (ISO) have published several recommendations on image and video compression, with a view to foster the adoption of digital imaging in multiple domains. In the field of medical imaging, such recommendations were validated by the Digital Imaging and Communications in Medicine (DICOM) committee that brings together a set of guidelines regarding storage, visualisation, and transmission of medical image and patient data.

The DICOM volume dedicated to data structures and encoding schemes provides a list of the data formats allowed to represent the compressed images and respective metadata. Typically, such formats encapsulate the definitions of some image compression standards in the form of the so-called transfer syntaxes.

Given that understanding how DICOM regulates image compression directly relates with a concise interpretation of the operating principles of such image compression standards, a brief review on the most important ones will follow.

4.1 JPEG-LS

The ISO 14495–1/ITU-T T.87 Lossless and near-lossless compression of continuous-tone still images[1], commonly referred to as JPEG-LS, is a method for coding continuous-tone, grey-scale, or colour digital still images, allowing lossless (bit-preserving) and near-lossless compression (where the error for each reconstructed sample is bounded by a pre-defined value). It is a very competitive compression algorithm given its characteristics of reduced memory footprint and fast execution. The JPEG-LS encoder implements a compression method based on context modelling of the input sample, prediction, and entropy coding. Firstly, for each input sample (e.g. pixel) a context is determined, where each context corresponds to a statistically homogeneous class. Then the current sample is predicted from a causal neighbourhood, and the prediction error is computed. In the case of near-lossless compression, the prediction error is also quantised.

The contexts modelling is based in gradient information of the input sample neighbourhood, resulting in a data model that enhances local compressibility, by allowing one to predict the probabilities of the local intensities with greater likelihood. Such an approach requires the information that characterises the adopted contexts (some of the local image intensities and the adopted partition schemes) to be known by the decoder afterwards.

The prediction of a sample is performed by using four neighbor samples: left, top, top-left, and top-right. The effectiveness of predictive coding largely depends on how well the adopted data model predicts the local intensity distributions. Regarding this, JPEG-LS provides two operating modes—regular and run-length—which are adaptively triggered based on the variability of the local image statistics. Such versatility constitutes a very useful feature since run-length encoding provides a much faster compression on images with homogeneous background and/or little detail, such as evidenced in some examples of fluorescence microscopy images. When the context information indicates that the neighborhood of the current pixel contains details, the regular operating mode is selected, where entropy coding is carried out using Golomb codes or arithmetic coding.

[1] ISO/IEC 14495–1 ITU-T Rec. T.87, Information technology – Lossless and near-lossless compression of continuous-tone still images: baseline, 1998. (Source code available at: http://libjpeg.sourceforge.net/)

4.2 JPEG 2000

Resulting from an effort of the JPEG committee to provide an irreversible-to-reversible compression alternative with features such as resolution scalability, partial decoding (random access, number of components), and region-of-interest coding, JPEG 2000 constitutes an ISO standardised transform-based coding system[2], built-upon the adoption of DWT as a decorrelation step. Furthermore, extensions such as the Multi-Component Transform (Part 2 of the ISO/IEC 15444 standard), which allows one to generalise the operating principle of the proposed single-component compression for multichannel or multi-frame images makes this coding system compatible with a wide range of image sequences, such as diffusion-weighted MRI. Another relevant extension[3] proposed by the standard JPEG 2000 coding system involves the volumetric expansion of the DWT to fully exploit the axial redundancy that some volumetric sequences may present, thus representing an advantage when compared to the per-image operation of the usual still-image encoders. This extension also supports the lossy-to-lossless coding functionality (quality scalability), resolution scalability and region-of-interest coding. In fact, this feature makes JPEG 2000 (Part 10 of the ISO/IEC 15444) to fit the requirements for optimal compression of inherently 3D single-channel data such as CT and MRI images, although any other type of volumetric datasets may benefit from the adoption of the 3D extension of the JPEG 2000 coding system. Furthermore, given that JPEG 2000 enables quality scalability through progressive-coding, combined with random access capabilities, decisively contributes for being the preferred choice for lightweight browsing of very high-resolution images like digital versions of pathology slides, in the same way it can be adopted to provide enhanced visualisation experience of high-resolution images such as high-resolution episcopic microscopy (HREM).

The JPEG 2000 coding pipeline is comprised of the following modules: multi-component transformation, tiling, wavelet-transform, quantiser, and entropy-coding. Finally, the rate allocator decides which compressed data is inserted in the bitstream. The multicomponent transformation module has reversible and irreversible modes, allowing for up to 16 384 components. In the tiling mode, each component can be divided into up to 65 535 rectangular tiles, and in the case of a single tile, its maximum resolution corresponds to the limit size of $2^{32} \times 2^{32}$. The sample bit depth can be represented by up to 38 bits/sample and can be different for each component. An L-Level wavelet transform is applied separately to each component (of each tile), generating a set of sub bands of wavelet coefficients. The reversible mode uses a (5,3) biorthogonal filter and the irreversible mode employs the (9,7) biorthogonal filter. The wavelet coefficients can be quantised differently in each sub band, but the step is one in the reversible mode. The wavelet coefficients are grouped into rectangular non-overlapped sets called code-blocks,

---

[2] ISO/IEC 15444–1 ITU-T Rec. T.800, Information technology—JPEG2000 image coding system: Core coding system, 2004. (Source code available at https://www.openjpeg.org)

[3] ISO/IEC 15444–10 ITU-T Rec. T.809, Information technology—JPEG2000 image coding system: Extensions for Three-dimensional Data, 2011. (Source code available at https://www.openjpeg.org)

which are entropy encoded employing a bit-plane-based binary arithmetic coding. Because it is possible to obtain the distortion reduction and the required bitrate during each coding pass, an optimal rate allocation is achievable if the whole image is encoded. This method also allows one to encode arbitrary shaped regions of interest (ROI), with higher quality than the background, transmitting them first. In JPEG 2000, the MAXSHIFT method is employed to identify the wavelet coefficients in the ROI and up-scaling them by a user pre-defined value. As ROI coefficients and background are in different bit-planes and have different quantisation indices, the rate allocator and the decoder can identify them. This feature is particularly important to encode bioimages since the region of interest and the background can be differently encoded.

4.3 H.265/HEVC

The H.265/HEVC (ISO/IEC 23008–2 Standard)[4] is a video coding standard named High Efficiency Video Coding, which inherently allows one to exploit both spatial and temporal redundancies in a lossless-to-lossy architecture. It was designed as a successor of the widely used Advanced Video Coding (AVC, H.264, or MPEG-4 Part 10), to support resolutions up to 8192×4320 pixels, and offering from 25% to 50% better data compression at the same level of video quality. This performance gain is achieved due to the use of additional prediction modes and partitioning schemes. In order to cope with the additional complexity, some tools have been introduced to take advantage of the parallel processing architectures. Besides exploiting spatial redundancy through intra-frame prediction techniques, the temporal redundancy of the sequence is also addressed through motion estimation and compensation techniques, commonly referred to as inter-frame prediction. The core system employs a hierarchical content-based image partitioning scheme, blocks from 4×4 to 64×64, prediction followed by DCT-based transformation and quantisation of the residuals and context-adaptive binary arithmetic coding, which refers to an efficient entropy coding technique.

To operate with the lossless profile, the HEVC main profile (HEVC) bypasses transform, quantisation, and in-loop filters. When compared to the non-lossless HEVC coding mode with the smallest quantisation parameter value (i.e., 0 for 8-b video and −12 for 10-b video), the lossless mode provides perfect fidelity and an average bitrate reduction of 3.2%-13.2%. In order to further improve the coding efficiency, a sample-based angular intra prediction (SAP) method is employed instead of the block-based one. This one uses adjacent neighbours for better intra prediction accuracy and performs prediction sample by sample. The experimental results reveal that the SAP provides an additional bit-rate reduction of 1.8%–11.8% on top of the HEVC lossless coding mode.

Analogously to the above-mentioned standard, some extensions have been included in the H.265/HEVC coding standard—Range Extensions (RExt) [6]—in

[4] ISO/IEC 23008–2 ITU-T Rec. H.265, Information technology—High efficiency video coding and media delivery in heterogeneous environments, 2013. (Source code available at: https://hevc.hhi.fraunhofer.de/)

order to provide alternative coding features and to make the coding framework compatible with superior image resolutions, bit-depths, as well as alternative colour encoding formats. The operation of H.265/HEVC system must be configured to match the characteristics of a given source image sequence, namely considering the spatial and temporal resolution, the number of components and the bit depth, among others. Also, the operating conditions must be taken into consideration regarding the visualisation and analysis systems. While the latter may require the original raw data, the former may work with visually lossless methods. In fact, some mobile devices currently provide H.265 decoders, however the limited availability of embedded profiles make them unsuitable for computer assisted diagnosis supported by image analysis, since such applications typically require the original raw data. In this context, H.265 also provides scalable encoding versions which have been adopted by DICOM, offering a more efficient way to store and manage image data. The targeted applications include using HEVC streams to store and exchange raw images but also to produce lossy content out of the archived lossless streams. The 'HEVC/H.265 Scalable Monochrome' enables the storage of monochrome sequences of still pictures with resolution up to 4096 × 2160, with a bit depth varying from 8 to 16 bits, at 50 Hz/60 Hz, in a lossless or lossy format and comprising a detachable lossy and lighter version. CT scanners typically produce such content. The 'HEVC/H.265 Scalable Main 4:4:4 Image Compression' tier to encode and decode RGB still pictures and video content with a resolution up to full HD with a bit depth of 8 bits per component. This will enable the storage of RGB video files and sequences of still pictures with a resolution of 1920 × 1080 and 8 bits per component, at 50 Hz/60 Hz, in a lossless or lossy format and comprising a detachable lossy and lighter version. Such content can typically be produced when rendering 3D views in 2D. Another relevant aspect is related to the encoding profiles more suitable to encode sequences of images, namely HEVC/HEVC-RA (random-access), where the third dimension is temporal (video) or spatial (volumetric). In such a case, a video encoder is able to exploit the redundancy between images along the sequence. To achieve higher compression ratios, the prediction techniques are applied to groups of images causing some dependencies between them, thus limiting the instantaneous (random) access to each image individually at the decoder. This means that, depending on how fine-grained the random access is required to be, one must select the appropriate number of reference images, which are used for predictive coding of the sequence, as well as the number of images between those that can be individually decoded.

In the following, we will revisit some up-to-date alternatives to the DICOM-compliant image compression algorithms provided that superior compression performance has been reported in the literature.

4.4 CALIC

Up to the emergence of JPEG-LS, the Context Adaptive Lossless Image Codec (CALIC) [7] was the best in-class lossless compression scheme given its compression efficiency. In practice, JPEG-LS shares some of the CALIC functionalities since both methods adopt context modelling. However, CALIC is built-upon a larger number of modelling contexts, which are used in prediction and in residual coding, making it a more complex system. The initial prediction employed in CALIC is based on the local gradient information that is subsequently refined when the prediction context and the associated residual bias are determined. Finally, after a process of contexts quantisation and modelling, the prediction residuals are entropically coded. In addition to the encoding mechanism for 2D images, some volumetric expansions of the CALIC coding system have been proposed.

4.5 MRP

The minimum rate predictors (MRP) is a state-of-the art lossless compression technique and here we refer to its volumetric extension—3D-MRP- which was proposed as a superior compression method for volumetric medical images [8]. Comparing to JPEG-LS, JPEG 2000, HEVC, and CALIC, the authors reported compression gains ranging from 12% to 15%. The 3D-MRP algorithm uses an adaptive block-based predictive scheme that dynamically assigns linear predictors accordingly to the local content of the image blocks. Unlike other techniques that address the problem of optimal design of predictors, the 3D-MRP implements a volumetric optimisation scheme aimed to iteratively design linear predictors that minimise the number of necessary bits to encode the resulting residues. Its adoption in the context of volumetric image compression allows one to reach increased compression performance, however at the cost of increased computational complexity.

4.6 JPEG-XL

The JPEG XL coding system currently constitutes a standardisation (ISO/IEC 18181)[5] effort to provide compression efficiency superior to existing standard codecs, while keeping a set of desirable features such as resolution scalability and support for images with up to 4096 channels. It also has support for: 16-bit integer and 16-bit float inputs, and outputs with corresponding precision, multi-channel lossless coding and 'visually lossless' coding (as defined in AIC-2 (ISO/IEC 29170–2)), without requiring mathematically lossless coding.

The codec architecture is backward-compatible with the legacy JPEG format and provides efficient lossy-to-lossless compression of high-quality imagery. In its core, by extending the fixed size DCT blocks implemented in the legacy JPEG into variable-size DCT blocks, combined with a progressive DCT-based decomposition, JPEG XL allows for both efficient and versatile image decorrelation and decoding. Furthermore, as opposed to the legacy JPEG format, JPEG XL implements an adaptive predictor that exploits the images bi-dimensionality in such way that it optimally updates the prediction mode, thus leading to storage savings. However, the most remarkable feature of JPEG XL consists of an alternative entropy coding

[5] ISO/IEC CD 18181 Information technology—JPEG XL image coding system, (under development). (Source code available at: https://gitlab.com/wg1/jpeg-xlm)

scheme, called Asymmetric Numeral Systems, that achieves significantly faster operation without sacrificing compression efficiency. For now, the group involved in the JPEG XL standardisation claims that its adoption can lead to storage savings up to approximately 60%, in common use at equivalent subjective quality [9]. This combined with the implemented group splitting principle, which allows for parallel decoding, makes JPEG XL a very convenient solution for biomedical image compression.

4.7 VVC

Versatile Video Coding (VVC) is currently under development by the Joint Video Experts Team (JVET) of ITU-T Video Coding Experts Group (VCEG) and ISO/IEC Moving Picture Experts Group (MPEG) to become the successor of H.265/HEVC as the next generation video coding standard [10]. The development of VVC intends to meet the requirements to ultra-high definition (4 K to 16 K) and 360-degree video computation, with a bit depth from 10 to 16 bits per component. It also has support for still picture coding, scalable video coding (temporal, spatial, and quality), as well as for lossless and subjectively lossless compression.

VVC inherits most of its coding architecture from HEVC/H.265 with the improvement of several techniques such as intra prediction with the addition of more directional modes, inter prediction by using neural networks, and focussing on affine models for motion compensation, the partition structure, in-loop filter, quantisation and entropy coding. It also uses new high complexity elements such as affine motion compensation, rectangular, and triangular image partitions, as well as some prediction in decoder side. With the following improvements VVC aims to achieve a 30%–50% bit-rate reduction over H.265/HEVC Main Profile for the same perceptual quality.

5 Comparative overview

The previously described image and video coding algorithms have been proposed to fulfil different image coding requirements, using the technology available at the time. Currently, some of these algorithms are no longer the state-of-the-art for natural images, but due to their functionalities they are particularly suited to encode certain types of images, as in the area of bioimaging.

In this context, table 1 shows the ability of these compression methods to encode volumetric data or to perform coding of ROIs. Additionally, a ranking of the

Table 1. Features comparison between coding algorithms.

	JPEG LS	JPEG 2000	H.265/HEVC	CALIC	3D MRP	JPEG XL
Volumetric coding	×	✓	✓	✓	✓	×
ROI coding	×	✓	✓	×	×	✓
Compression efficiency	6	5	2	4	1	3
Low complexity	2	3	4	5	6	1

methods is presented (1-best and 6-worse), to classify each method in terms of compression efficiency and coding complexity. As can be seen, it may happen that the most efficient algorithm presents the highest complexity. Therefore, for each application it is necessary to evaluate the operating conditions.

6 Conclusions

The massive adoption of digital imaging in research and medical practice currently constitutes a data bottleneck for institutions, which increasingly demand compression systems for efficient storage and distribution. In this context, the adoption of image compression standards (some included in the DICOM standard) has enabled superior compatibility and interoperability between picture archiving and communication systems. Nevertheless, besides achieving state-of-the-art compression efficiency, alternative compression methods must be considered for specific applications, given the richer set of features that may provide.

We focussed on the most relevant lossless coding schemes by describing their principles and highlighting core functionalities, but also on recent research aiming to improve the modest compression performance provided by lossless techniques. Concurrently, even though no mature guideline exists for regulating the use of lossy compression in clinical settings, the adoption of visually lossless techniques, capable of delivering high-quality visual contents at reduced bit rates, constitutes a natural trend, particularly suitable for supervised image inspection purposes and semi-automated analysis. Furthermore, the growing adoption of hybrid multimodal imaging systems such as CT/PET and MRI/PET, combined with the advent of deep learning, has been opening new research paths to adopt cross-modality image synthesis as a tool to encode complementary, yet correlated, datasets in a more compact format.

Combined with the wide range of preclinical imaging modalities, image-based applications such as quantification and segmentation can constitute important research platforms for quantitatively characterising the impact of data compression on specific routine protocols. In fact, in the short term, image compression for preclinical imaging might involve data-specific coding schemes (not necessarily related to the human visual system) to optimise the resulting bitstreams against objective quality metrics determined by post-processing applications.

References and further reading

[1] Han L and Yin Z 2018 *Mach. Vision Appl.* **29** 1257–67

[2] Dong X, Wang T, Lei Y, Higgins K, Liu T, Curran W, Mao H, Nye J and Yang X 2018 *Phys. Med. Biol.* **64** 215016

[3] Armanious K, Jiang C, Abdulatif S, Küstner T, Gatidis S and Yang B 2019 *Proc. of EUSIPCO 2019—27th European Signal Proc. Conf. 2019* 09

[4] Nadarajan G and Doyle S 2020 Medical imaging 2020: digital pathology *Proc. of SPIE* 11320

[5] Ben-Cohen A, Klang E, Raskin S, Soffer S, Ben-Haim S, Konen E, Amitai M and Greenspan H 2019 *Eng. Appl. Artif. Intel.* **78** 186–94

[6] Flynn D, Marpe D, Naccari M, Nguyen T, Rosewarne C, Sharman K, Sole J and Xu J 2016 *IEEE Trans. Circuits Syst. Video Technol.* **26** 4–19
[7] Wu X and Memon N 1997 *IEEE T. Comm.* **45** 437–44
[8] Lucas L F R, Rodrigues N M M, da Silva Cruz L A and de Faria S M M 2017 *IEEE Trans. Med. Imaging* **36** 2250–60
[9] Alakuijala J *et al* 2019 Applications of digital image processing XLII *Proc. of SPIE* 11137
[10] Bross B, Chen J and Liu S 2019 *JVET-N1001: Versatile Video Coding (Draft 5), Joint Video Experts Team (JVET), 14th Meeting* (Geneva, CH, Technical Report)

IOP Publishing

Imaging Modalities for Biological and Preclinical Research: A Compendium, Volume 2

Parts II–IV: *In vivo* preclinical imaging, multimodality imaging and outlook

Andreas Walter, Julia G Mannheim and Carmel J Caruana

Chapter III.4.g

Computer-assisted analysis of multimodal image data—perspectives and conclusion

Nataša Sladoje, Joakim Lindblad and Perrine Paul-Gilloteaux

Modern image processing methods, heavily relying on the latest machine and deep learning techniques, are rapidly finding increased use in the study of complex, large, and heterogeneous multimodal image data, making the comprehensive analysis of such complex data feasible. In this chapter we give a general view on computer assisted image processing in a multimodal context. We discuss the main challenges of the field, some popular approaches in addressing these challenges, and ongoing community efforts to facilitate communication, collaboration, and knowledge exchange in this highly multidisciplinary, and rapidly developing field.

1 Challenges in a multimodal setting

There are many challenges in the highly multidisciplinary field of computerised image processing. The development of new algorithms and software relies on *substantial cross-expertise*; in the physics of the imaging modality, the biochemistry of stains, markers and tracers, and the knowledge of the biological model and/or of the disease and organs studied, all brought into the fields of computer science and mathematics. When the task is analysis of multimodal image data, this plurality is getting even more extreme. Knowledge of the specificities of several, often very different, modalities has to be combined to reach the sought goals, while the challenges of different sample preparation techniques, and imaging conditions required for the diverse modalities multiply. The probability of successful acquisition of high-quality data by several involved complex imaging processes rapidly approaches zero. We may quickly realise that it is relatively easy to generate *non-manageable amounts of data*, but only a small portion of that amount is of a quality which enables reliable information extraction, by a human expert or by a machine.

And yet, the same small portion may still be far too big and complex for manual analysis.

Intelligent automated approaches are essential for efficient extraction of the most valuable information from the *complex data*, while at the same time reducing its dimensionality to a manageable and comprehensible level. Available multimodal image data analysis methods tend to be highly optimised for a particular combination of two modalities, or even a particular type of imaged objects (e.g., anatomical structures). They are often prohibitively slow in modern, often terabyte sized, imaging scenarios. Generality and applicability to new, or additional, modalities, or scenes/objects tend to be very limited. Only very recent learning-based approaches [1] enable the accurate alignment of 3D medical volumes within minutes, rather than hours. The aim towards modality-agnostic, generally applicable, but still powerful, fast, and easy to use techniques and tools, is long-term.

Integration and analysis of the heterogeneous information of multimodal data is a demanding task. Images are often of very *different dimensions* (2D, 3D, 4D, multiplexed or spectral data with hundreds of measurements for one spatial localisation), and establishing correspondences between them may be very difficult. Furthermore, it may also be of interest to combine image information with non-imaging data, such as -omics data. Considering that monomodal image analysis tools are still much more common and available in greater variety and abundance, integration of the different types of information is often performed at the very end of the analysis, after being extracted from each of the modalities independently. However, such an approach greatly limits the available complementary information during processing, often leading to reduced quality of results. The full potential of multimodal information is reached only if the joint cross-modal content and the modality-specific content are considered together and optimally combined throughout the complete processing chain.

Finally, an important challenge faced within computer assisted image analysis, in particular when concerning life science applications, is the *evaluation* of the developed methods, essential for their confident usage and further improvements. In a multimodal setting, this challenge becomes even bigger, due to additional complexity of acquiring ground truth (the correct/desired result of the data analysis), most often based on human expertise and manual analysis. Alternative approaches, e.g., based on automated or semi-automated procedures, as well as suitable, possibly indirect, quantitative, and/or qualitative evaluation criteria, are proposed, enabling the use of larger data and facilitating benchmarking and stimulating further improvements.

2 Principles and main approaches

There exists a great variety of approaches for addressing the different tasks and challenges of multimodal image processing and analysis. We highlight a few general topics which influence the design choices made.

2.1 Model-based versus learning-based approaches

Knowledge about the physics and the mathematical model of the imaging technique, as well as the biology of the observed specimen, provide valuable information on how to solve the different tasks of multimodal image processing. Such domain specific knowledge is traditionally inbuilt in model-based image processing and analysis algorithms, designed to address particular analysis tasks in a sequence of standard steps (often referred to as a workflow). Numerous algorithms are developed to solve specific tasks, such as image denoising, image registration, segmentation, object description and feature extraction, classification, and image interpretation. These algorithms in general require some level of parameter tuning to perform at their best for a particular dataset. This does require some skill, but the number of parameters is usually relatively small. The algorithmic approaches may differ significantly and should be carefully selected for each application. Such *engineered, model-based*, approaches have the appealing property of being intuitive and transparent (at least for the educated professional). For the appropriate task, model-based methods reliably deliver top-class performance. However, that performance may easily be degraded by unsuitable parameter selection, as well as imperfections of the model's ability to fit the particular situation (including unforeseen imaging variations) and can therefore become sub-optimal.

The rapid increase of both computing power and the amount of available data has brought into focus *data-driven, learning-based methods.* These methods 'learn' to estimate a function which maps the input image data directly to the desired analysis output, based on a number of annotated examples (in that sense, solving a nonlinear regression problem). The learned function is then applied to new, unseen, data of a similar type, to predict the intended output. Best known and currently very popular are *deep convolutional neural networks (DCNNs)*, for which the learnt model (function) is defined by a very large number of parameters (often millions). A very large number of annotated images, and often a large amount of time, are needed to automatically optimise these parameters and thereby 'learn' the model. Once defined (trained), such models may, at a high speed, exhibit outstanding performance on a very large variety of image analysis tasks in all application fields. An outstanding property of DCNNs is their general applicability; whereas model-based approaches often function well only for one particular tightly specified task, deep learning approaches, all relying on the same building blocks and the same learning algorithms, can with little effort be tuned to almost every image processing and analysis problem (given a sufficient amount of training data). This versatility, together with high performance, make them an often preferred choice, in spite of some clear disadvantages: need for large quantities of annotated training data; slow and compute-intensive training; highly complex and non-transparent (black box) processing, to name a few. Expert annotation of medical and biomedical data, e.g., in the form of manually drawn regions, in the amount required for training of high performing models, is extremely expensive. The highly complex non-linear combination of millions of tuned parameters make DCNN models extremely difficult to interpret. Their adjustment to new datasets often requires above-average computer science competence. Their complexity and non-intuitive design raise questions of reliability and trustworthiness of the solutions, which may cause hesitation regarding their use in, e.g., medicine and healthcare. These issues are all the subject of intensive research: transfer learning, few-shot or zero-shot learning techniques aim to reduce the need for annotated data, whereas a whole new field of explainable artificial intelligence (XAI) [2] has emerged, aiming to provide approaches to visualise and explain the networks' decision making process.

Whether to go for a model-based or a learning-based approach for a specific problem depends on many factors such as access to data, variability of the problem, need for fast results, available know-how and computer resources, etc. In many situations, hybrid techniques—combining fast and flexible deep learning together with the reliability, steerability, and interpretability of model-based approaches—may offer the best of two worlds. Several current trends in method development follow this direction.

2.2 Moving information between modalities

The true power of multimodal data is in the heterogeneity of the information it captures. This enables supporting the processing of one modality (e.g., lacking annotated data, being noisy, and/or with only sparse and unreliable labels provided) by utilising knowledge from another modality, not burdened with similar challenges. This general approach of transfer of information is among the essential ones in multimodal data analysis. If utilised in a learning-based setting, this is referred to as *co-learning*. An important property is that the supporting modality may be excluded when appropriate, such that, at inference time, the trained/developed model may operate on the supported, resource-poor, modality solely.

A technique of growing popularity is to utilise one modality to provide ground truth annotations of the other. For example, let us assume availability of images acquired by two modalities, out of which one is slow and difficult to acquire, but allows efficient automated segmentation of objects of interest, whereas the other is, e.g., heavily corrupted by noise and very difficult to (automatically) segment. Transfer of the segmentation masks from the first modality (after the corresponding images are aligned) to the images acquired by the second modality enables detection and delineation of the objects in it. This approach is often cheaper, faster, and more reliable than to depend on human annotation of the second modality.

A recent popular approach is to train deep neural networks on (aligned) multimodal image data to construct inference models to generate images of one particular modality given the other, thereby possibly reducing the need for the actual imaging in one of the modalities. The approach is to learn characteristic features of appearance of images of one modality and how to transform them to appear as coming from the other. The approach is referred to as *image synthesis* (cross-modal image translation); cycle-consistent generative adversarial networks (GANs) enable one to circumvent the need for aligned training images. Examples include a deep learning-based approach to generate PET images from MRI; similar classification performance is demonstrated using the generated PET images as for the true PET images on diagnosing Alzheimer's disease. The same principle has also been applied to microscopy images. One example [3] suggests a cross-modality translation approach to perform *virtual staining* of wide-field autofluorescence images of tissue sections, reaching the appearance of bright-field histologically stained images. Another proposes to predict fluorescent labelling from transmitted light images [4].

Image translation methods perform well if the difference between modalities is not too big. With a sufficient amount of annotated well aligned data, their performance will further improve, and the number of imaging modalities required to address a particular research question could possibly be reduced. Ongoing efforts towards collecting and sharing such repositories of annotated and well aligned data, as well as public archives, such as the *BioImage Archive*, are therefore of highest importance. However, if the modalities are very different, image translation exhibits a main downside: one modality may simply not provide enough information to generate the corresponding image from the other. This means that the method (often a GAN) has to fill in the missing information by guessing what should be there; although good at guessing, the network often creates structures which look plausible but deviate from the truth. Such guessed information may rather disturb the subsequent processing than be of help.

2.3 Images versus their representations

Image data is often overwhelmingly rich with information, in particular when the data originates from multiple and diverse modalities. For subsequent tasks, it may be beneficial to summarise and simplify the original (multimodal) data and reduce its complexity, (optimally) combining the complementarity and redundancy of the multiple modalities. Methods to transform images into such suitable *representations* depend on both the data and the task to be solved in the subsequent analysis utilising these representations; content relevant to be preserved for, e.g., classification, may not be relevant for registration, and vice versa. The heterogeneity of multimodal data makes it challenging to construct such representations, in particular if a desire is to reduce the need for domain knowledge and instead aim for general techniques. Most successful are learning-based methods; a recent survey [5] summarises existing solutions and open challenges in this direction. Approaches differ in the extent to which the learnt representations are constrained to be shared between the modalities. Recent works promote learning *disentangled representations*, to enable differentiating modality-specific and cross-modality shared information. Learnt representations are proposed for (joint) segmentation in MRI and CT images [6]. In many cases, the proposed approaches are optimised for particular pairs of modalities; the development of more generally applicable methods is of high interest.

2.4 Fusion—early, late, or middle

Integration of information available from multiple modalities to perform a highly informed analysis is referred to as *fusion*. Three main approaches can be distinguished. *Early fusion* integrates information from the individual modalities, in the form in which they are acquired. In the case of neural networks, for example, this reduces to a direct usage of the initial multimodal images as a concatenated (stacked) input to the network. The subsequent processing is performed on the joint (fused) information. A typical application example is analysis (e.g., segmentation) of multiparametric MRI images. Early fusion yields a truly multimodal feature representation, since the features are integrated from the start. An added advantage is the requirement of only one processing pipeline; for a learning-based method this means one learning phase only. However, it turns out that early fusion often struggles to reach good performance. This comes as a consequence of the difficulty to combine very different types of information directly as delivered from the imaging devices, where each sensor has its specific properties. Moreover, the more subtle modality specific properties may easily be lost in early fusion.

Late fusion instead integrates the information resulting from the analysis performed on the individual representations. The modalities are to a high extent processed separately, and joint cross-modal information is extracted and used only towards the end of the process. A downside of late fusion is that the processing, which takes place before the fusion, is not able to exploit the multimodal information, which may lead to a suboptimal result. Another drawback is the need for parallel processing pipelines (often requiring separate learning stages) for the non-fused data.

An appealing third alternative is to aim somewhere in between the two extremes. A *mid-fusion* approach allows distinct processing of sensor data, to reach an information format which is more uniformly shared between the modalities. At that stage, the information is easier to combine, whereas the subsequent processing can harvest on access to both modality-specific and cross-modality shared information, maximising the benefits of multimodality. This approach clearly relates to the notion of representation learning; deep learning-based methods gradually remove the dividing line between multimodal representation and fusion. Multimodal information is integrated through joint learnt representations targeted to optimise ‘end-to-end’ learning towards the final analysis goal (e.g., classification, segmentation) [6].

Regardless of the approach used, the quality of alignment of the image data strongly influences the quality and reliability of the fused output.

2.5 Registration for segmentation or segmentation for registration

Essentially each one of the multimodal image analysis tasks heavily relies on good alignment and accurate spatial correspondence established between the images. This is achieved through *image registration.*

Image registration methods are typically based on either comparison of large-scale intensity patterns and optimisation of a suitable distance/similarity measure between them, or on establishing correspondence between a number of salient feature points, to estimate a linear or non-linear transformation which matches the detected pairs of such points. The latter, feature-based approach is popular in computer vision primarily due to high speed. However, feature-based methods tend to be less resilient to noise, and they often fail to robustly detect cross-modal correspondences. Whereas medical image registration heavily relies on persistent visibility of anatomical structures (e.g., organ contours), a lack of salient correspondences is typical for biomedical images. This makes region based methods more suitable for such data.

Deep learning-based approaches aim to learn, directly from data, either image alignment, or a similarity measure, or both [7]. The requirement for a large number of aligned (training) image pairs is prohibitive for applications in biomedicine, due to complexity of imaging and cost of manual data annotation. Approaches which require very few, or even no annotated (training) examples to learn transformations exist and are attracting a lot of attention; they are known as *few-shot learning* and *unsupervised learning*. However, these solutions come with increased complexity, and most often inferior performance compared to fully supervised approaches.

Interestingly, registration is one of the few sub-topics of image processing where deep learning-based approaches do not outperform the model-based ones. State-of-the-art methods for image alignment combine 'the best of the two worlds' where deep learning is utilised as, e.g., an optimiser, or a tool to speed up the overall registration procedure.

Registration of distinctly different modalities remains a very challenging, and still only rarely addressed task. Registration may, through the transfer of information approach mentioned above, enable multimodal segmentation, where a segmented image in one modality is aligned to a non-segmented image of another modality, and the segmentation from the first is transferred to the second. Reversely, the registration may also be aided by successful single modality analysis: an example of such a method, which improves registration by utilising segmentation, is given in [1]. Another recent example is a challenging registration of bright-field and SHG microscopy in histopathology [8], which relies on knowledge about the biological structures present in the modalities; these structures are segmented in one modality to guide the matching with the content of the other.

One interesting recent approach, taken in [9], utilises an intermediate modality as an 'interpolator' to gradually reduce the difference between the data of primary interest. If our particular problem does not provide such an 'intermediate' modality, we may instead aim to synthesise one, utilising the above-mentioned image synthesis/translation approaches. Monomodal registration, often being much easier, can then be performed between one original and one synthesised image, where the derived transformation is further used to align the two input multi-modal images. Another possibility is to transform the input images into modality-independent representations and then optimise (mono-modal) similarity between the representations [10]. This approach, based on the above discussed representation learning, is not forced to go all the way from one modality into the other. With less need for information extrapolation, it may provide more reliable data for the subsequent mono-modal registration.

3 Examples of application—available software and how to find the right one

The high multidisciplinarity of the field and the expertise required to address life science research questions by computational methods and automated (correlative multimodal) image data analysis place high demands on communication channels and accessibility of software tools. Tools do exist in large amounts and variety, but are not always easy to find, select, implement, adjust, and confidently apply; some are offered as an open source and are freely available, others are provided within commercial software packages, often integrated with the imaging equipment. The required understanding of the biological/medical problem, physics of the imaging techniques used, as well as skills in computer science and image analysis, for a confident selection of the most appropriate methods, are rarely found in one person.

With an aim to provide an easy-to-access 'bridge' between biomedical questions and the image analysis toolbox, the *BioImage Informatics Index*, *biii.eu*, a web-based platform developed by the *Network of European Bioimage Analysts—NEUBIAS*, has recently been offered to the wider communities of both users and developers. The platform supports searching for solutions to a biological problem (using the terminology of life sciences), for a particular method (using the terminology of computer science), or for a particular ready-to-use implementation of a method (referring to, e.g., known software packages), accommodating multiple views on the same problem. The BioImage Informatics Index is a continuously updated community effort, fostering knowledge exchange and collaboration through crowdsourcing. Rather than listing a few of the numerous software tools and packages developed for particular tasks and types of images, we encourage the readers to explore, and perhaps themselves contribute to *biii.eu*.

Another recent initiative, taken within the *COMULIS* (*Correlated Multimodal Imaging for Life Sciences*) Network, is aimed at presenting a preprocessed view of *biii.eu* to particularly meet the needs of researchers working with multimodal image data. A dynamic overview of available methods for multimodal image registration, segmentation, and visualisation, with a particular focus on applications within life sciences, can be accessed from the COMULIS webpage.

One more community driven initiative is the *Scientific Community Image Forum* at *forum.image.sc*, which provides a place for everyone to ask questions related to software-oriented aspects of scientific imaging, particularly (but not limited to) image analysis, processing, acquisition, storage, and management of digital scientific images. The forum is aimed to support beginners as well as experts, both life scientists and computer scientists, offering to everyone practical advice and theoretical solutions to scientific problems with images. The goal is inclusivity and discussion on any software package in the imaging ecosystem is equally welcome. More than 30 open source software projects have pledged to use the *image.sc* forum as their recommended discussion channel.

The selection of software should most often be based on its performance on a particular task. The top performing reported methods, e.g., winners of international challenges, are usually highly optimised for particular data, and a particular task, and will often underperform when applied to other data and different combinations of modalities. The evaluation of multimodal methods is, however, a very challenging and open problem, primarily due to difficulty of manual annotation/solutions to provide *ground truth* data. The development of well performing automated analysis methods also heavily relies on availability of *ground truth* data. The above-mentioned modality-transfer, semi-automated solutions, self-supervised and unsupervised learning methods, and crowdsourcing, are some of the approaches taken to generate ground truth data. Given a data set with an established ground truth, evaluation criteria can be defined to enable benchmarking of different solutions and thereby facilitate reliable selection of suitable methods for a particular task. A recent result of the *NEUBIAS* community is *BIAFLOWS* [11], an open-source *Bio Image Analysis workflows benchmarking platform*, which aims to enable reproducible benchmarking of bioimage analysis methods coming from any software ecosystem. Through a web interface, workflows can be launched and assessed remotely, with respect to a range of standard benchmark metrics. If extended with multimodal image datasets, corresponding ground truth, and suitable evaluation metrics, *BIAFLOWS* has the potential to become a highly valuable tool for fast and reliable evaluation and benchmarking of multimodal image analysis methods.

4 Conclusion

Automated analysis of multimodal image data is a necessity—the data is simply too large and too complex to be analysed manually. A great variety of methods exist; they can reach results of accuracy comparable with human experts, but most often incomparably faster. To bring the methods closer to the wide multidisciplinary community of researchers and practitioners relying on multimodal image analysis, joint efforts are invested into creating and maintaining open software platforms which provide overview, quick access, implementations, example applications, as well as evaluation of a number of popular multimodal image analysis methods; some are mentioned above.

With no intention, nor chance, to provide a complete and thorough presentation of the field of computer assisted multimodal image analysis, the sections included in this chapter rather aim at giving a flavour of its potential, and indicate some directions for its further development. Possibilities are, with no doubt, endless, and so are the rewards!

References and further reading

[1] Balakrishnan G, Zhao A, Sabuncu M R, Guttag J and Dalca A V 2019 *IEEE Trans. Med. Imaging* **38** 1788–800

[2] Lapuschkin S, Wäldchen S, Binder A, Montavon G, Samek W and Müller K R 2019 *Nat. Comm.* **10** 1–8

[3] Rivenson Y *et al* 2011 *Nat. Biomed. Eng.* **3** 466

[4] Christiansen E M *et al* 2018 *Cell* **173** 792–803

[5] Baltrušaitis T, Ahuja C and Morency L P 2018 *IEEE Trans. Pattern Anal. Mach. Intell.* **41** 423–43

[6] Dou Q, Liu Q, Heng P A and Glocker B 2020 *IEEE Trans. Med. Imaging* **39** 2415–25
[7] Yang X, Kwitt R, Styner M and Niethammer M 2017 *NeuroImage* **158** 378–96
[8] Keikhosravi A, Li B, Liu Y and Eliceiri K W 2020 *Biomed. Opt. Express* **11** 160–73
[9] de Vos B D, Berendsen F F, Viergever M A, Sokooti H, Staring M and Išgum I 2019 *Med. Image Anal.* **52** 128–43
[10] Pielawski N 2020 *NeurIPS* https://proceedings.neurips.cc/paper/2020/hash/d6428eecbe0f7dff83fc607c5044b2b9-Abstract.html
[11] Rubens U *et al* 2020 *Cell Patterns* **1** 100040

Part IV

Future of imaging in biological and preclinical research

IOP Publishing

Imaging Modalities for Biological and Preclinical Research:
A Compendium, Volume 2
Parts II–IV: *In vivo* preclinical imaging, multimodality imaging and outlook
Andreas Walter, Julia G Mannheim and Carmel J Caruana

Chapter IV.1

Emerging technologies and outlook

Andreas Walter, Arrate Munoz-Baruttia, Martin Jones, Julia G Mannheim, Lucy Collinson, Marco Andreana, Angelika Unterhuber and Paul Verkade

Following the eras of biochemistry in the first half and molecular biology in the second half of the 20th century, the 21st century started off as the age of -omics. Recent technical developments such as super-resolution fluorescence light microscopy and cryo-electron microscopy demonstrate that we have now entered the age of imaging. These imaging technologies, some even while still under development, allow us to make breakthrough discoveries. New and even more advanced imaging technologies are still being developed and becoming available to the wider scientific community. In the last part of this ebook we will highlight a number of cutting-edge imaging technologies for biological and biomedical research that we believe will become essential tools to advance our quest for knowledge and help cure and prevent diseases.

1 Introduction

The field of bioimaging has been experiencing several major boosts in recent years: advancements in light microscopy now make it possible to routinely resolve subcellular structures both below the diffraction limit using super-resolution fluorescence light microscopy (commonly referred to as super-resolution microscopy) as well as in ***thick tissue with increased penetration depth and lateral resolution*** ((lattice) light sheet microscopy, multiphoton microscopy or adaptive optics (compare chapters I.2.a, I.2.b, I.2.c and I.2.e)). This applies both to biological and (pre)clinical studies. A high-resolution optical glimpse into (human or animal) thick tissue without the need for surgery is nowadays also possible using optical coherence tomography (OCT) or photoacoustic tomography (PAT) with penetration depths of up to 40 mm (PAT) and resolutions down to a few micrometers (OCT) (compare chapters II.2 and II.3).

doi:10.1088/978-0-7503-3747-2ch31

Similarly, the resolution revolution in cryo-electron microscopy (cryo-EM) led to an exponential increase in resolved single particle structures and is accompanied by a 'volume revolution' that allows imaging of relatively large volumes at high resolution using serial block face scanning electron microscopy (SBF SEM), focused ion beam scanning electron microscopy (FIB/SEM), array tomography, and a rapidly evolving set of new imaging tools including plasma FIBs and multibeam SEMs, which together are termed 'volume EM' (compare chapter I.5.d).

Further trends in bioimaging that will significantly enrich biomedical research include the integration of localised spectroscopic information that can profile a sample. ***Spectroscopic imaging*** is usually label-free, provides previously inaccessible information at high spatial resolution about the chemical (e.g. X-ray fluorescence spectroscopy) or molecular (e.g. mass spectrometry imaging) composition of the sample (compare chapters I.9.f and I.9.g). Other modalities give information about the biophysical and mechanical properties of the sample (such as Brillouin spectroscopy, Raman spectroscopy or atomic force microscopy—compare chapters I.9.c, I.9.b and I.6). Hence, they complement the structural and functional information acquired by conventional imaging modalities.

The most-used preclinical *in vivo* imaging modalities to visualise entire organs and even whole bodies are optical (bioluminiscence and fluorescence), magnetic resonance imaging (MRI) and positron emission tomography (PET) (compare chapters II.4.a–b, II.7.a, and II.10). Besides these imaging capabilities, multimodality PET/CT (X-ray computed tomography) instruments are widely present in preclinical research laboratories (compare chapter III.2.a). More in-depth information about the sample composition is obtained by emerging modalities such as phase-contrast X-ray imaging and spectral CT. ***Macroscopic preclinical imaging*** has demonstrated its tremendous impact over the last decades as a pioneer of hardware development (e.g. the combination of PET with magnetic resonance imaging (MRI) to enable simultaneous PET/MR imaging—compare chapter III.2.c), as well as new biological targets (such as tracers, drugs, contrast agents, etc), on their way into clinical routine diagnostics. Furthermore, preclinical imaging is experiencing a boost with recent developments of nanoprobes with targeting, diagnostic and theranostic capabilities.

Additionally, the frontiers of bioimaging are currently being pushed towards the integration and correlation of several modalities to tackle biomedical research questions holistically and across multiple scales via ***correlated multimodality imaging (CMI)*** (compare chapter III.3). CMI is specifically moving towards the integration of *in vivo* imaging with subsequent *ex vivo* imaging modalities to bridge the gap between preclinical and biological imaging. The first European Initiatives confirm this trend, such as the COST Action Correlated Multimodal Imaging in Life Sciences (COMULIS, CA17121), which aims at fueling urgently needed collaborations in the field of CMI, promoting and disseminating its benefits through showcase pipelines, and paving the way for its technological advancement and implementation as a versatile tool in biological and preclinical research [1].

Often these advanced imaging setups are custom-built and require expert knowledge and tedious optimisation. Challenges to face include, for instance, which

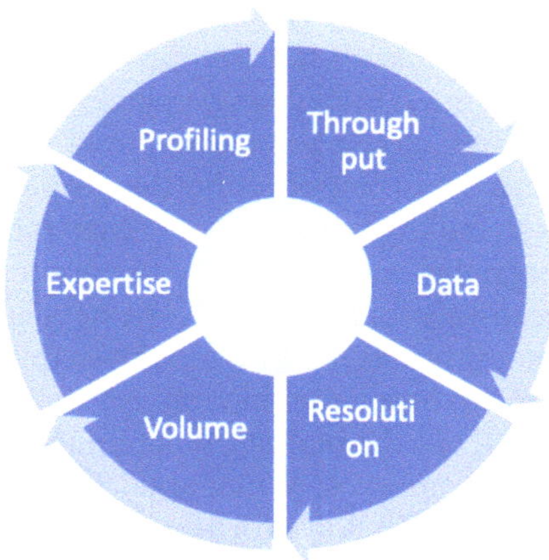

Figure 1. CMI is reliant on six pillars that both drive but are also the bottlenecks in the CMI technology.

probes to use to visualise the sample, how to prepare a good sample, and which strategy to use to best move a sample between different modalities (figure 1). There is, however, no 'one-size-fits-all' solution and most biological questions need customised workflows, which generally will come at the expense of throughput. To achieve meaningful statistics, throughput and quantitative imaging are crucial to biomedical research. Although the amount of imaging data acquired has increased massively, throughput and ***automation of microscopy*** will be crucial for mainstream biomedical research and quantitative conclusions. Automation will be accompanied by an additional boost in imaging data.

Consequently, ***big data in bioimaging*** will be a substantial challenge for the community. It is already becoming routine to collect raw images of hundreds of gigabytes per imaging session. Recent advances in cryo-EM, including the advances in detector technology with faster frame rates, have substantially increased data generation with a typical single-particle imaging session acquiring data in the giga to terabyte range, and lattice light-sheet microscopy, for example, acquires data in the range of terabytes for a single 4D data set. Additionally, the biggest bottleneck of CMI is currently data handling and storage due to the plethora of complex, multimodal, time-varying and diverse volumetric imaging data (figure 1).

2 Emerging technologies

2.1 (*In vivo*) imaging of thick tissue at high resolution

Imaging of thick tissue at high resolution allows looking at details of biological systems in their 3D environment in their context as an organ, a body part or as a complete organism. In any tissue, organ or body part, cells are arranged in specific ways, forming complex, functional structures. They respond to each other, and in some cases, even move together. Specially designed optical microscopy technologies allow scientists to record, track and study how these processes happen both for *in vivo* and *ex vivo* specimens.

Light sheet fluorescence microscopy (LSFM) refers to a set of techniques which image by sweeping a tissue sample with a thin layer of light. Images obtained are specific to the fluorescence emitted in that single plane, excluding any signal coming from the rest of the sample and allowing visualisation of inner structures without the need for physical sectioning. Most importantly, this technique avoids any unnecessary excitation of the surrounding planes and thus reduces phototoxicity and premature fluorophore degradation. The ability to capture fine details in addition to the capacity to study relatively large samples for extended periods of time, similar to optical projection tomography (OPT) (compare chapter I.2.f), have enabled a boost of discoveries in the developmental biology research in the last years. LSFM gained prominent attention from the biological community after it was successfully used to image living zebrafish embryos. When applied to previously cleared *ex vivo* animal or human samples, LSFM allows for the in-depth study of features such as the characterisation of biological structures, disease localisation and quantification or drug behavior screening. This information can serve as a tool to develop new treatments or diagnostic procedures.

A step forward towards the challenge to observe cells without affecting their behavior is the use of lattice light sheet fluorescence microscopy. Like LSFM, it uses ultrathin sheets of light derived from 2D optical lattices to illuminate the sample. By simultaneously implementing either super-resolution structured illumination or by dithering the lattice to create a uniform light sheet, it is possible to capture previously unseen dynamic biological phenomena in three dimensions, in multiple colours, often at sub-second intervals, for hundreds to thousands of time points at the diffraction limit and beyond [2].

Multi-view selective plane illumination microscopy (SPIM) utilises multiple views of the sample to reconstruct a 3D volume. This technique achieves improved illumination homogeneity, larger sample sizes, increased axial resolution and enhanced signal-to-noise response. For example, the dual excitation method reduces striping artefacts caused by shadows from punctate high attenuation regions.

Another interesting recent development is the Mesolens [3] that helps to overcome a typical limitation of current optical microscope objectives of low magnification (e.g. low numerical aperture and therefore, too little resolution and discrimination deeper inside the tissue). The Mesolens offers both a relatively large field of view and a relatively high resolution allowing one to capture single-shot images of whole mouse embryos or explants from the brains of rat embryos both with cellular resolution.

The prevailing biomedical challenge today lies in the ability to accurately identify and locate lesions (pathologies) for immediate intervention and early disease diagnosis. Advanced optical imaging offers a unique potential to meet this challenge.

Optical coherence tomography (OCT) has been one of the most innovative and rapidly emerging optical imaging modalities of the last decades because of its capability to non-invasively exploit morphological, functional and microangiographic tissue information within the first few millimetres of organs. OCT has achieved widespread clinical acceptance (such as in ophthalmology, cardiology and dermatology) due to its real-time imaging capabilities without the need of exogenous contrast agents at micrometer resolution.

Nonlinear optical imaging (NLOI), including multiphoton excitation and coherent Raman spectroscopy, has recently shown great potential for clinical applications, but its use is still mainly limited to optical laboratories due to the optical requirements in terms of signal excitation. NLOI provides fast, cellular resolution images with molecular contrast, allowing imaging depths up to 1 mm, which is much deeper than classical confocal microscopy. Despite impressive technical advances and progress in NLOI over the last years, it is still challenging to transfer these technologies in a compact and robust way to the clinical routine.

Miniaturisation of these optical imaging methods into an endoscope along with laser development will enable inspection of the inner (human or animal) tissues/organs, paving the way for clinical acceptance. Moreover, the synergetic combination of different imaging technologies in multimodal settings, will facilitate pathological evaluation, to optimally adopt and monitor the efficacy of applied therapies in the clinics.

As described above, each of these techniques has its unique qualities and is suited for certain biological applications. With these specifics also come limitations. In table 1 we have selected a few of the imaging modalities described and highlighted some of their parameters, including what they are best suited for and their limitations. This table is a very limited selection of imaging modalities and we refer to [4] for a more comprehensive overview.

2.2 Spectroscopic label-free imaging

Spectroscopic imaging adds localised spectroscopic information to the acquired structural and functional parameters without the need of labelling the sample with probes or contrast agents. Current leading imaging modalities include atomic force microscopy, mass spectrometry imaging (MSI), X-ray fluorescence, Brillouin, Raman imaging, nanoSIMS, energy dispersion spectroscopy (EDS), and energy electron loss spectroscopy (EELS) [5]. They complement traditional imaging approaches by providing localised chemical (elemental or molecular identification for a broad variety of biomolecules like lipids, proteins, amino acids or metabolites, and drug compounds) or biophysical information (vibrational modes or mechanical properties such as elasticity or stiffness) to characterise—at high spatial resolution—the biomechanics and tissue properties beyond visualisation of a specific resolution range.

Spectroscopic imaging techniques bear the potential to aid in the detection of biomarkers and to screen the states of live cells in a label-free manner. Spectroscopic imaging has led to a substantial impact in clinical research by revealing biomolecular changes associated with disease and providing low-cost imaging of drugs. Many pathological conditions, such as cancer, neurodegenerative, cardiovascular or bone diseases, have been shown to be closely related to mechanical changes at the cellular level [6].

Research focused on elemental and molecular information that play crucial roles in both physiological and pathological metabolic processes has also seen a recent increase in interest. The new and still evolving field of molecular histology integrates insights from tissue MALDI MSI or infrared spectroscopy into classical histomorphology

Table 1. Comparison of some of the common microscopy techniques in biological and pre-clinical research across the resolution scales. Adapted from [4].

Modality	TEM	Super-resolution FM	LSFM	NLO (CRS, TPM, SHG, THG)	OCT	PAT
Contrast mechanism	Electrons interact with the sample. Heavy atoms deviate electrons more from their path and generate contrast. Heavy metals are generally used to enhance contrast in biological samples.	Uses the physical properties of statistically photoactivatable fluorophores (PALM) or selective deactivation of fluorophores (STED). Contrast: quantum efficiency of fluorophore, labelling density.	LSFM scans a thin slice of the sample—optical sectioning—using a plane of light instead of a point.	Fluorescence. Nonlinear susceptibility. Molecular vibration.	Optical scattering light resulting from refractive index changes (speckle and/or phase decorrelation).	Endogenous and/or exogenous optical absorption. Photoacoustic effect.
Penetration [mm]	<0.001	Technology dependent: ~0.01	⩾0.1	~0.5	~1-2	~10
Axial resolution [μm]	>0.002	Technology dependent: <0.250	⩾0.4	⩾0.3 (diffraction limited)	⩾0.5 (light source dependent)	~80 (implementation dependent)
Lateral resolution [μm]	<0.001	Technology dependent: >0.005	⩾0.500 (objective dependent)	⩾1 (diffraction limited)	⩾1 (diffraction limited)	~40 (photoacoustic wave)
Speed	Slow	Slow	Video rate	Slow	Video rate	Slow

ROI [mm^2]	<0.01 × 0.01	<0.1 × 0.1	~0.2-1 (limitation is the production of a uniform, thin light sheet over the volume)	0.3 × 0.3	10 × 10	150 × 150
Advantages	Ultrastructural information. Reference space in relation to structure of interest. Very high resolution.	Highest resolution light microscopy techniques. Co-localisation of single molecules.	Live imaging. Less photo damage to sample (longer imaging). *In vivo* volumetric imaging.	Subcellular 3D section capability. Biochemical contrast. Metabolic information. Morphologic contrast. Label free / non-invasive.	*In vivo*. Fast. Non-invasive. Label-free. Depth information. Morphology. Optical biopsy. Optical angiography (quantification of blood flow, microvascularisation).	*In vivo*. Deep tissue imaging. Endogenous and exogenous contrast. Non-invasive. Non-destructive.
Limitations	No live imaging. Usually elaborate sample processing (time consuming).	System complexity. Phototoxicity associated with intense quenching beam. No real time imaging. Limited to nearly immobile molecules. Mostly not very suited for live imaging.	Extra optics to generate the light sheet. Special holders (old architectures). Lower resolution (improved with the new lattice light sheet microscopes).	Superficial. Reduced field of view. Speed	Limited molecular information. Lack of chemical selectivity. Broadband light sources. Resolution Limited by speckle.	Resolution. Speed. Structural contrast. Couplant (water).

(*Continued*)

Table 1. (*Continued*)

Modality	TEM	Super-resolution FM	LSFM	NLO (CRS, TPM, SHG, THG)	OCT	PAT
Examples biomedical applications	Visualisation of viruses attached to and inside cells. Loss or defects in intracellular organelles caused by mutations. Uptake of drugs into biological systems	Image fluorescent molecules inside cells at the high resolution. Understanding molecular interactions and dynamics. Study the fundamentals of biology through single-molecule fluorescence.	Developmental biology studies, e.g. development of embryos over a long-time course. Mesoscopic imaging on cleared samples.	Early detection of cancers. Monitoring of the effect of various agents on the skin. Determination of atherosclerotic plaque composition. Rapid identification of pathogenic microorganisms.	Study of flow within the vascular structures of the eye. Skin imaging. Cardiac imaging.	Anatomy and physiology of small animals. Oxygenation status. Blood flow. Functional vasculature features. Skin burns. Vulnerable plaques in the coronary arteries.

TEM = Transmission Electron Microscopy, LSFM = Light Sheet Fluorescence Microscopy, NLO = Non-Linear Optical techniques, CRS = Coherent Raman Scattering, TPM = Two-Photon Microscopy, SHG = Second Harmonic Generation Microscopy, THG = Third Harmonic Generation Microscopy, OCT = Optical Coherence Tomography, PAT = Photo Acoustic Tomography, PALM = Photo Activated Localisation Microscopy, STED = STimulated Emission Depletion.

without the need of prior knowledge about the presence of specific molecules. Since spectroscopic imaging technologies rely on intrinsic contrast, they make the application of dyes or fluorescent proteins unnecessary—circumventing the possible interference with function often seen with fluorescent markers. Markers are typically limited to 3–4 colours due to spectral overlap, which makes it difficult to discriminate between the labelled structures or cells. As a disadvantage, many spectroscopic imaging approaches cannot be applied to living tissue *in vivo* since they often do not preserve the tissue. Although these imaging technologies have great diagnostic potential, clinical applications are still in their infancy due to the absence of suitable, reliable, and clinically approved methodologies. Specifically, the field needs to perform large-scale multicenter validation studies to demonstrate its clinical utility and significance. Standardised methods need to be developed to improve its reproducibility at the sample preparation level. It will be essential to increase the spatial resolution and develop streamlined data analysis and quantitation methods that can handle the large amounts of data being produced.

Future approaches will include imaging mass cytometry (compare chapter I.9.h), which enables panels of proteins to be simultaneously imaged in clinical biopsies using highly multiplexed immunohistochemistry based on lanthanide-labelled probes and elemental mass spectrometry. This technique will additionally help to identify molecular signatures for diseases and allow stratifying patients. Additionally, the integration of '-omics' bears huge potential to complement imaging data since it can help to reveal relationships between genotype and phenotype [7]. It can thus contribute to better understand the onset and progression of a variety of diseases and to identify new diagnostic and prognostic biomarkers. The integration and combination of imaging and -omics data is challenging due to the heterogeneity of the sources and the high dimensionality of the resulting feature space [8] and will require interoperable data formats.

2.3 Correlated multimodality imaging

CMI is an advanced imaging approach that allows researchers to gain both holistic and multiscale information about a single specimen. We believe that it will be the method of choice to explore cells, organisms and diseases holistically by deciphering molecular mechanisms in structural context across spatial and temporal scales [9]. CMI needs to span relevant resolution ranges, from millimeters to nanometers, whilst providing complementary information about dynamics, function, morphology and localisation within a close-to-native 3D ultrastructural context. It aims at combining high lateral resolution with high penetration depths and scanning a wide field of view in a reasonable time to allow the correlation of complementary parameters across the entire sample.

Humans are very diverse and react differently to challenges such as viruses (e.g. the recent Covid-19 outbreak). There is an increasing need to extract as much information as possible out of a single sample, and profile it on an individual level. This biological and clinical need is the drive to generate a wide variety of complementary combinations of imaging technologies. CMI is a highly dynamic field which will continue to push the frontiers of imaging towards more and more complex and advanced imaging combinations beyond correlative microscopy, such as CLEM (compare III.1.a), or hybrid imaging in (pre)clinical research and diagnostics. These novel CMI approaches include advanced (currently) non-commercial, custom-built setups, such as correlative soft X-ray tomography, and range from advanced super-resolution cryo-CLEM to novel CMI pipelines that specifically bridge preclinical (*in vivo*) imaging and biological (*ex vivo*) microscopy and zoom in from a living sample to individual cellular structures. So far, novel CMI pipelines with more than two modalities beyond hybrid (pre)clinical setups or CLEM are still in their infancy since they require substantial optimisation and method development to tackle: (1) compatibility of sample preparation across modalities without compromising data quality; (2) soft- and hardware solutions to re-identify the region of interest after relocation between imaging platforms; and (3) solutions to correlate complex, multiscale, diverse volumetric imaging data including data handling and visualisation.

For the future, the main aims for CMI will be to (1) integrate preclinical *in vivo* imaging and biological (*ex vivo*) microscopy to zoom in from a living sample to individual close-to-native ultrastructure and (2) adding localised spectroscopic or chemical information to the acquired structural and functional parameters. CMI will play a vital role in the future of bioimaging and in life sciences.

2.4 Macroscopic preclinical imaging

Over the last decade, macroscopic preclinical *in vivo* imaging has served as a groundbreaking tool to study pathological diseases, monitor disease progression and develop new therapeutics. Tested in preclinical phase, some of these techniques and substances have made their way into clinical routine diagnostics. There is continuous development ongoing in regard to hard- and software design, e.g., to enhance the resolution and sensitivity of imaging devices (such as PET or single photon emission tomography (SPECT)), develop new data analysis techniques or new MR sequences. Furthermore, new probes, tracers, contrast agents and therapeutics are being developed to reveal comprehensive molecular and biological information of underlying diseases.

There is a contiguous shift towards precision medicine, in which therapy strategies are tailored to each patient based on e.g. genomic data, to maximise treatment outcome [10]. One challenge, however—that is already being tackled and will continue to be an active field of research—is the integration of multiparametric data across scales. Especially the combination and interlink of multiparametric data, such as macroscopic imaging data with -omics or pathological data, has been proven to be challenging but nevertheless of highest importance to reveal the full mechanisms of diseases [11, 12]. Advanced data analysis approaches based on, e.g., deep learning or neural networks, are being implemented to combine these parameters.

2.5 Automation in bioimaging

In time-lapse fluorescence imaging—to get reproducible, unbiased data—environmental conditions such as light, temperature, and CO_2 must be controlled. Severe limitations in accuracy, reproducibility and throughput are caused by the involvement of humans in all steps of an imaging workflow. The situation is even more critical since multiple runs of experiments or complex assays need to be performed in life science research. Automation of the workflows can help to alleviate those issues.

A number of initiatives to tackle them have been proposed. For example, the micropilot software (EMBL) automatically detects cells of interest and launches complex imaging experiments, including three-dimensional multicolour time-lapse or fluorescence recovery after photobleaching in live cells. In three independent experimental setups, this allowed the developers to statistically analyse biological processes in detail and demonstrate the power for systems biology. Micro-manager (https://micro-manager.org/) is a Java-based software package for control of automated microscopes. Together with the image processing application ImageJ, micromanager provides a comprehensive, freely available, imaging solution. Python's microscope package (https://www.python-microscope.org/) offers the ability to automate microscope experiments, as well as to control local and remote microscope devices, and to aggregate them into complex setups.

Novel solutions are also being proposed for the automation of correlated multimodal imaging approaches. The automation of the CLEM workflow, autoCLEM is a successful example. The correlation of fluorescence light microscopy signals to localise the regions of interest under cryogenic conditions has the limitations typical of manual setups in microscopy as described above and is even aggravated as the multiple correlation steps are at different scales. AutoCLEM is composed of a high-speed confocal imaging module to automatically acquire a fluorescence light microscope grid atlas that is linked to the cryoEM grid atlas. The fluorescence light microscopy coordinates of the targeted areas are automatically converted to cryoEM coordinates and refined using fluorescent fiducial beads. The autoCLEM workflow thus improves the efficiency and throughput of CLEM methods.

The resolution revolution of cryo-EM has encouraged the development of modelling tools able to produce high-quality atomic models from cryo-EM density maps. To facilitate the use of these tools by non-experts, there is a need to create solutions for inter-package integration in the model building area and full tracking of the complete workflow, from image processing to structure validation such as the recently presented extension of the image processing framework Scipion [13].

3 Big data in bioimaging

Across all types of imaging and microscopy, technological advances have meant that the acquired data are getting more complex and measured at increasing rates, pushing us well-and-truly into the Big Data regime. Since these imaging advances are occurring at a faster rate than the corresponding improvements in computational infrastructure, this produces a significant bottleneck in our ability to analyse the data in a timely fashion. Correlative and multimodal imaging provides a set of mechanisms to leverage the intrinsic benefits of different imaging modalities, such as large-volume imaging of intact specimens in X-ray microscopy, specificity in fluorescence microscopy and ultrastructural resolution in electron microscopy. To reduce the downstream computational burden, for example, the 'finding-a-needle-in-a-haystack' class of problems in large volumes, high-resolution imaging can be efficiently tackled by pinpointing structures of interest with targeted labelling in a lower resolution modality and thereby reducing the overall data load. Furthermore, object identification by specific labelling at lower resolution can aid interpretation when morphological analysis alone is insufficient for positive classification in higher-resolution imaging.

Apart from interventions *during* the imaging process, new tools for subsequent analysis are also necessary since, once the images have been acquired, even with highly optimised acquisition strategies, the data can still be too large for traditional analysis methods to cope with. A field that has provided significant advances in many large-scale image analysis problems is artificial intelligence, in particular the area of deep learning. This family of methods includes both supervised and unsupervised approaches, whereby the algorithm iteratively learns how to solve the problem at hand. The most widely used supervised methods, such as convolutional neural networks, are capable of extraordinary levels of performance, but at the cost of requiring significant amounts of annotated data for the training process. Once bootstrapped beyond this initial phase, however, pre-trained networks can be readily fine-tuned to generalise across different data, a problem that is notoriously difficult in traditional image analysis methods.

4 Other trends

4.1 Zooming in

Following in the footsteps of the 'resolution revolution' in single particle cryo-EM, the field of structural cell biology is now aiming to place molecules and protein complexes in their native cellular context. Pioneered by the group of Baumeister, identifying 'organelles' (70s ribosomes with an approximate diameter of 20 nm) in bacteria using template matching [14], the field has now progressed to identifying individual protein complexes inside cells based on template matching [15]. This effort is not only aided but strongly guided and dependent on correlative methods to pinpoint the exact location of the structure of interest, ideally down to (tens of) nanometers. The latest in these developments are so-called bar-coding and super-resolution volume CLEM. In the bar-coding approach, Briggs and coworkers tackle the issue of throughput by labelling samples with multiple labels, expressed at different levels, giving each sample a unique signal that can be correlated with its underlying ultrastructure [16]. The latest in the field of CLEM is the integration of super-resolution cryo-fluorescence microscopy to increase the precision of relocation of structures of interest for structural cell biology. This technology is currently spearheaded by researchers from Janelia farm [17]. By using fixation through cryo-vitrification, GFP-expressing cells retain their fluorescence. This fluorescence can be analysed using super-resolution light microscopy where the stage and sample are kept frozen. To be able to analyse the whole cell, rather than analysing the sample in the cryo-TEM, the sample was freeze substituted and the area of interest retraced and imaged using FIB-SEM. Each of the individual steps of this workflow are not widely available to most labs yet, but with the field expanding rapidly, such methodologies will soon be commonplace, possibly run as centralised centers of excellence.

4.2 Zooming out

At the other end of the spectrum, the need to understand how organelles and cells interact within complex three-dimensional tissues and organs has generated a 'quiet revolution' in volume imaging. As discussed above, whilst light microscopy is being developed to penetrate deeper into tissue volumes using luminescence, light sheets, tomography and episcopic techniques, X-rays enable a virtual view inside the sample to image and locate regions of interest, and electron imaging depth is being extended by serial slicing and imaging of samples using diamond knives, ion beams and plasma milling. Volume imaging has arguably had the greatest impact in developmental biology (light sheet microscopy, optical projection tomography and high resolution episcopic microscopy (compare chapter I.2.g, *High-resolution episcopic microscopy*) of embryonic development in a series of model organisms) and neuroscience (FIB/SEM, serial block face SEM (SBF-SEM) and array tomography of tissue from model organisms and post-mortem human brains) —compare the individual chapters for a description and application of the various technologies. Indeed, synchrotron hard X-ray microscopy (compare chapters I.4.a and I.4.b) is now in use as an imaging tool in its own right, with resolutions that approach SBF-SEM at far greater imaging speeds. Linking volume imaging techniques together into CMI workflows is usually a game of 'finding the needle in the haystack', with the problem becoming harder as the sample size increases and the event of interest becomes rarer in space and time. The key is to image the intact sample non-destructively at low resolution, and to gradually whittle the sample down to a manageable scale to image destructively at the resolution required to image the fine detail of the structure of interest. It is notable that much of this final high-resolution imaging is performed on resin-embedded metal-stained tissues since the original samples are too large to be properly vitrified or cryo-protected intact. It is now possible to map (parts of) the brain and look at interconnectivity of neurons over large distances. Using a correlative approach, individual synapses deep inside a tissue can be followed over time and subsequently retraced and analysed at high resolution using the sectioning and imaging capabilities, respectively of SBF-SEM [18].

5 Conclusions

There is great excitement entering the era of imaging. The most recent developments in imaging technology have pushed the boundaries of resolution, volume, and analysis capabilities. Those developments will aid us in our basic understanding of biology but also in tackling diseases in a clinical setting. By combining both fields using CMI techniques, we believe that ultimately a seamless bidirectional workflow will be generated where basic science will support clinical research and vice versa. In conclusion, the clinic will steer basic researchers into new research areas and function as the ultimate test of basic research.

References and further reading

[1] Walter A, Kleywegt G J and Verkade P 2021 *Methods Cell Biol.* https://doi.org/10.1016/bs.mcb.2020.12.010

[2] Chen B C *et al* 2014 *Science* **346** 1257998

[3] McConnell G, Tragardh J, Amor R, Dempster J, Reid E and Amos W B 2016 *eLife* e18659

[4] Walter A *et al* 2020 *Front. Phys.* **8** 47

[5] Jiang H *et al* 2014 *Methods* **68** 317–24

[6] Huang *et al* 2004 *J. Cell Sci.* 2014 **127** 4186–200

[7] Vergara F M *et al* 2020 BioRxiv doi: https://doi.org/10.1101/2020.02.26.961037

[8] Hériché J K, Alexander S and Ellenberg J 2019 *Ann. Rev. Biomed. Data Sci.* **2** 175–97

[9] Keuenhof K *et al* 2020 *Methods Cell Biol.* 2020 https://doi.org/10.1016/bs.mcb.2020.10.002

[10] Ginsburg G S and Phillips K A 2018 *Health Affairs* **37** 694–701

[11] Disselhorst J A, Krueger M A, Ud-Dean S M M, Bezrukov I, Jarboui M A and Trautwein C *et al* 2018 *PNAS* **115** E2980–7

[12] Lambin P *et al* 2017 *Nat. Rev. Clin. Oncol.* **14** 749–62

[13] Martinez M 2020 *J. Chem. Inf. Model.* **60** 2533–40

[14] Ortiz J O, Förster F, Kürner J, Linaroudis A A and Baumeister W 2006 *J. Struct. Biol.* **156** 334–41

[15] Mahamid J, Pfeffer S, Schaffer M, Villa E, Danev R, Cuellar L K, Förster F, Hyman A A, Plitzko J M and Baumeister W 2016 *Science* **351** 969–72

[16] Bykovm Y S, Cohen N, Gabrielli N, Manenschijn H, Welsch S, Chlanda P, Kukulski W, Patil K R, Schuldiner M and Briggs J A G 2019 *J. Cell Biol.* **218** 2797–811

[17] Hoffman D P *et al* 2020 *Science* **367** eaaz5357

[18] Lees R M, Peddie C J, Collinson L M, Ashby M C and Verkade P 2017 *Methods Cell. Biol.* **140** 245–76

www.ingramcontent.com/pod-product-compliance
Ingram Content Group UK Ltd.
Pitfield, Milton Keynes, MK11 3LW, UK
UKHW051907100926
8556IPUK00005B/790